OXFORD LOGIC GUIDES

Series Editors

D.M. GABBAY
A.J. MACINTYRE
D. SCOTT

OXFORD LOGIC GUIDES

Available books in the series:

10. Michael Hallett: *Cantorian set theory and limitation of size*
17. Stewart Shapiro: *Foundations without foundationalism*
18. John P. Cleave: *A study of logics*
21. C. McLarty: *Elementary categories, elementary toposes*
22. R.M. Smullyan: *Recursion theory for metamathematics*
23. Peter Clote and Jan Krajícek: *Arithmetic, proof theory, and computational complexity*
24. A. Tarski: *Introduction to logic and to the methodology of deductive sciences*
25. G. Malinowski: *Many valued logics*
26. Alexandre Borovik and Ali Nesin: *Groups of finite Morley rank*
27. R.M. Smullyan: *Diagonalization and self-reference*
28. Dov M. Gabbay, Ian Hodkinson, and Mark Reynolds: *Temporal logic: Mathematical foundations and computational aspects, volume 1*
29. Saharon Shelah: *Cardinal arithmetic*
30. Erik Sandewall: *Features and fluents: Volume I: A systematic approach to the representation of knowledge about dynamical systems*
31. T.E. Forster: *Set theory with a universal set: Exploring an untyped universe, second edition*
32. Anand Pillay: *Geometric stability theory*
33. Dov M. Gabbay: *Labelled deductive systems*
35. Alexander Chagrov and Michael Zakharyaschev: *Modal logic*
36. G. Sambin and J. Smith: *Twenty-five years of Martin-Löf constructive type theory*
37. María Manzano: *Model theory*
38. Dov M. Gabbay: *Fibring Logics*
39. Michael Dummett: *Elements of intuitionism, second edition*
40. D.M. Gabbay, M.A. Reynolds and M. Finger: *Temporal Logic: Mathematical foundations and computational aspects, volume 2*
41. J.M. Dunn and G. Hardegree: *Algebraic methods in philosophical logic*
42. H. Rott: *Change, choice and inference: A study of belief revision and nonmonotoic reasoning*
43. Johnstone: *Sketches of an elephant: A topos theory compendium, volume 1*
44. Johnstone: *Sketches of an elephant: A topos theory compendium, volume 2*
45. David J. Pym and Eike Ritter: *Reductive logic and proof search: Proof theory, semantics and control*
46. D.M. Gabbay and L. Maksimova: *Interpolation and definability: Modal and intuitionistic logics*
47. John L. Bell: *Set Theory: Boolean-valued models and independence proofs, third edition*
48. Laura Crosilla and Peter Schuster: *From sets and types to topology and analysis: Towards practicable foundations for constructive mathematics*
49. Steve Awodey: *Category theory*
50. Roman Kossak and James Schmerl: *The structure of models of Peano Arithmetic*

The Structure of Models of Peano Arithmetic

ROMAN KOSSAK
City University of New York
JAMES H. SCHMERL
University of Connecticut, Storrs

CLARENDON PRESS • OXFORD
2006

OXFORD
UNIVERSITY PRESS

Great Clarendon Street, Oxford OX2 6DP

Oxford University Press is a department of the University of Oxford. It furthers the University's objective of excellence in research, scholarship, and education by publishing worldwide in

Oxford New York

Auckland Cape Town Dar es Salaam Hong Kong Karachi Kuala Lumpur Madrid Melbourne Mexico City Nairobi New Delhi Shanghai Taipei Toronto

With offices in

Argentina Austria Brazil Chile Czech Republic France Greece Guatemala Hungary Italy Japan Poland Portugal Singapore South Korea Switzerland Thailand Turkey Ukraine Vietnam

Oxford is a registered trade mark of Oxford University Press in the UK and in certain other countries

Published in the United States by Oxford University Press Inc., New York

First published 2006

British Library Cataloguing in Publication Data
Data available

Library of Congress Cataloging in Publication Data
Data available

Typeset by Newgen Imaging Systems (P) Ltd., Chennai, India
Printed in Great Britain
on acid-free paper by
Biddles Ltd., King's Lynn, Norfolk

ISBN 0–19–856827–4 978–0–19–856827–8

1 3 5 7 9 10 8 6 4 2

PREFACE

In the 1930's, nonstandard models of arithmetic were introduced into mathematics by Thoralf Skolem in two papers [188] and [189]. Even though the logic community was slow in recognizing the importance of Skolem's contribution of the method of definable ultrapowers (see [207]), it now seems almost obligatory to include nonstandard models in introductory courses on mathematical logic. The point that they help to emphasize is the limited expressive power of first order logic: there are mathematical structures (one of them being the most classical of mathematical structures—the standard model of arithmetic) which are indistinguishable with respect to their first-order properties but whose isomorphism types are dramatically different. The drama is personified by nonstandard elements. However, the discussion at the introductory level usually ends here, leaving out the complex picture obtained by a closer scrutiny of the spectrum of isomorphism types of nonstandard models. Once we know that nonstandard models exist, it is very natural to ask how different they are from the standard one and also from each other. In other words, we would like to know to what extent the first-order theory of a model of arithmetic determines properties that are not first-order expressible. A priori, there is no guarantee that the possible answers would be relevant to other developments in model theory. Even a quick initial glance at a nonstandard model reveals a very rich structure. It could be that the diversity among nonstandard models is so vast that no coherent picture in the form of a relative classification can emerge. In fact, this might be the the state of affairs for the spectrum of all nonstandard models. Fortunately, when we consider some well defined and important subclasses of nonstandard models, a more attractive picture can be painted. This is the subject of this book.

There was little progress in the model theory of arithmetic between Skolem's discovery and two important developments that took place at the end of 1950's. In the first of these, Stanley Tennenbaum proved in a famous unpublished paper [205] that in no nonstandard model can either addition or multiplication have a recursive presentation. This result pointed to an essential difficulty in constructing nonstandard models (there are no such objects in the world of constructive mathematics!). The second of these was the fundamental theorem of Robert MacDowell & Ernst Specker [123]. Skolem had proved that the standard model has an elementary end extension. The MacDowell–Specker Theorem involves a refinement of Skolem's method to show that *every* model of Peano Arithmetic (PA) has an elementary end extension. Refining this further, Haim Gaifman (in several papers, but most importantly in [45]) developed a technology of iterating elementary end extensions along linear orders to obtain models having additional interesting features. Thus, Tennenbaum says: there are no effective constructions; but the response from MacDowell, Specker, and Gaifman is: some interesting set-theoretic constructions are easily available.

A very important concept that emerged from the proof of Tennenbaum's theorem is that of the standard system of a nonstandard model. This is the collection of sets of natural numbers coded in the model. The standard systems are precisely the ω-models of the fragment second-order arithmetic known as WKL_0. Having been studied by Dana Scott [180] such models are also called Scott sets. There are $2^{\aleph_0}$ countable Scott sets, and each of them is the standard system of a model of PA. Under the Continuum Hypothesis, every Scott set is the standard system of a model of PA. Every countable nonstandard model has $2^{\aleph_0}$ distinct initial segments which are themselves models of PA. The standard system of the model puts some restrictions on the possible complete theories of these initial segments (including a restriction on their theories). Still, each countable Scott set $\mathfrak{X}$ is the standard system of $2^{\aleph_0}$ elementarily inequivalent models of PA, and also for each model M having a standard system of $\mathfrak{X}$, there are $2^{\aleph_0}$ pairwise nonisomorphic countable models elementarily equivalent to M having $\mathfrak{X}$ as their standard systems. Considering the complexity of both classes of objects involved, this is a mess! However, for us this is only a point of departure. Suppose the complete theory of a model and its standard system are given. What else can be said about the model? The theorem of MacDowell and Specker suggests that it might be fruitful to consider elementary submodels. For any model M, the family of elementary submodels of M forms a lattice $\mathrm{Lt}(M)$, with naturally defined operations $\wedge$ and $\vee$. What are the lattices which can be represented as $\mathrm{Lt}(M)$ for some model of arithmetic M? This is one of the main questions we will consider. Some answers are given in Chapter 3 and much of the material from the previous chapters, especially from Chapter 3 on types, is developed with an eye on applications to the lattice problem. While, all distributive lattices satisfying an obvious necessary condition (they must be algebraic) can be represented as $\mathrm{Lt}(M)$, many questions concerning nondistributive lattices are open. In particular, we do not know if there is a finite lattice which cannot be represented as $\mathrm{Lt}(M)$.

The MacDowell–Specker Theorem has an interesting feature. It is almost independent of the language in which arithmetic is formalized. Let $\mathcal{L}$ be any countable language extending the language of PA, and let T be a theory in $\mathcal{L}$ that extends PA proves the scheme of induction for all formulas of $\mathcal{L}$. It turns out that most arguments concerning models of PA apply without modifications to models of T. We address this by formulating our results for PA^* rather than PA, where PA^* is any T as above. Actually, many results carry over to uncountable languages as well, with one notable exception, the MacDowell–Specker Theorem. There are models of PA^* in a language of cardinality $\aleph_1$ with no elementary end extensions. This result is due to George Mills, and it uses forcing in models of arithmetic. The purpose Chapter 6 is to give a proof of Mills' theorem and to show how forcing can be applied to construct interesting models of PA^*.

While we do not know exactly which lattices can be represented by lattices of elementary submodels of models of PA, the analogous question concerning automorphism groups has a complete answer, which we give in Chapter 5. For every infinite linearly ordered structure $(A, <, \dots)$, there is a model M of PA such that the automorphism groups of $(A, <, \dots)$ and M are isomorphic. Moreover, one can obtain such an M as an elementary end extension of any model of PA. Conclusion: nothing special here about arithmetic (but, it should be noted that the proof uses the full power of arithmetic and involves a formalized Ramsey style theorem of Nešetřil and Rödl). So we learn that in general there is no connection between the standard system of a model or its theory and its automorphism group. But this is not the end of the story. In Chapters 8 and 9, we discuss countable recursively saturated models of PA and show that something special *is* happening in this class of models. Most of the results there are formulated in terms of automorphisms and automorphism groups. In particular, many properties characterizing the important class of arithmetically saturated models involve automorphisms; for example, a countable recursively saturated model M of PA is arithmetically saturated iff it has an automorphism moving all undefinable elements, and this happens iff the automorphism group of M has uncountable cofinality. The automorphism group of a countable arithmetically saturated model, considered as an abstract group, determines the standard system of the model.

Some important results on automorphism groups of models of PA are included in Kaye and Macpherson's volume on automorphisms of first-order structures[77]. Here we concentrate on the results obtained after the volume was published, although we do include the complete proof of the theorem of Daniel Lascar on the small index property of countable arithmetically saturated models of PA.

The aim of Chapter 10 is to present some exotic species of models with properties dramatically contrasting those of the countable ones. In particular we construct a recursively saturated rather classless model using $\diamondsuit$ (a result due to Matt Kaufmann [67]) and then again using weak $\diamondsuit$. We do it despite the fact that Saharon Shelah has already proved that the existence of such models is a theorem of ZFC (and we explain why). Other topics in this chapter include nonisomorphic, but still very similar, models. Previous constructions of such models used extra set-theoretic assumptions. Here it is done in ZFC. Rigid recursively saturated models are also constructed in this chapter and as are models of Peano Arithmetic in the language with Ramsey quantifier and with the stationary quantifier.

One of the topics we neglect in this book is reducts. Any classification of models of PA must include subclassifications of their natural reducts. For a model M, let $(M, +)$, $(M, \times)$, and $(M, <)$ denote, respectively, the reducts of M to, respectively, $+$, $\times$, and $<$. It turns out that in the countable case all these reducts are nicely classifiable. All nonstandard countable models share the same order type $(\omega^* + \omega)\rho$, where ρ is the order type of the rationals. For every countable the isomorphism type if its reducts to $+$ and $\times$ is determined uniquely by the standard system of M. Consequently, for any countable models M and N of

PA, $(M,+) \cong (N,+)$ iff $(M,\times) \cong (N,\times)$. In Chapter 10 we prove that this is not the case for models of cardinality $\aleph_1$. Very little is known about order types of uncountable models. We honor this important and rather unexplored topic in Chapter 11 where we give proofs of two striking theorems concerning order types, one due to Jean-François Pabion [142] on κ-saturated reducts $(M,<)$ and one due to Shelah [185] on the existence of (κ,κ)-gaps.

There is extensive literature on models of PA, but essentially there are only two books: Kaye's [71] and Hájek & Pudlák's [50]. It has been often noted that bookwriting in this area is not easy. Here is what Laurie Kirby wrote in his review of Kaye's book [82]

> Our vocabulary lacks a term to denote a person whose calling is the study of models of arithmetic. Model theorists, topologists, even functional analysts can identify themselves succinctly, but we have to resort to such locutions as "I'm in models of arithmetic." And the name of the field itself—"models of arithmetic"—also seems to bespeak an insecurity about whether it is a field at all: the objects of study are baldly named without any pretensions to a grand theory or -ology. Models of arithmetic certainly is a bona fide field. It has its own meetings, folklore, and stars. It has built up a coherent body of knowledge relevant to some of the central problems of modern logic. But it has never sat comfortably within the traditional fourfold division of logic, it is sparsely populated and has been known to lie dormant for decades, and it has never had a "bible." Access to this difficult terrain has been daunting to outsiders.

These remarks were appropriate in 1992 and are still appropriate today. Kaye's book is still the "bible," while Hájek and Pudlák's book has become a standard reference in the model theory of fragments of arithmetic. Before 1991, the only, and not easily available, source was the excellent notes of Craig Smoryński [191] from his lectures on nonstandard models at the University of Utrecht in 1978. The notes were an inspiration for many of us who studied models of PA in the 1980s.

While writing this book we assumed that the reader is familiar with Kaye's book, and we tried to avoid repetitions. There is some overlap in the Chapter 7 on cuts. In this chapter we discuss strong cuts and their various characterizations. Strong cuts play an important role in arithmetic saturation—one of the main themes in this book.

Most of the results presented here have been previously published, but many of the proofs are new. Some proofs are simpler, or at least much shorter, than the original ones. This is a great advantage of having the whole range of techniques presented in a unified way in one place. Some results appear here for the first time. In the Remarks and References we made an effort to include accurate credits and references. Anyone who has worked in this or any area understands well that this was not an easy job, and certainly there will be errors and omissions.

Exercises are an integral part of the book. To measure the difficulty of exercises, we have designed a suitable ranking system. Exercises marked with ♣ are the easiest. These are traditional exercises for practice. Usually they just involve going through the definitions of the concepts involved. The highest rank is ♠. We have used it sparingly, as it is reserved for those exercises we could not do. Exercises marked with ♦ and ♥ are in between. Included in the ♦ and ♥ categories are many lemmas, propositions, and theorems we took from the literature. We recommend that the reader at least read them, as they help to provide a relatively complete account of what is known about the subject. One should note that the rankings of all these exercises is quite subjective; others may rank them differently, and we probably would too if we were to do it again. The rankings are also relative, as they are based on the material developed in the book. As stand alone exercises, many would have to be ranked higher.

Throughout the book, the reader will find numerous instances of the (Do it!) command. As in most mathematical texts, it is assumed that the reader will fill in the more routine parts of the arguments. There will be many statements starting with "Clearly, ...," or "One can verify ...," or "A similar argument shows ...." These are treacherous points. It is always good for the reader to pause for a moment and to verify whether she or he really believes the author(s). The role of the (Do it!)'s is to provide alert stopmarks. They remind you not to read too fast.

ACKNOWLEDGMENTS

For many years, jointly and separately, we have enjoyed working together with Ali Enayat, Richard Kaye, and Henryk Kotlarski on many projects related to the subject of this book. We would like to thank them here. We also want to thank Andreas Blass, Haim Gaifman, Matt Kaufmann, Laurie Kirby, Julia Knight, Kenneth McAloon, Angus Macintyre, David Marker, Jeff Paris, Stephen Simpson, Stuart Smith, Craig Smoryński, and Alex Wilkie. They all have made significant contributions and influenced us either by their published work or/and personal communication.

We worked on this project in intervals of various lengths over the last ten years. The first author did most of the work during his sabbatical leave in 2003–2004. He would like to thank Juliette Kennedy and Jouko Väänänen of the University of Helsinki and Costas Dimitracopoulos of the University of Athens for their great hospitality, interest, and support. Both authors spent the month of May of 2004 at the Mathematics Research Institute in Oberwolfach supported by a Research in Pairs grant. It was a very productive month (and we both gained a few pounds). The first author would like to acknowledge the CUNY Professional Staff Congress grants in the years 1998–2001 which provided him with valuable release time in the early stages of this work. The final manuscript was prepared at the University of Connecticut in Storrs in the summer of 2005. We thank the UConn Mathematics Department Head Miki Neumann for his support.

Finally, to our wives, Sue (Jim's) and Wanda (Roman's): we could not have done it without you. Thanks for your encouragement, tolerance, and abiding love.

CONTENTS

1

BASICS

It is perhaps a bit embarrassing that a discipline as fundamental as mathematical logic, after more than a century of intensive investigations, has not arrived at a consistent and generally accepted notational and terminological conventions. However, it is quite remarkable that, apparently due to the robustness of the main concepts, the diversity of styles and conventions has never been a serious obstacle in studying model theory. Below we give an account of our own choices (arrived at with some effort). We also give proofs of some basic results, including Ehrenfeucht's Lemma, and Blass' theorem on intersections of finitely generated submodels, and a special case of Friedman's Embedding Theorem.

The recent years have brought the decline and fall of "Recursion" Theory and emergence of "Computability" Theory. Whatever was recursive in the twentieth century is computable now. For us this creates a problem. Large parts of this book are devoted to recursively saturated models of PA, and *computably saturated* just does not sound right. For this reason we are sticking with the old terminology.

On many occasions we define sequences, terms, types, and other objects *recursively*. Sometimes this is done formally within PA, sometimes in the real world. Often such constructions are not effective; hence we have decided to call them *inductive*, rather than recursive. So, we will be constructing objects by induction or formal induction.

1.1 Notation and basic definitions

The language of Peano Arithmetic (PA) consists of the symbols : $+, \times, \leq, 0, 1$ and is denoted by $\mathcal{L}_{\mathsf{PA}}$. A formula of $\mathcal{L}_{\mathsf{PA}}$ is a first-order formula in this language. The well-known axioms declare that M is a model of PA iff M is the nonnegative part of a discretely ordered ring and satisfies the *least number principle*: every nonempty definable subset of M has a least element. The least number principle is equivalent to Peano's induction schema and to other well-known principles we freely use throughout the book.

Even though our main concern is with models of PA, it often convenient to consider models of arithmetic in a more general setting. Let $\mathcal{L}$ be any language extending $\mathcal{L}_{\mathsf{PA}}$. Then $\mathsf{PA}^*(\mathcal{L})$ is the $\mathcal{L}$-theory consisting of the basic axioms of PA and the induction schema for all formulas of $\mathcal{L}$. We are studying $\mathsf{PA}^*(\mathcal{L})$, especially when $\mathcal{L}$ is countable. We make the following convention: if $\mathcal{L}$ is countable, we often suppress specific reference to $\mathcal{L}$ and then write simply PA^*. Whenever we refer to "model" without additional explanation, we mean a model of $\mathsf{PA}^*(\mathcal{L})$, where $\mathcal{L}$ is some given countable language extending $\mathcal{L}_{\mathsf{PA}}$. Sometimes, we may want to restrict $\mathcal{L}$ further. In situations where we want to allow for uncountable $\mathcal{L}$, then we specifically say that $\mathcal{L}$ may be uncountable and then refer to models of $\mathsf{PA}^*(\mathcal{L})$. In cases where we want $\mathcal{L}$ to be just $\mathcal{L}_{\mathsf{PA}}$, then we refer to models of PA. This is done when what is being discussed might not apply to models of PA^* but also on some other occasions for notational convenience and with a warning for the reader. Letters M and N, and sometimes also K and L, will be used for models of $\mathsf{PA}^*(\mathcal{L})$. For general first-order structures we will use $\mathfrak{A}$ and $\mathfrak{B}$.

When dealing with recursively saturated models of PA^*, we always assume that $\mathcal{L}$ is finite. With some extra care we could also work with recursive languages.

If M is an $\mathcal{L}$-structure for some language $\mathcal{L}$, then by $\mathcal{L}(M)$ we denote the language $\mathcal{L}$ augmented with constants for each element of M.

For a formula $\varphi(x)$ and a model M, we denote by $\varphi(M)$ the set $\{x \in M : M \models \varphi(x)\}$.

We denote the set of all formulas of $\mathcal{L}$ by $\mathsf{Form}_{\mathcal{L}}$ or just by Form if it is clear what $\mathcal{L}$ is. If $\mathcal{L}$ is finite, identify the formulas of $\mathcal{L}$ with their codes via some standard arithmetic coding of finite sequences. For a finite $\mathcal{L}$, let $\mathsf{Form}_{\mathcal{L}}(x)$ be an an arithmetical formula representing $\mathsf{Form}_{\mathcal{L}}$ in PA. Then for every model $M \models \mathsf{PA}$

$$\mathsf{Form}_{\mathcal{L}} = \mathbb{N} \cap \{x \in M : M \models \mathsf{Form}_{\mathcal{L}}(x)\}.$$

If M is nonstandard, $\mathsf{Form}_{\mathcal{L}}(M)$ contains nonstandard elements. This is a simple consequence of the overspill principle (Proposition 1.1.1). The elements of $\mathsf{Form}_{\mathcal{L}}(M)$ are referred to as formulas in the sense of M.

We do not use separate symbols for models and their universes.

We use ω to denote the set of natural numbers (as well as its order type). The *standard* model of PA is $\mathbb{N} = (\omega, +, \times, \leq, 0, 1)$. By a standard model of PA^* we mean any expansion of the standard model. We let TA, standing for *True Arithmetic*, be the complete theory of $\mathbb{N}$. A model of PA^* is nonstandard if it is not isomorphic to a standard model. The standard model $\mathbb{N}$ is isomorphic to a proper initial segment of any nonstandard model. Hence, we assume that $\mathbb{N}$ is an initial segment of every model of PA^*.

To express that n is a natural number we usually write $n < \omega$; however, the notation $n \in \mathbb{N}$ is sometimes used when we want to emphasize that n is a standard element of some nonstandard model.

The set of subsets of a model M which are definable with parameters in M is denoted by $\mathrm{Def}(M)$. If a set is definable without parameters, we say that it is 0-definable.

By Σ_n and Π_n and Δ_n we denote the usual levels of the arithmetic hierarchy of formulas, with $\Sigma_0 = \Pi_0 = \Sigma_0$ defined as the class of formulas all of whose quantifiers are bounded. If Γ is a class of formulas, then $\mathrm{Def}_\Gamma(M)$ is the collection of sets definable in M by the formulas in Γ.

For every formula $\varphi(\bar{u}, v)$ of $\mathcal{L}$ we define the *Skolem term*

$$t_\varphi(\bar{u}) = \begin{cases} \min\{v : \varphi(\bar{u}, v)\}, & \text{if } \exists v\ \varphi(\bar{u}, v) \\ 0 & \text{otherwise} \end{cases}$$

$\mathsf{Term}_\mathcal{L}$ denotes the set of all Skolem terms of $\mathcal{L}$.

When expansions of a model are considered, we use the same symbols for elements and subsets of models and their formal names. For example, if M is a model for a language $\mathcal{L}$ and $X_0, X_1, \ldots$ is a sequence of subsets of M, then $(M, X_0, X_1, \ldots)$ is a first-order structure for the language $\mathcal{L} \cup \{X_0, X_1, \ldots\}$. This convention allows us to use alternatively expressions of the form $x \in X_i$ and $X_i(x)$. The sequence $X_0, X_1, \ldots$ does not have to be countable.

If $X \subseteq M$ is such that $(M, X) \models \mathsf{PA}^*$, then we say that X is an *inductive* subset of M.

Some results are more conveniently stated and proved in the context of second-order arithmetic and its fragments, in particular ACA_0. If $\mathfrak{X} \subseteq \mathcal{P}(M)$, then $(M, \mathfrak{X})$ is a second-order structure in which second-order variables range over $\mathfrak{X}$. Then $(M, \mathfrak{X})$ is a model of ACA_0 if for each $X \in \mathfrak{X}$, $(M, X) \models \mathsf{PA}^*$, and $\mathfrak{X}$ is closed under arithmetical definability. This means that for all $X_0, \ldots, X_{n-1} \in \mathfrak{X}$, all $\bar{a} \in M$, and every first-order formula $\varphi(x, \bar{y}, X_0, \ldots, X_{n-1})$ the set

$$\{x \in M : (M, X_0, \ldots, X_{n-1}) \models \varphi(x, \bar{a}, X_0, \ldots, X_{n-1})\}$$

is in $\mathfrak{X}$.

For every model M, $(M, \mathrm{Def}(M)) \models \mathsf{ACA}_0$. This implies that ACA_0 is a conservative extension of PA.

Another important theory is WKL_0. It is a fragment of second-order arithmetic consisting of the induction schema for Σ_1 formulas, Δ_1 comprehension schema, and Weak König's Lemma. Let us say that a *binary tree* is a subset B of the set $2^{<\omega}$ of finite 0–1 sequences such that if $\sigma : \{0, \ldots, n-1\} \longrightarrow \{0, 1\}$ is in B, then so is the restriction of σ to each set $\{0, \ldots, k-1\}$ for each $k < n$. A tree is B unbounded if it contains sequences of any finite length. Weak König's Lemma says that every unbounded binary tree has an unbounded branch. Later in this chapter we discuss coding of finite sequences in arithmetic. Weak König's Lemma is easily formalizable and provable in ACA_0.

We use the notation $I \subseteq_{\mathsf{end}} M$ if I is an initial segment of M. We call an initial segment I a *cut* of M if $I \neq \emptyset$ and is closed under successor. A *proper cut* is a cut which is a proper subset. If I is a cut of M, then we say that $X \subseteq I$ is *cofinal* in I if for every $x \in I$ there is $y \in M$ such that $x < y$. We say that $Y \subseteq M \setminus I$ is *coinitial* in $M \setminus I$ if for every $x > I$ there is $y \in Y$ such that $x > y$.

The following proposition is a direct consequence of the induction schema.

Proposition 1.1.1 *No proper cut of a nonstandard model is definable.* □

Proposition 1.1.1 implies the following, frequently used, Overspill Principle and its flip-side, the Underspill Principle

Proposition 1.1.2 (Overspill) *Let I be a proper cut of a nonstandard model M, and suppose that $D \subseteq M$ is definable. If $I \cap D$ is cofinal in I, then for each $d > I$, there is j such that $I < j < d$ and $j \in D$.* □

Proposition 1.1.3 (Underspill) *Let I be a proper cut of a nonstandard model M, and suppose that $D \subseteq M$ is definable. If $(M \setminus I) \cap D$ is coinitial in $M \setminus I$, then for each $d \in I$, there is $j \in I$ such that $d < j$ and $j \in D$.* □

For $a \in M$, a_M denotes the set $\{x \in M : M \models x < a\}$. This notation is particularly useful in situations when $a \in M$, M is cofinal in N, and a_M is a proper subset of a_N. We also use the interval notation $[a, b]_M = \{x \in M : a \leq x \leq b\}$, or $[a, b]$ when it is clear what M is. In particular, we often write $[0, a-1]$ rather than a_M. If $a \in M$ and X and Y are subsets of M, then we write $X < a$ if $\forall x \in X\ (x < a)$, and $X < Y$ if $\forall x, y\ (x \in X \wedge y \in Y \longrightarrow x < y)$. We abuse notation if $X = I$ and $Y = J$ are initial segments. In this case $I < J$ means that I is a proper initial segment of J. If $A \subseteq M$, then:

$$\inf(A) = \{x \in M : \forall y \in A\ (x \leq y)\},$$
$$\sup(A) = \{x \in M : \exists y \in A\ (x \leq y)\}.$$

1.2 Skolem closures

For a structure M and $A \subseteq M$, $\mathrm{Scl}^M(A)$ denotes the *Skolem closure* of A in M, that is the set of elements of M which are definable in M with parameters from A, or, equivalently

$$\mathrm{Scl}^M(A) = \{t(a_0, \ldots, a_{n-1}) : a_0, \ldots, a_{n-1} \in A \text{ and } t \in \mathsf{Term}\}.$$

Since PA has a definable pairing function, in the definition of Skolem closure we could just use binary terms or even just unary terms and the pairing function. If $M = \mathrm{Scl}^M(A)$, then we say that M is *generated* by A. If A is finite we say that M is *finitely generated*. If $A = \{a_0, \ldots, a_{n-1}\}$ for some $n < \omega$, then $\mathrm{Scl}^M(A) = \mathrm{Scl}(\langle a_0, \ldots, a_{n-1}\rangle)$, where $\langle a_0, \ldots, a_{n-1}\rangle$ is an element of M coding $\{a_0, \ldots, a_{n-1}\}$. See Section 1.4 below for details on coding. Hence, every finitely generated model of PA^* is generated by a single element.

The Skolem closure of A is the smallest elementary substructure of M which contains A (DO IT!). If there is no danger of confusion we drop the superscript M in $\mathrm{Scl}^M(A)$.

Let N be an elementary extension of a model M, and let a be an element of $N \setminus M$. Then $M(a) = \mathrm{Scl}^N(M \cup \{a\})$ is an elementary extension of M. The isomorphism type of $M(a)$ over M is determined uniquely by the type of a over M (DO IT!); hence, we can speak of $M(a)$ without referring to N.

If T is a completion of PA^* and $M \models T$, then let $M_T = \mathrm{Scl}^M(0)$. For every $N \models T$, $M_T \cong N_T$; hence M_T is the *prime* model of T. Notice that M_{TA} is $\mathbb{N}$.

1.3 End extensions and cofinal extensions

We say that a model N is an *end* extension of a model M if M is a submodel and a cut of N. If N is an end extension of M and the extension is elementary, then we say that M is an elementary cut of N, and we denote this by $M \prec_{\mathsf{end}} N$.

We say that N is a *cofinal* extension of M if M is a submodel of N and for each $x \in N$ there is $y \in M$ such that $N \models x \leq y$. This is denoted by $M \prec_{\mathsf{cof}} N$. In these definitions the extensions are not required to be proper; sometimes to emphasize this possibility we use the symbols $\preccurlyeq_{\mathsf{end}}$ and $\preccurlyeq_{\mathsf{cof}}$.

If Γ is a class of formulas and M is a submodel of N, then we write $M \prec_\Gamma N$ if for all $\bar{a}$ in M and all φ in Γ,

$$M \models \varphi(\bar{a}) \Longleftrightarrow N \models \varphi(\bar{a}).$$

If in addition $M \prec_{\mathsf{end}} N$, we write $M \prec_{\mathsf{end},\Gamma} N$. We could also introduce $\prec_{\mathsf{cof},\Gamma}$, but as the results below indicate, this would not be used often.

Theorem 1.3.1 *Suppose $M \models \mathsf{PA}^*$, N is a cofinal extension of M, and $M \prec_{\Sigma_0} N$. Then $M \prec N$ and, in particular, $N \models \mathsf{PA}^*$.* □

If in Theorem 1.3.1 M and N are models of PA, then the assumption $M \prec_{\Sigma_0} N$ can be eliminated. This follows from the MRDP theorem, whose formalized version says:

Theorem 1.3.2 (MRDP Theorem) *For every Σ_1 formula $\varphi(\bar{x})$ of $\mathcal{L}_{\mathsf{PA}}$, there is a quantifier-free formula $\psi(\bar{x}, \bar{y})$ such that*

$$\mathsf{PA} \vdash \forall \bar{x}(\varphi(\bar{x}) \longleftrightarrow \exists \bar{y} \psi(\bar{x}, \bar{y})).$$

□

Thus, for models of PA, $M \subseteq N$ implies $M \prec_{\Sigma_0} N$ (DO IT!). Hence, we have the following theorem.

Theorem 1.3.3 *Let M and N be models of PA. If M is cofinal in N, then $M \prec N$.* □

The following proposition has a straightforward proof by induction on the complexity of Σ_0 formulas.

Proposition 1.3.4 *If N is a model of PA^* and M is a cut of N, then $M \prec_{\Sigma_0} N$.* □

Suppose M and N are models of PA^* and $M \subseteq N$. Let K be $\sup(M)$ in N. Then $K \subseteq_{\mathsf{end}} N$, and it follows from Proposition 1.3.4 that $K \prec_{\Sigma_0} N$.

Theorem 1.3.5 (Gaifman's Splitting Theorem) *If M and N are models of* PA, *M is a submodel of N, and $K = \sup(M)$ in N; then $M \preccurlyeq_{\mathsf{cof}} K \subseteq_{\mathsf{end}} N$.*

Proof We can assume that $K \neq N$. By Theorem 1.3.2 we have $M \preccurlyeq_{\Sigma_0} N$ and by Proposition 1.3.4, $K \preccurlyeq_{\Sigma_0} N$. Hence, $M \preccurlyeq_{\Sigma_0} K$, and the result follows by Theorem 1.3.1. □

Since the proof of Gaifman's Splitting Theorem uses the MRDP Theorem, it cannot be generalized directly to models of PA^*. For models of PA^* we have the following variant (Do it!).

Theorem 1.3.6 *If $M \preccurlyeq N \models \mathsf{PA}^*$ and $K = \sup(M)$ in N, then $M \preccurlyeq_{\mathsf{cof}} K \preccurlyeq_{\mathsf{end}} N$.* □

Gaifman's Splitting Theorem explains why we usually consider end extensions and cofinal extensions separately. Each of these two types of extensions requires a different approach. For example, it follows directly from the Compactness Theorem and the results we already mentioned, that every nonstandard model has a proper elementary cofinal extension. While it is also true that every model of PA has an elementary end extension, this is much more difficult to prove.

The following property of cofinal extensions is frequently used.

Theorem 1.3.7 *If $M \prec_{\mathsf{cof}} N$ and $M^* \models \mathsf{PA}^*$ is an expansion of M, then there is a unique expansion N^* of N such that $M^* \prec N^*$.*

Proof (Sketch) Suppose $M^* = (M, X_0, X_1, \dots)$. For each i and each $a \in M$, $X_i \cap a_M$ is coded by an element of M (see the next section for conventions about coding). Let $b_{i,a}$ be such a code. Then let

$$Y_i = \bigcup_{a \in M} \{x \in N : N \models x \in b_{i,a}\}.$$

Now let $N^* = (N, Y_0, Y_1, \dots)$. It is easy to verify that $M^* \prec_{\Sigma_0} N^*$, and then the result follows from Theorem 1.3.1. □

If $M \prec N$, then we say that N is a *minimal* extension of M if for each M', $M \preccurlyeq M' \preccurlyeq N$ implies that either $M' = M$ or $M' = N$. By Theorem 1.3.6, every minimal extension is either an elementary end extension or a cofinal extension.

1.4 Coding bounded sets and classes

For simple coding tasks we use *Cantor's pairing function*:

$$\langle x, y \rangle = \frac{1}{2}[(x+y)^2 + 3x + y].$$

For every model M, Cantor's pairing function establishes a one-to-one correspondence between M^2 and M. We use it as a convenient way for partitioning unbounded sets into unboundedly many unbounded subsets. Let X be an unbounded definable subset of a model M, and let $f : M \longrightarrow X$ be the definable function enumerating X in increasing order. Then for each $a \in M$, let $X_a = \{f(\langle a, i\rangle) : i \in M\}$. In other words, $x \in X_a$ iff x is the $\langle a, i\rangle$th element of X for some $i \in M$. Clearly, each X_a is unbounded.

By iterating Cantor's function, for each $n < \omega$, we get a definable one-to-one correspondence between M^n and M. We also need a uniform coding scheme for arbitrary definable bounded sets and sequences. Definable bounded subsets of a model M are called *M-finite.* There are many ways to code M-finite sets. To be specific we fix one coding method. Informally, if A is M-finite, the *code* of A in M is $\Sigma_{a\in A}2^a$. We designate 0 to be the code of the empty set. More formally, there is a formula $\varphi(x, y)$ of $\mathcal{L}_{\mathsf{PA}}$, expressing that the xth bit in the binary representation of y is 1. Then one can define $x \in y$ to be $\varphi(x, y)$ and prove that $(M, \in)$ is a model of all the axioms of ZF, with one exception: $(M, \in)$ satisfies the negation of the axiom of infinity. The full strength of PA is not necessary here; $I\Sigma_0 + \mathsf{exp}$ suffices. A complete proof is given, for example, in [50].

Our method of coding could be replaced by any other definable coding everywhere in this book except for Chapter 11, where some specific properties of the binary coding are used. In that chapter the number $\Sigma_{a\in A}2^a$ is called the *canonical code* of A.

It is a common practice to identify M-finite sets with their codes. We frequently take advantage of this. If $x, y \in M$, then the cartesian product $x \times y$ is the set $\{\langle x', y'\rangle : x' \in x \wedge y' \in y\}$. Clearly, $x \times y$ is M-finite. Then we define functions and sequences in the usual way. In particular, for an M-finite A, $\mathrm{card}^M(A)$ denotes the cardinality of A in M, that is the unique $c \in M$ such that there is an $f \in M$ which is a one-to-one and onto function $f : c_M \longrightarrow A$. Sometimes $\mathrm{card}^M(A)$ is referred to as the *internal cardinality of* A. For the real world cardinality of A, we use $|A|$.

Let us fix some notation. If $s \in M$ codes a sequence, then $\ell(s)$ is the length of s, and for $i < \ell(s)$, $(s)_i$ is the ith term of s. We need some estimates on the size of codes. Directly from the definitions one can obtain the following.

Proposition 1.4.1 *If $\ell(s) = x$ and for all $i < x$, $(s)_i < y$, then $s \leq 2^{(x+y+1)^2}$.* □

Let $M^{<M}$ be the set of codes of all sequences coded in M. We define the partial ordering $\trianglelefteq$ on $M^{<M}$ so that if $\sigma, \tau \in M^{<M}$, then $\sigma \trianglelefteq \tau$ iff (the sequence coded by) σ is an initial segment of (the sequence coded by) τ. We write $\sigma \triangleleft \tau$ if $\sigma \trianglelefteq \tau$ and $\sigma \neq \tau$. We use ˆ for concatenation. By definition, $\sigma\hat{}\tau$ is the element of $M^{<M}$ which codes the concatenation of σ with τ.

Let I be a cut of a model M. We say that $X \subseteq I$ is *coded* in M if $X = I \cap Y$ for some $Y \in \mathrm{Def}(M)$. The collection of all subsets of I coded in M is denoted by $\mathrm{Cod}(M/I)$. Notice that if $I \subseteq_{\mathsf{end}} M \subseteq_{\mathsf{end}} N$, then $\mathrm{Cod}(M/I) = \mathrm{Cod}(N/I)$.

We say that a subset X of M is a *class* of M if for every $a \in M$, $X \cap a_M \in \mathrm{Def}(M)$. The set of classes of M is denoted by $\mathrm{Class}(M)$.

The next proposition can be easily proved by induction on the complexity of Σ_0 formulas (DO IT!).

Proposition 1.4.2 *A subset X of a model M is a class iff $(M, X) \models I\Sigma_0$.* □

One can show, and we ask the reader to do so in the exercises at the end of this chapter, that every subset of a model M which is coded in an end extension must be a class of M and that every countable model has classes which cannot be coded in end extensions. It is easy to see that every countable model has undefinable classes (DO IT!). This is not so for some uncountable models. A model M is *rather classless* if $\mathrm{Class}(M) = \mathrm{Def}(M)$. We construct rather classless models in the next chapter.

1.5 Standard systems

Let M be a nonstandard model. The *standard system* of M is the collection

$$\mathrm{SSy}(M) = \mathrm{Cod}(M/\mathbb{N}) = \{X \subseteq \mathbb{N} : \exists Y \in \mathrm{Def}(M)[X = Y \cap \mathbb{N}]\}.$$

Let T be a theory in some extension $\mathcal{L}$ of $\mathcal{L}_{\mathsf{PA}}$. Let X be a subset of $\mathbb{N}$. We say that T *represents* X iff there is a formula $\varphi(x)$ of $\mathcal{L}$ such that for each $n < \omega$,

$$n \in X \Longleftrightarrow T \vdash \varphi(n)$$

and

$$n \notin X \Longleftrightarrow T \vdash \neg\varphi(n).$$

We denote by $\mathrm{Rep}(T)$ the collection of sets represented by T. Peano Arithmetic represents every recursive set and, in particular, represents itself.

We use the following version of Gödel's First Incompleteness Theorem.

Theorem 1.5.1 *Any consistent theory $T \supseteq \mathsf{PA}^*$ which represents itself is incomplete.*

Proof Suppose that $\theta(x)$ represents T in T. By Gödel's Diagonalization Theorem, there is a sentence ψ be such that

$$\mathsf{PA}^* \vdash \psi \longleftrightarrow \neg\,\Theta(\psi).$$

Then neither ψ nor $\neg\psi$ is in T. □

Corollary 1.5.2 *If $T \supseteq \mathsf{PA}^*$ is consistent and complete, then $T \notin \mathrm{Rep}(T)$.* □

If $\mathsf{PA}^* \subseteq T$ and $M \models T$ is nonstandard, then $\mathrm{Rep}(T) \subseteq \mathrm{SSy}(M)$. If $T \supseteq \mathsf{PA}^*$ is complete and its prime model M_T is nonstanard, then $\mathrm{Rep}(T) = \mathrm{SSy}(M_T)$.

A *Scott set* is an ω-model of WKL_0. In other words, a Scott set is a nonempty collection $\mathfrak{X}$ of subsets of $\mathbb{N}$ which satisfies the following conditions:

(1) $\mathfrak{X}$ is closed under Boolean operations;
(2) If $A \in \mathfrak{X}$ and $B \subseteq \mathbb{N}$ is recursive in A, then $B \in \mathfrak{X}$;
(3) If $T \in \mathfrak{X}$ is an infinite binary tree, then T has an infinite path $P \in \mathfrak{X}$.

The following theorem summarizes some basic properties of Scott sets.

Theorem 1.5.3 (1) *For each model M, $\mathrm{SSy}(M)$ is a Scott set.*
(2) *If $\mathfrak{X}$ is a countable Scott set, $T \supseteq \mathsf{PA}^*$ is complete, and $\mathrm{Rep}(T) \subseteq \mathfrak{X}$, then there is a model M of T such that $\mathrm{SSy}(M) = \mathfrak{X}$.*
(3) *Let $T \supseteq \mathsf{PA}^*$ be a theory which represents itself, let $\mathfrak{X}$ be a countable Scott set, and suppose that $T \in \mathfrak{X}$. Then there are continuum many completions $\overline{T} \supseteq T$ such that $\mathfrak{X} = \mathrm{Rep}(\overline{T})$.* □

The next two results will be used in Chapters 8 and 9.

Theorem 1.5.4 *Let $\mathfrak{X}$ be a countable Scott set. Then there are countable Scott sets $\mathfrak{X}_0, \mathfrak{X}_1$ extending $\mathfrak{X}$ such that $(\mathbb{N}, \mathfrak{X}_0) \models \mathsf{ACA}_0$ and $(\mathbb{N}, \mathfrak{X}_1) \not\models \mathsf{ACA}_0$.*

Proof (Sketch) To get $\mathfrak{X}_0$ use a countable chain argument (or downward Skolem–Löwenheim Theorem) to obtain a Scott set extending $\mathfrak{X}$ which is closed under arithmetical definability. To get $\mathfrak{X}_1$, start with the set A such that $\mathfrak{X} = \{A_n : n < \omega\}$, where $A_n = \{x : \langle x, n\rangle \in A\}$. Then, using Henkin's construction, build a nonstandard model M of PA, which is Δ_2 from X and such that $X \in \mathrm{SSy}(M)$. Let $\mathfrak{X}_1 = \mathrm{SSy}(M)$. It follows that every set in $\mathfrak{X}_1$ is Δ_2 definable from X; hence, $(\mathbb{N}, \mathfrak{X}_1) \not\models \mathsf{ACA}_0$. □

The last theorem of this section can be proved by combining Theorem 1.5.3, Corollary 1.11.3, and Friedman's Embedding Theorem (Theorem 1.13.1).

Theorem 1.5.5 *Let M be a countable model. Then for every countable Scott set $\mathfrak{X} \supseteq \mathrm{SSy}(M)$, there is a cofinal elementary extension N of M such that $\mathrm{SSy}(N) = \mathfrak{X}$.* □

1.6 Types

An n-type $p(x_0, \ldots, x_{n-1})$ is a collection of formulas in the variables $x_0, \ldots, x_{n-1}$ in a given language $\mathcal{L}$. If M is an $\mathcal{L}$-structure, then p^M is the set of all n-tuples which realize $p(x_0, \ldots, x_{n-1})$ in M.

For a consistent theory T and $n < \omega$, we let $S_n(T)$ be the set of complete n-types which are consistent with T. If M is a model of PA^* and T is the theory of the expanded structure $(M, a)_{a \in M}$, then we may write $S_n(M)$ instead of $S_n(T)$.

Suppose that $M \prec N$, $a \in N$, and $N = \mathrm{Scl}(M \cup \{a\})$. Suppose further that a realizes the type $p(x) \in S_1(M)$. Then we write $N = M(a)$ and say that

$M(a)$ is a *$p(x)$-extension* of M. Notice that Ehrenfeucht's Lemma (Theorem 1.7.2 below) implies that a is the only element in p^N. If M is the prime model of T, $p(x) \in S_1(T)$, and $M(a)$ is a $p(x)$-extension of M, then we will say that $M(a)$ is a *prime model* of $p(x)$. If there is no need to refer to a specific element realizing $p(x)$, then the $p(x)$-extension of M will be denoted by $M(p)$.

Definable coding of finite sequences allows to reduce most considerations about n-types of completions of PA^* to 1-types.

A type $p(v) \in S_1(M)$ is *bounded* if $(v \leq a) \in p(v)$ for some $a \in M$; otherwise $p(v)$ is *unbounded.* If $p(v)$ is a pure type extending a completion T of PA^*, then $p(v)$ is bounded (unbounded) if it is bounded (unbounded) over the prime model of T.

For $\bar{a} = (a_0, \ldots, a_n) \in M$, the type of $\bar{a}$ in M is

$$\mathrm{tp}(\bar{a}) = \{\varphi(v_0, \ldots, v_n) : M \models \varphi(a_0, \ldots, a_n)\}.$$

If there is a danger of confusion, we might write $\mathrm{tp}^M(\bar{a})$ instead $\mathrm{tp}(\bar{a})$.

The set $S_n(T)$ is equipped with a topology. The basic open sets are the sets of the form $\{p \in S_n(T) : \varphi \in p\}$. Each of these sets is also closed. It is easy to see that $S_n(T)$ is a Hausdorff space. The Compactness Theorem shows that $S_n(T)$ is compact. The set of bounded types in $S_1(T)$ is open. The unbounded types form a closed set, and if the language of T is countable, this set, with the induced topology, is homeomorphic to the Cantor set.

1.7 Blass–Gaifman and Ehrenfeucht lemmas

One of the very useful features of PA^* is its ability to define functions by induction. Because of this feature, iterates of definable functions are definable. Let us make that a little more precise. Consider a Skolem term $t(x)$. The yth iterate of $t(x)$ is the Skolem term $t^{(y)}(x)$ for which the following two sentences are consequences of PA^*:

$$\forall x[t^{(0)}(x) = x],$$
$$\forall x \forall y[t^{(y+1)}(x) = t(t^{(y)}(x))].$$

For a simple specific instance of this, if $t(x)$ is the term $x+1$, then $t^{(y)}(x)$ is the term $x+y$.

The use of iterates is essential in the proofs of the next two lemmas which state fundamental properties concerning the nature of definable functions in models of PA^*.

Lemma 1.7.1 (Blass–Gaifman Lemma) *Let $a, b \in M \models \mathsf{PA}^*$ and $t(x)$ be a Skolem term such that $M \models a < b \leq t(a)$. Then there is a Skolem term $s(x)$ such that $M \models a < b \leq s(a) = s(b)$. Moreover, we can also require that* $\mathsf{PA}^* \vdash \forall x \forall y[x < y \longrightarrow x \leq s(x) \leq s(y)]$.

Proof We can assume that $\mathsf{PA}^* \vdash \forall x \forall y[x < y \longrightarrow x < t(x) \leq t(y)]$ for if this is not so, then replace $t(x)$ with the term $t'(x)$ defined inductively by

$$t'(0) = 1 + t(0) \text{ and } t'(x+1) = \max(1 + t'(x), t(x+1)).$$

Then $M \models \forall x \exists y[t^{(y)}(0) > x]$, so we can let $r(x)$ be term denoting the least such y. Clearly, $r(a) \leq r(b) \leq r(a) + 1$. If $r(b)$ is even, then let $e(y)$ be the Skolem term such that $e(y)$ picks out the even one of y and $y + 1$. If $r(b)$ is odd, then $e(y)$ picks out the odd one. In either case $r(a) \leq r(b) \leq e(r(a)) = e(r(b))$. Then let $s(x)$ be the term $t^{(e(r(x)))}(0)$. □

Theorem 1.7.2 (Ehrenfeucht's Lemma) *Let $a, b \in M \models \mathsf{PA}^*$ and $t(x)$ be a Skolem term such that $M \models a \neq b = t(a)$. Then $\mathrm{tp}(a) \neq \mathrm{tp}(b)$.*

Proof We define a Skolem term $r(x)$ in each of two cases.

Case 1: $b < a$. We can assume that $M \models \forall x[t(x) \leq x]$ for, if this is not so, then replace $t(x)$ with the Skolem term $\min(x, t(x))$. Then

$$M \models \forall x \exists y[t^{(y)}(x) = t^{(y+1)}(x)],$$

so we can let $r(x)$ be the term denoting the least such y.

Case 2: $a < b$. We can assume that $M \models \forall x[t(x) > x]$ for, if this is not so, then replace $t(x)$ with the Skolem term $\max(x + 1, t(x))$. Since $a < b = t(a)$, we can find a term $s(x)$ as in the Blass–Gaifman Lemma. Then

$$M \models\models \forall x \exists y[t^{(y)}(x) \geq s(x)],$$

so we can let $r(x)$ be the term denoting the least such y.

In either case, $r(a) = r(b) + 1$. Thus, $r(a)$ is even iff $r(b)$ is odd and, therefore, $\mathrm{tp}(a) \neq \mathrm{tp}(b)$. □

See the last three exercises in this chapter for an alternative proof of Ehrenfeucht's Lemma involving graph coloring.

The standard model, or more generally, any prime model of PA^*, is *rigid*, that is it does not have nontrivial automorphisms. Ehrenfeucht's Lemma implies an even stronger statement. A model N is *finitely generated over a model M* if there is $a \in N$ such that $N = \mathrm{Scl}^N(M \cup \{a\})$. Notice that if N is finitely generated over M, then N is a simple extension of M.

Lemma 1.7.3 *Suppose N is finitely generated over a model M. If $f : N \longrightarrow N$ is an elementary embedding such that $f(x) = x$ for all $x \in M$, then f is the identity function.*

Proof Consider PA^* in the language with constants for all elements of M. In this language, each element of N is of the form $t(a)$ for some Skolem term t. Hence, according to Ehrenfeucht's Lemma, for each $b \in N$, if $b \neq a$, then $\text{tp}(b) \neq \text{tp}(a)$ and the result follows. □

1.8 Recursive saturation and arithmetic saturation

Recursive saturation plays an important role in this book. When discussing recursive saturation of models of PA^* we assume that the language is finite. Then we can identify syntactic objects with their Gödel numbers. Under this convention, if $\bar{b}$ is a tuple of elements of a model M and $p(v,\bar{b})$ is a type, then $p(v,\bar{b})$ is recursive (arithmetic, etc.) if the set $\{\varphi(v,\bar{w}) : \varphi(v,\bar{b}) \in p(v,\bar{b})\}$ is recursive (arithmetic, etc.).

A first-order structure $\mathfrak{A}$ is *recursively saturated* if for each tuple $\bar{b}$ in $\mathfrak{A}$ and every recursive type $p(v,\bar{w})$, if $p(v,\bar{b})$ is finitely realizable, then $p(v,\bar{b})$ is realized in $\mathfrak{A}$.

Proposition 1.8.1 *Let M and N be recursively saturated models. Then $M \equiv_{\infty\omega} N$ iff $M \equiv N$ and* $\text{SSy}(M) = \text{SSy}(N)$. □

It follows that each countable recursively saturated model is uniquely determined up to isomorphism by its complete theory and standard system.

Proposition 1.8.1 is a particular instance of the back-and-forth characterization of $\mathcal{L}_{\infty\omega}$-equivalence. If M and N are recursively saturated elementarily equivalent models, then the back-and-forth system for M and N is

$$\mathcal{I} = \left\{(\bar{a},\bar{b}) \in M^{<\omega} \times N^{<\omega} : \text{tp}(\bar{a}) = \text{tp}(\bar{b})\right\}.$$

If $(\bar{a},\bar{b}) \in \mathcal{I}$, then for each $a \in M$ there is $b \in N$ such that $(\bar{a}\hat{\ }a, \bar{b}\hat{\ }b) \in \mathcal{I}$, and for each $b \in N$ there is $a \in M$ such that $(\bar{a}\hat{\ }a, \bar{b}\hat{\ }b) \in \mathcal{I}$ (DO IT!). In particular, if M and N are both countable, then $M \cong N$.

Let $\mathfrak{X}$ be a collection of subsets of ω. We say that a first-order structure $\mathfrak{A}$ is *$\mathfrak{X}$-saturated*

(1) If for each tuple $\bar{a}$ in $\mathfrak{A}$, $\text{tp}(\bar{a}) \in \mathfrak{X}$.
(2) If $\bar{b}$ is a tuple in $\mathfrak{A}$, $p(v,\bar{w})$ is a type in $\mathfrak{X}$ and $p(v,\bar{b})$ is finitely realizable in $\mathfrak{A}$, then $p(v,\bar{b})$ is realized in $\mathfrak{A}$.

Proposition 1.8.2 *A model M is recursively saturated iff it is* $\text{SSy}(M)$-*saturated.*

Proof Since PA represents every recursive set, $\text{SSy}(M)$-saturation implies recursive saturation. To see the converse, let $\langle \varphi_n(v,\bar{w}) : n < \omega \rangle$ be a recursive enumeration of all formulas in the displayed variables, and suppose that for some $\bar{b} \in M$ and $A \in \text{SSy}(M)$, the type

$$p(v,\bar{b}) = \left\{\varphi_n(v,\bar{b}) : n \in A\right\},$$

is finitely realizable in M. Let $a \in M$ be such that for all $n < \omega$, $n \in A$ iff $M \models n \in a$. Now consider the recursive type

$$q(v, a, \bar{b}) = \{n \in a \longrightarrow \varphi_n(v, \bar{b}) : n < \omega\}.$$

Clearly, $q(v, a, \bar{b})$ is just $p(v, \bar{b})$ in disguise. Since $q(v, a, \bar{b})$ is finitely realizable in M, it is realized in M and so is $p(v, \bar{b})$.

To see that for all $\bar{a} \in M$, $\mathrm{tp}(\bar{a})$ is in $\mathrm{SSy}(M)$, consider the recursive type $r(w)$ consisting of formulas of the form $\varphi(\bar{v}) \in w \longleftrightarrow \varphi(\bar{a})$, where $\varphi(\bar{v})$ ranges over all formulas with the variables $\bar{v}$. □

A first-order structure $\mathfrak{A}$ is *arithmetically saturated* iff for all $\bar{a}$, $\bar{b}$ in $\mathfrak{A}$, each finitely realizable type $p(v, \bar{b})$ which is arithmetic in $\mathrm{tp}(\bar{a})$ is realized in $\mathfrak{A}$.

Arithmetic saturation is a robust notion. We say much more about it in Chapter 8. Now let us just note the following proposition.

Proposition 1.8.3 *Let M be a recursively saturated model. The following conditions are equivalent:*

(1) *M is arithmetically saturated;*
(2) *$(M, \mathrm{SSy}(M)) \models \mathsf{ACA}_0$;*
(3) *The standard cut is strong in M.* □

The equivalence of the first two conditions follows directly from Proposition 1.8.2. A cut I of M is *strong* in M if for each $a \in M$, there is a $c > I$ such that for all $i \in I$, $(a)_i > I$ iff $(a)_i > c$. The equivalence of (2) and (3) for all models M, not just for recursively saturated ones, is established in Chapter 7.

If $\mathbb{N}$ is not strong in M, then there is $a \in M$ coding a sequence of a nonstandard length such that for every nonstandard $e \in M$, there is $i < \omega$ such that $\mathbb{N} < (a)_i < e$. If $\mathbb{N}$ is not strong in M and M is recursively saturated, then there is an a with the additional property that all for all $i < \omega$, if $(a)_i$ is nonstandard, then it is undefinable. To see this consider an $a \in M$ such that for every nonstandard $e \in M$ there is $i < \omega$ such that $\mathbb{N} < (a)_i < e$. By recursive saturation, there is $b \in M$ such that $\mathrm{Scl}(0) = \{(b)_i : i \in \mathbb{N}\}$. If a does not already have the additional property, then replace it with an element realizing the recursive type:

$$\{(v)_i = \min\{x : \forall j \leq \ell(b)[j \leq (a)_i \longrightarrow (x \neq (b)_j)]\} : i < \omega\}.$$

Thus, we have proved the following proposition:

Proposition 1.8.4 *If M is recursively saturated and $\mathbb{N}$ is not strong in M, then there is $a \in M$ such that for every nonstandard $e \in M$ there is $i < \omega$ such that $\mathbb{N} < (a)_i < e$ and for all $i < \omega$, if $(a)_i > \mathbb{N}$, then $(a)_i \notin \mathrm{Scl}(0)$.* □

1.9 Satisfaction classes and resplendency

Nonstandard satisfaction classes extend the notion of satisfaction to nonstandard formulas. Let M ba a model for a language $\mathcal{L}$. The standard satisfaction class $S_{\mathsf{st}}(M)$ is the set

$$\{\langle\varphi, a\rangle \in M : M \models \varphi(a)\}.$$

A subset $S \subseteq M$ is usually defined to be a nonstandard satisfaction class for M if $S_{\mathsf{st}}(M) \subseteq S$, S contains pairs $\langle\varphi, a\rangle$ for all (or some) nonstandard $\varphi \in \mathsf{Form}_{\mathcal{L}}(M)$ and S satisfies Tarski's inductive conditions for satisfaction. For a precise definition, see [71]. For applications in this book, we can use a simplified definition.

Definition 1.9.1 Let S be a subset of a model M. We say that S is a partial inductive satisfaction class, abbreviated $(M, S) \models \mathsf{Sat}(S)$, if S is inductive and

$$\langle\varphi, a\rangle \in S \Longleftrightarrow M \models \varphi(a). \qquad \square$$

Our definition of a partial inductive satisfaction class does not mention Tarski's inductive conditions for satisfaction. However, if M is nonstandard and $(M, S) \models \mathsf{Sat}(S)$, then, by overspill, there is a nonstandard e such that S satisfies Tarski's inductive conditions for all Σ_e formulas in the sense of M.

We say that $S \subseteq M$ is a *full satisfaction class* for M if for all $\varphi \in \mathsf{Form}(M)$ and all $a \in M$, either $\langle\varphi, a\rangle \in S$ or $\langle\neg\varphi, a\rangle \in S$, and S satisfies Tarski's inductive conditions. (See [71] for details). Notice that we do not assume that S is inductive. By a theorem of Kotlarski, Krajewski, and Lachlan [115], every countable recursively saturated model has a full satisfaction class. It is easy to prove that if M has a full inductive satisfaction class, then $M \models \mathrm{Con}(\mathsf{PA})$. Hence, not every countable recursively saturated model has a full inductive satisfaction class. Let us also note an important theorem of Lachlan [119].

Theorem 1.9.2 *If a countable model M has a full satisfaction class, then M is recursively saturated.* $\square$

A first-order structure $\mathfrak{A}$ is *resplendent* iff for every tuple $\bar{a}$ of $\mathfrak{A}$ and every Σ^1_1-sentence $\exists X \Psi(X, \bar{a})$, if $\mathrm{Con}(\mathrm{Th}(\mathfrak{A}, \bar{a}) + \exists X \Psi(X, \bar{a}))$, then $(\mathfrak{A}, X, \bar{a}) \models \Psi(X, \bar{a})$ for some $X \subseteq \mathfrak{A}$. A first-order structure $\mathfrak{A}$ is *chronically resplendent* if for each sentence $\exists X \Psi(X, \bar{a})$ such that $\mathrm{Con}(\mathrm{Th}(\mathfrak{A}, \bar{a}) + \exists X \Psi(X, \bar{a}))$, there is $X \subseteq \mathfrak{A}$ such that $(\mathfrak{A}, X, \bar{a}) \models \Psi(X, \bar{a})$ and $(\mathfrak{A}, X)$ is resplendent.

For $\mathfrak{X} \subseteq \mathcal{P}(\omega)$, we say that $\mathfrak{A}$ is $\mathfrak{X}$*-resplendent* iff for every tuple $\bar{a}$ of $\mathfrak{A}$ and every set T of sentences in some finite extension of the language of $\mathfrak{A}$ with an additional predicate symbol, if T is in $\mathfrak{X}$ and $\mathrm{Th}(\mathfrak{A}, \bar{a}) + T$ is consistent, then $(\mathfrak{A}, \bar{a})$ has an expansion satisfying T. It is well-known that every resplendent model is *Rec*-resplendent, where *Rec* is the collection of all recursive sets.

Theorem 1.9.3 (1) *Every countable recursively saturated first-order structure is chronically resplendent.*

(2) *Every countable recursively saturated model M of* PA^* *is* $\mathrm{SSy}(M)$-*resplendent.* □

It is an open problem whether there is a resplendent first-order structure which is not chronically resplendent.

The following simple proposition has many applications.

Proposition 1.9.4 *Let M be a nonstandard model of* PA^*.

(1) *If M has a partial inductive satisfaction class, then M is recursively saturated.*

(2) *If M is resplendent, then M has a partial inductive satisfaction class.*

Proof For the proof of (1), let S be a partial inductive satisfaction class for M, and let $\langle \varphi_n(v,w) : n < \omega \rangle$ be a recursive enumeration of all formulas of $\mathcal{L}$. Let $A \in \mathrm{SSy}(M)$ and suppose that the type $p(v,a) = \{\varphi_n(v,a) : n \in A\}$ is finitely realizable in M. Let $A' \in \mathrm{Def}(M)$ be such that $A = A' \cap \mathbb{N}$. Then for each $n < \omega$,

$$(M,S) \models \exists x \forall i < n\ [i \in A' \longrightarrow \varphi_i(x,a) \in S].$$

By overspill, this is also true for some nonstandard n. It follows that $p(v,a)$ is realized in M. Notice that this argument shows that M is $\mathrm{SSy}(M)$-saturated.

To prove (2), let us assume that M is resplendent. Consider the recursive theory:

$$T(S) = \{\forall x\ (\varphi(x) \in S \longleftrightarrow \varphi(x)) : \varphi \in \mathcal{L}\} + \mathsf{PA}^*.$$

Every finite fragment of $T(S)$ has a model of the form (M,D), where D is a definable subset of M (Do IT!). Hence, $T(S)$ is consistent, and the result follows. □

Corollary 1.9.5 *No model M has a definable partial inductive satisfaction class.*

Proof By passing to an elementary extension if necessary, we can assume that M is nonstandard. Suppose $S \in \mathrm{Def}(M)$. Let $K = \mathrm{Scl}^M(a)$, where a is a nonstandard element of M and a is such that S is definable from parameters in $\mathrm{Scl}^M(a)$. Then $S \cap K$ is definable in K, and since $K \prec M$, $S \cap K$ is a partial inductive satisfaction class for K. But K is not recursively saturated (Do IT!), and we get a contradiction with Proposition 1.9.4. □

Corollary 1.9.5 is a general version of Tarski's Theorem on Undefinability of Truth which says in its simplest form that Th($\mathbb{N}$) is not definable in $\mathbb{N}$.

Here is another corollary of Proposition 1.9.4.

Theorem 1.9.6 *Every cofinal extension of a recursively saturated model is recursively saturated.*

Proof Let M be a countable recursively saturated model. By Proposition 1.9.4, M has a partial inductive satisfaction class S. Let N be a cofinal extension of M. Use Theorem 1.3.7 to get $\overline{S}$ such that $(M, S) \prec (N, \overline{S})$. Then $\overline{S}$ is a partial inductive satisfaction class for N, and the result follows again by Proposition 1.9.4.

The uncountable case now follows by a Skolem–Löwenheim type argument (Do it!). □

Corollary 1.9.7 *Every countable recursively saturated model M has cofinal extensions M_0 and M_1 such that M_0 is arithmetically saturated and M_1 is not.*

Proof Directly from Theorems 1.5.4, 1.9.6, and Proposition 1.8.3. □

To simplify some statements, we will use a hierarchy of formulas whose levels are closed under negation. For $n < \omega$ let Q_n be the closure of the set of Σ_n formulas of $\mathcal{L}$ under negation, conjunction, and bounded quantification. For $e \in M$, $Q_e(M)$ is the set of Q_e formulas of $\mathcal{L}(M)$ in the sense of M. We say that a partial inductive satisfaction class S is an *Q_e-class*, if S satisfies Tarski's inductive conditions of satisfaction for all formulas in $Q_e(M)$. If S is an Q_e-class for every $e \in M$, then S is full and we will say that S is an Q_∞-class. We have shown that a countable model is recursively saturated iff it has an Q_e-class for some nonstandard e. Notice that, since partial satisfaction classes are inductive, if a partial inductive satisfaction class is not full, then there is a largest $e > \mathbb{N}$ such that S is a Q_e-class.

Let us finish this section with a useful lemma. It can be easily proved by induction on complexity of formulas. If $S \subseteq M$ is an Q_e-class and $d < e$, then $S|_d$ denotes the restriction of S to $Q_d(M)$ sentences.

Lemma 1.9.8 *If S is an Q_e-class, D is an Q_d-class, $d \leq e$, and $(M, S, D) \models$ PA^*, then $D = S|_d$.* □

1.10 Cuts and gaps in recursively saturated models

A nonstandard model can be partitioned into convex subsets in many natural and useful ways. For example, every model M is partitioned into $\mathbb{Z}$-blocks, where the $\mathbb{Z}$-block of every standard n is $\mathbb{N}$ and, for nonstandard $a \in M$, the $\mathbb{Z}$-block of a is

$$\{a + k : k \in \mathbb{Z}\}.$$

This idea can be generalized as follows. Let $\mathcal{F}$ be a set of definable functions $f : M \longrightarrow M$ for which $x \leq f(x) \leq f(y)$ whenever $x < y$. There is a partition of M into sets, which we call *$\mathcal{F}$-gaps*. For any $a \in M$, $\text{gap}_{\mathcal{F}}(a)$ is the smallest set C such that $a \in C$ and whenever $b \in C$, $f \in \mathcal{F}$, and $b \leq x \leq f(b)$ or $x \leq b \leq f(x)$, then $x \in C$.

Thus, $\mathbb{Z}$-blocks are the successor gaps, that is the $\mathcal{F}$-gaps, where $\mathcal{F}$ consists of the successor function. Similarly, one can define polynomial gaps, exponential gaps, etc. The *gap* of $a \in M$, denoted by $\text{gap}(a)$, is the $\mathcal{F}$-gap of a, where $\mathcal{F}$ is the set of all definable functions $f : M \longrightarrow M$ for which $x \leq f(x) \leq f(y)$ whenever $x < y$. In the literature, gaps are often called *skies*.

Every model M has the *least gap*, the gap of 0. If for some $a \in M$, $M = \sup(\text{gap}(a))$, then we call $\text{gap}(a)$ the *last gap* of M. A model with a last gap is called *short*. A model which is not short is *tall*.

Gap terminology is particularly useful in the study of recursively saturated models. Here are some examples. All statements below are good exercises (DO IT!). Assume that M is recursively saturated.

For any $a \in M$, the type consisting of formulas $t(a) < v$, where t ranges over all Skolem terms, is realized in M. Hence M is tall.

Let $a, b \in M$ be such that $\text{gap}(a) < b$. Then $I = \sup(\text{gap}(a))$ is a short elementary cut such that $a \in I < b$. Another elementary cut is $J = \inf(\text{gap}(b))$. Moreover, $I < J$; I is short, hence it is not recursively saturated; J is tall and is recursively saturated. In the literature on recursively saturated models of PA, I is often denoted by $M(a)$ and J by $M[b]$. We do not follow this tradition here, as it clashes with other standard conventions we have adopted.

If $K \prec_{\text{end}} M$ is not recursively saturated, then there is a such that $K = \sup(\text{gap}(a))$.

If M is countable and a, b, I, J are as above, then there are continuum many elementary cuts K such that $I < K < J$. Countably many of these are short. Using independent minimal types, discussed in Chapter 3, one can prove that there are countably many nonisomorphic such short cuts. All other cuts K are recursively saturated, and in fact, are isomorphic to M.

We say that $a \in M$ codes an *ascending sequence of gaps*, abbreviated $a \in \text{ASG}(M)$, if $\ell(a)$ is nonstandard and for each $i < \ell(a)$, $\text{gap}((a)_i) < (a)_{i+1}$.

Theorem 1.10.1 *For any model M the following are equivalent:*

(1) *M is recursively saturated.*

(2) *For all $a \in M$ there is $b \in \text{ASG}(M)$ such that $(b)_0 > a$.*

(3) *For all $a \in M$ there is $b \in \text{ASG}(M)$ such that $\ell(b) > a$.* □

While all tall elementary cuts K in a countable recursively saturated model M are isomorphic to M, there are many nonisomorphic pairs (M, K). We use the following result in Chapter 10.

Theorem 1.10.2 *Every countable recursively saturated model M has continuum many elementarily inequivalent pairs of the form (M, K), where K is a recursively saturated elementary cut.* □

1.11 Truth definitions and restricted saturation

While satisfaction over any model of PA is undefinable, and only recursively saturated models have partial inductive satisfaction classes, satisfaction for Σ_n formulas is definable, and this has important consequences. This is true for $\mathsf{PA}^*(\mathcal{L})$ for any finite $\mathcal{L}$. Recall that we identify formulas with their Gödel numbers. Also, if a is element of a model M and $\varphi(x)$ is a formula of $\mathcal{L}$, then we can regard $\varphi(a)$ as an element of M which codes the substitution of the numeral of a for x in $\varphi(x)$.

Let $\mathcal{L}$ be a finite extension of $\mathcal{L}_{\mathsf{PA}}$.

Theorem 1.11.1 *There is a Σ_1 formula $\mathsf{Tr}_0(x, y)$ of $\mathcal{L}$ such that for every Σ_0 formula $\varphi(y)$ of $\mathcal{L}$,*

$$\mathsf{PA}^*(\mathcal{L}) \vdash \left[\forall y(\varphi(y) \longleftrightarrow \mathsf{Tr}_0(\varphi, y))\right]. \qquad \square$$

The construction of Tr_0 is relatively straightforward along the lines: "it is snowing" is true iff it is snowing. Chapter 9 of [71] provides all details.

Corollary 1.11.2 *For each $n > 0$ there is a Σ_n formula $\mathsf{Tr}_{\Sigma_n}(x, y)$ of $\mathcal{L}$ such that for every Σ_n formula $\varphi(x)$ of $\mathcal{L}$,*

$$\mathsf{PA}^*(\mathcal{L}) \vdash \left[\forall y(\varphi(y) \longleftrightarrow \mathsf{Tr}_{\Sigma_n}(\varphi, y))\right]. \qquad \square$$

We call $\mathsf{Tr}_{\Sigma_n}(x, y)$ the *universal* Σ_n formula. Similarly, for each $n > 0$ there is a universal Π_n formula $\mathsf{Tr}_{\Pi_n}(x, y)$.

Corollary 1.11.3 *For any nonstandard model M and any $n < \omega$*

$$\{\varphi \in \Sigma_n : M \models \varphi\} \in \mathrm{SSy}(M). \qquad \square$$

Corollary 1.11.4 *Let $M \models \mathsf{PA}^*(\mathcal{L})$ be nonstandard. Let $p(v, \bar{b})$ be a type coded in M. If there is $n < \omega$ such that all formulas in $p(v, \bar{b})$ are Σ_n and $p(v, \bar{b})$ is finitely realizable in M, then $p(v, \bar{b})$ is realized in M.* □

The existence of universal Σ_n truth formulas implies immediately that the arithmetical hierarchy is proper. It follows from Corollary 1.11.2 and Tarski's theorem on the undefinability of truth (Do it!).

The following simple exercise has applications in Chapter 6.

Exercise 1.11.5 Let M be a model of $\mathsf{PA}^*(\mathcal{L})$. Then for $n > 1$, every Δ_{n+1}-definable subset of M is Δ_2-definable in $(M, \mathsf{Tr}_{\Sigma_n}(M))$.

Let M be a model of $\mathsf{PA}^*(\mathcal{L})$. The set $\mathsf{Tr}_{\Sigma_n}(M)$ is called the Σ_n-*complete* set of M.

1.12 Arithmetized Completeness Theorem

Let M be a model, let $\mathcal{L} \subseteq M$ be a definable language in the sense of M and let $T \subseteq M$ be a definable $\mathcal{L}$-theory such that $M \models \mathrm{Con}(T)$. Then Henkin's construction performed in M gives a definable $\mathcal{L}$-structure $\mathfrak{A}$ and a Δ_2-formula H which defines in M a full satisfaction class for $\mathfrak{A}$ making all sentences of T true. In particular, for all standard formulas $\varphi(x_0, \ldots, x_{n-1})$ of $\mathcal{L}$ and all $a_0, \ldots, a_{n-1} \in \mathfrak{A}$

$$\mathfrak{A} \models \varphi(a_0, \ldots, a_{n-1}) \Longleftrightarrow M \models H(\varphi, \langle a_0, \ldots, a_{n-1} \rangle),$$

hence $\mathfrak{A}$ is a model of $T \cap \mathbb{N}$. This discussion is an outline of the proof the following nonstandard version of the Hilbert–Bernays Completeness Theorem:

> THE ARITHMETIZED COMPLETENESS THEOREM: Let M be a model. If $T \in \mathrm{Def}(M)$ and $M \vdash \mathrm{Con}(T)$, then there is a model $\mathfrak{A}$ having a definable full satisfaction class S such that $T \subseteq S$ and, moreover, S is Δ_2 in T.

The Arithmetized Completeness Theorem is a convenient tool for constructing end extensions of models of PA. In order to obtain end extensions which are models of PA, we need to know that for every nonstandard model M, there is a $C \in \mathrm{Def}(M)$ such that $\mathsf{PA} \subseteq C$ and $M \models \mathrm{Con}(C)$. A Theorem of Mostowski [135] implies that such a set C always exists.

A theory T is *reflexive* if $T \vdash \mathrm{Con}(T')$ for every finite $T' \subseteq T$; T is *essentially reflexive* if every consistent extension S of T in the same language is reflexive. Mostowski [135] proved that PA is essentially reflexive. The proof gives a stronger principle which we will need in the next chapter.

> MOSTOWSKI'S REFLECTION PRINCIPLE: For any model M of $\mathsf{PA}^*(\mathcal{L})$ for a finite $\mathcal{L}$ and any $n < \omega$, $M \models \forall \sigma[\mathsf{Tr}_{\Sigma_n}(\sigma) \longrightarrow \mathrm{Con}(\sigma)]$.

Let M be a model and let $C \in \mathrm{Def}(M)$ be such that $\mathsf{PA} = C \cap \mathbb{N}$ and $M \models \mathrm{Con}(C)$. Then let $\mathrm{ACT}(M, C)$ be the model obtained by applying the Arithmetized Completeness Theorem to M and C. There is always a choice for $\mathrm{ACT}(M, C)$, depending on the selection of the formulas defining the universe $\mathrm{ACT}(M, C)$ and its full satisfaction class. For applications discussed in this book, those differences do not matter.

By induction, M is isomorphic (via a definable isomorphism) to a cut of $\mathrm{ACT}(M, C)$ (DO IT!). We identify M with this cut. The following properties of $\mathrm{ACT}(M, C)$ are not difficult to prove (DO IT!). (HINT FOR PART (1): use Corollary 1.9.5.)

Theorem 1.12.1 *Let C be a definable subset of a model M such that $\mathsf{PA} \subseteq C$ and $M \models \mathrm{Con}(C)$. Let $N = \mathrm{ACT}(M, C)$. Then:*

(1) *N is not an elementary extension of M.*
(2) *If M is nonstandard, then N is recursively saturated.*
(3) *For every $A \in \mathrm{Def}(N)$, $A \cap M \in \mathrm{Def}(M)$.* □

The Arithmetized Completeness Theorem is a powerful tool, but it is not used much in this book. One important application is Theorem 2.4.2. A connection between the Arithmetized Completeness Theorem and the Low Basis Theorem is explored further in the chapter on Generics and Forcing.

1.13 Friedman's Embedding Theorem

In this book we are chiefly concerned with elementary extensions; thus considerable amount of material on arbitrary initial segments, nonelementary extensions, and models of fragments of arithmetic are left out. Still, we need to mention one basic result. It is a theorem of Friedman which characterizes fully those models of PA that are initial segments of a given countable model. We formulate the theorem in full generality and we give a sketch of a proof of a special case.

For a class Γ of formulas and a model M, let

$$\mathrm{Th}_\Gamma(M) = \{\varphi \in \Gamma : M \models \varphi\}.$$

Recall that $M \prec_{\mathsf{end},\Gamma} N$ means that for all $\varphi(\bar{x}) \in \Gamma$ and all $\bar{a}$ in M, $M \models \varphi(\bar{a})$ iff $N \models \varphi(\bar{a})$.

Theorem 1.13.1 (Friedman's Embedding Theorem) *Let M and N be countable models of $\mathsf{PA}^*(\mathcal{L})$ for a finite $\mathcal{L}$. The following are equivalent:*

(1) *There is a cut $I \prec_{\mathsf{end},\Sigma_n} N$ such that $M \cong I$;*
(2) *$\mathrm{SSy}(M) = \mathrm{SSy}(N)$ and $N \models \mathrm{Th}_{\Sigma_{n+1}}(M)$.*

Proof Let us sketch the proof of the case of $n = 0$. Let K be an elementary end extension of M, and let $a \in K \setminus M$. Using the Σ_1 universal truth formula Tr_{Σ_1}, and the fact that $N \models \mathrm{Th}_{\Sigma_1}(K)$, one can show that there is $b \in N$ such that $\mathrm{Th}_{\Sigma_1}(K, a) = \mathrm{Th}_{\Sigma_1}(N, b)$. Then it follows that $(a_K, +, \times) \equiv (b_N, +, \times)$, where $+$ and $\times$ are the graphs of addition and multiplication restricted to a_K and b_N, respectively. Since a satisfaction relation can be defined in M for both these structures, they are recursively saturated. Also, $(a_K, +, \times)$ and $(b_N, +, \times)$

have the same standard system. Now, a back-and-forth argument can be used to show that $(a_K, +, \times) \cong (b_N, +, \times)$, and the result follows. □

Corollary 1.13.2 *Every countable nonstandard model of* PA *is isomorphic to a proper cut of itself.* □

One can use the universal Σ_n formulas to derive the full version of Theorem 1.13.1 from the special case we proved (Do it!).

Friedman's Embedding Theorem has many refinements and extensions for fragments of arithmetic and other theories. One of its consequences is the following result, due to Jensen & Ehrenfeucht [60].

Theorem 1.13.3 *Each nonstandard model has* $2^{\aleph_0}$ *many pairwise elementarily inequivalent cuts which are models of* PA.

1.14 Exercises

♣**1.14.1** Let M be a nonstandard model. For any $a \in M$, sup(gap(a)) is the smallest elementary cut of M which contains a, and inf(gap(a)), if nonempty, is the largest elementary cut of M which does not contain a.

♣**1.14.2** If $M \subseteq_{\mathsf{end}} N$ and $X \subseteq M$ is coded in N, then X is a class of M.

♦**1.14.3** Every countable model M has classes which cannot be coded in any end extension of M.

♣**1.14.4** Let Ind(M) be the set of those $X \subseteq M$ for which $(M, X) \models \mathsf{PA}^*$. For every model M, $\mathrm{Def}(M) \subseteq \mathrm{Ind}(M) \subseteq \mathrm{Class}(M)$. For countable models the second inclusion is always proper. For countable models the first inclusion is also always proper, this, however, is a more difficult exercise. A solution is given in Chapter 6. The reader might want to look for another proof (♠).

♦**1.14.5** We say that a sequence $\langle a_n : n < \omega \rangle$ is *recursively definable* in a model M if there are $c \in M$ and a recursive sequence of terms $\langle t_n(x) : n < \omega \rangle$ such that for all $n < \omega$, $a_n = t_n(c)$. A tall model M is recursively saturated iff for every recursively definable sequence $\langle a_n : n < \omega \rangle$ of M there is $b \in M$ such that for all $n < \omega$, $a_n = (c)_n$.

♣**1.14.6** If a model M has a full, inductive satisfaction class, then $M \models$ Con(PA).

♦**1.14.7** Let S be a Q_e-class for a model M, and suppose that $e' + \mathbb{N} < e$. If $S' = S|_{e'}$ is the restriction of S to $Q_{e'}(M)$ sentences, then (M, S') is recursively saturated.

Hájek [49] defined a model M to be *thrifty* if $\mathrm{SSy}(M) = \mathrm{Rep}(\mathrm{Th}(M))$. Clearly, every elementary end extension of a prime model is thrifty.

♦**1.14.8** There are thrifty models which are not elementary end extensions of their prime elementary submodels.

The following exercises give an alternative proof of Ehrenfeucht's Lemma using graph coloring.

Definition 1.14.9 $G = (V, E)$ is a definable (0-definable) graph in a model M, if G is a graph whose set of vertices V and set of edges E are definable (0-definable) in M. The chromatic number of G in M, $\chi_M(G)$ is the smallest $k \in M$ for which there exists a definable k-coloring of G, that is a definable $\alpha : V \longrightarrow \{0, \ldots, k-1\}$ such that for all $\{a, b\} \in E$, if $a \neq b$, then $\alpha(a) \neq \alpha(b)$. If G has no k-coloring for any $k \in M$, then $\chi_M(G) = \infty$. □

♦**1.14.10** The chromatic number of a definable graph is well-defined. Moreover, if G is 0-definable in a model M and $\chi_M(G) = k$, then $k \in \mathrm{Scl}(0)$ and there exists a 0-definable k-coloring of G.

♦**1.14.11** If $G = (V, E)$ is a 0-definable graph in a model M such that for some $\{a, b\} \in E$, $a \neq b$, and $\mathrm{tp}(a) = \mathrm{tp}(b)$, then for each $n < \omega$, $\chi_M(G) > n$.

♦**1.14.12** Derive Ehrenfeucht's Lemma as a corollary of the preceding two exercises, that is prove that if $a, b \in M$, $a \neq b$, and $t(a) = b$ for some Skolem term t, then $\mathrm{tp}(a) \neq \mathrm{tp}(b)$. (HINT: consider the graph (M, E), where xEy iff $t(x) = y$.)

1.15 Remarks & References

The material summarized in this chapter covers many decades of work. It would be a separate project (and an interesting one too) to give the full account of who did what, when, and why. We limit ourselves to a few comments and a list of references. Smoryński's articles [196] and [197] are informative and entertaining surveys covering most of the material in this chapter.

Gaifman's Splitting Theorem was proved by Gaifman in [44]. It should be noted that the proof of Theorem 1.3.3 uses the Matiyasevich–Robinson–Davis–Putnam Theorem in an essential way. In fact, one can prove that Theorem 1.3.3 is in a sense equivalent to MRDP (see [70] Exercise 7.6). Theorem 1.3.7 was proved independently by Kotlarski [109] and Schmerl [164].

While MRDP is provable in PA (see [46] for a proof in $I\Sigma_0 + \exp$), MRDP is not valid for some models of PA*. There are an $X \subseteq \mathbb{N}$ and a bounded formula of $\mathcal{L}_{\mathsf{PA}} \cup \{X\}$ which is not equivalent in $(\mathbb{N}, X)$ to any existential formula. An example was given by Michael Weiss in his unpublished Ph.D. thesis [208].

A hierarchy of classes with regard to codability in end extensions of countable models was defined and studied by Kossak & Paris [105]. In particular, it is shown there that every countable model has classes which can be coded in some end extensions but not in any elementary end extensions.

Scott sets were introduced by Scott [180] who proved that countable Scott sets are exactly the families of sets of natural numbers representable in completions of PA. Scott sets are also called c-closed families. Standard systems were defined by Friedman [41]. They proved to be of fundamental importance in the study of nonstandard models of first-order and second-order arithmetic and set theory. Sometimes the standard system of a model M is referred to as the set of reals of M. Scott sets are important in the study of Turing degrees of nonstandard models. This topic is not dealt with in this book. For a comprehensive survey see [88]. Theorem 1.5.3 was generalized by Knight & Nadel [85] who proved that every Scott set of cardinality $\aleph_1$ is the standard system of a model of PA. It is a long standing open problem whether in ZFC one can prove that every Scott set is the standard system of a nonstandard model.

The equivalence (2) $\Longleftrightarrow$ (3) in Proposition 1.8.3 is due to Kirby & Paris [83].

The gap terminology was adopted in [75] motivated by some results of Kotlarski from [111].

The Blass–Gaifman Lemma was proved independently by Blass [12] and Gaifman [45]. Ehrenfeucht's Lemma is from [28].

Recursive saturation is one of the main themes in this book. The concept was introduced and studied by Barwise & Schlipf in [5] and [6]. Similar (in fact equivalent) notions were considered in explicit or implicit forms by others: Jensen & Ehrenfeucht [60], Wilmers in his Ph.D. thesis [209] and in [213], and Hamid Lessan in his Ph.D. thesis [121]. For the developments discussed in this book articles of Schlipf [160], Jensen & Ehrenfeucht [60], and a series of Smoryński's articles starting with [193] were the most influential. In particular, Smoryński introduced the notion of ascending sequences of skies, which we renamed to ascending sequences of gaps. The characterization of recursively saturated models in Theorem 1.10.1 was proved by Smoryński & Stavi in [199].

Resplendency was defined by Barwise and Schlipf. It is a robust notion which can be defined in many alternative ways. See [193] and the chapter on recursive saturation in [77] for historical notes and further discussion. Theorems 1.9.3 and 1.10.2 are due to Smoryński, the first having been proved in [193], and the second in [192].

Basic model theory of recursively saturated models and their standard systems can be viewed as a special case of the theory of recursively saturated models of rich theories. A first-order theory T in a recursive language is *rich* iff there is a recursive sequence $\langle \varphi_n(x) : n < \omega \rangle$ of formulas such that for all disjoint finite sets $X, Y \subseteq \omega$,

$$T \vdash \exists x \big[\bigwedge_{n \in X} \varphi_n(x) \wedge \bigwedge_{m \in Y} \neg \varphi_m(x) \big].$$

Then for a recursively saturated model M of T, the standard system of M is defined as the collection of sets of the form $\{n < \omega : M \models \varphi_n(a)\}$ for $a \in M$. See [196] or [71] for details.

The study of satisfaction classes of nonstandard models was initiated by A. Robinson in [157]. The theme was picked up by Geiser [47], but it was not until much later when satisfaction classes became an independent object of study and a model-theoretic tool. The first systematic study was done by Krajewski [117].

Theorem 1.9.6 is due to Smoryński & Stavi [199]. Their method also shows that cofinal extensions preserve $\aleph_0$-saturation. This cannot be improved to $\aleph_1$-saturation, as shown by Kotlarski in [109]. On the other hand, every *simple* cofinal extension of an $\aleph_1$-saturated model is $\aleph_1$-saturated. If a model is not recursively saturated, then none of its elementary end extensions is recursively saturated. It turns out, however, that there are models which are not recursively saturated but have simple cofinal extensions which are. The study of such models has led to a discovery of a whole spectrum of weak notions of saturation, which were studied in detail by Kaufmann & Schmerl in [68] and [69] . The second paper proves, among many other results, that, assuming $\diamondsuit$, there is a model which is not recursively saturated, but every proper, simple cofinal extension of which is $\aleph_1$-saturated. Exercise 1.14.5 is from [69].

Kotlarski & Kaye in [76] study extensions of the form $\mathrm{ACT}(M, C)$. In particular, they consider and partially answer the question: given $a \in M$, is there a $C \subseteq M$ such that a is definable in $\mathrm{ACT}(M, C)$?

Friedman's theorem was proved by Friedman in [41], the proof sketched in this chapter is due to Dimitracopoulos & Paris [26]. The result has been generalized by Tanaka [203] who proved that every countable nonstandard model of WKL_0 is isomorphic to a proper cut of itself. Theorem 1.13.3 is due to Jensen & Ehrenfeucht [60].

2

EXTENSIONS

Where do nonstandard models arithmetic come from? By Tennenbaum's Theorem, we cannot hope to construct such models by effective algebraic methods. Instead, we usually start with one model M, given, for example, by the Completeness Theorem, and then we either analyze submodels generated by various subsets of M, or we build extensions. In this chapter we discuss elementary extensions. By Gaifman's Splitting Theorem each elementary extension splits into a cofinal extension followed by an elementary end extension. We discuss methods of constructing such extensions. We give a proof of the theorem of MacDowell and Specker saying that every model of PA^* in a countable language has an elementary end extension. We also discuss minimal and superminimal extensions and use them to construct Jónsson models of PA^*. In general, models of PA^* do not amalgamate easily; we give an example of it in Section 2.3, followed by several positive results on amalgamations involving conservative extensions. In particular we give a proof of a theorem of Blass characterizing conservative extensions in terms of amalgamations. We finish the chapter with a brief discussion of nonelementary extensions. *Recall that in this and in the following chapters, unless otherwise stated, a model means a model of* PA^* *in a countable language.*

2.1 Simple extensions

A *simple* extension is an extension generated by one element. More formally: if $M \prec N$, we say that N is a *simple* extension of M if there is $a \in N$ such that $N = \mathrm{Scl}(M \cup \{a\})$.

Suppose that $M \prec N$, $a \in N$, and $N = \mathrm{Scl}(M \cup \{a\})$. Suppose further that a realizes the type $p(x) \in S(M)$. Then we write $N = M(a)$ and say that $M(a)$ is a *$p(x)$-extension* of N. Notice that Ehrenfeucht's Lemma 1.7.2 implies that a is the only element of N that realizes the type $p(x)$.

Recall that M is *short* if M is a cofinal extension of a simple extension of its prime submodel, and M is *tall* iff it is not short.

We begin with a rather unexpected property of short models. It is used in later chapters.

Theorem 2.1.1 *Suppose that M is a short model and M_1, M_2 are cofinal, elementary submodels of M. Then $M_1 \cap M_2$ is a cofinal submodel of M.*

Proof Let $d \in M$ be such that $\mathrm{Scl}(d)$ is cofinal in M. Since each of M_1 and M_2 is an elementary cofinal submodel of the short model M, there are elements

$a \in M_1$ and $b \in M_2$ such that $a, b > d$. Then $\mathrm{Scl}(a)$ and $\mathrm{Scl}(b)$ are cofinal in M. Hence, there is a Skolem term $t(x)$ such that $M \models b < t(a)$. By Blass–Gaifman Lemma 1.7.1, there is a Skolem term $s(x)$ such that $M \models a < b \leq s(a) = s(b)$. Then for $c = s(a) \in \mathrm{Scl}(a) \cap \mathrm{Scl}(b)$. Thus, $\mathrm{Scl}(c) \preccurlyeq \mathrm{Scl}(a) \cap \mathrm{Scl}(b) \preccurlyeq M_1 \cap M_2$, and since $s(c) \geq a$ and $\mathrm{Scl}(a)$ is a cofinal substructure of M, then $\mathrm{Scl}(c)$ is also a cofinal substructure of M. □

Every finitely generated model is a simple extension of its prime submodel. Every model M can be represented as $M = \bigcup_{\alpha<\kappa} M_\alpha$, for $\kappa = |M|$, where $\{M_\alpha : \alpha < \kappa\}$ is a continuous chain, and $M_{\alpha+1}$ is a simple extension of M_α, for each $\alpha < \kappa$, and M_0 is the prime submodel of M. Hence, to understand how models of arithmetic are built, we need to take a look at simple extensions.

A typical construction of a simple extension is based on an inductive construction of a nonprincipal type. Usually this is done by defining some "largeness" property and then proving appropriate theorems about this property. When doing this, you think about the proofs in the standard model (i.e. in the real world) and the proofs routinely carry over. Often, but not always, the important feature of this approach is the definition of "large" and not the lemmas which are often easy.

To construct the required type, we first enumerate all formulas of $\mathcal{L}(M)$ with one variable, $\varphi_0(v)$, $\varphi_1(v), \ldots$. Here we need the language and the model M to be countable. We select a definable large subset X_0 of M, and we inductively construct a descending sequence of large definable subsets of M, $X_0 \supseteq X_1 \supseteq \cdots$, in such a way that, for every n, either $X_{n+1} \subseteq \varphi_n(M)$ or $X_{n+1} \subseteq \neg\varphi_n(M)$. The sequence $X_0, X_1, \ldots$ determines a type in $S_1(M)$, namely the type

$$p(v) = \{\varphi(v) : \exists n < \omega \; [X_n \subseteq \varphi(M)]\}.$$

If $M(a)$ is the p-extension of M, then for every $\varphi(v) \in \mathcal{L}(M)$ we have

$$M(a) \models \varphi(a) \quad \text{iff } \exists n < \omega \; [X_n \subseteq \varphi(M)].$$

This allows one to determine some properties of $M(a)$ by the appropriate choice of the sequence $X_0, X_1, \ldots$. For example, if for some $n < \omega$, the set X_n is bounded in M, then $p(v)$ is bounded, and it follows that $M \prec_{\mathsf{cof}} M(a)$ (Do it!).

If each set X_n is unbounded in M and $M(a)$ is the p-extension of M, then $M < a$. This, however, is not enough to conclude that $M(a)$ is an end extension of M. To guarantee this, for every term $t(v)$ of $\mathcal{L}(M)$, if $t(a)$ is not in M, then we must have $t(a) > M$. This can be achieved by defining $X_0, X_1, \ldots$ so that each Skolem term t is either constant or unbounded on some set X_n (Do it!). Moreover, since every unbounded definable subset of a model M can be partitioned into two disjoint definable unbounded sets, we can obtain continuum different types $p(v)$, for which $M(a)$ is an elementary end extension of M.

Let us summarize the above discussion with a proposition.

Proposition 2.1.2 *Every countable model has an elementary end extension.* □

Proposition 2.1.2 is also true for uncountable models and the idea behind the proof is similar; however, instead of enumerating all formulas of $\mathcal{L}(M)$, we enumerate only formulas $\varphi(v, w)$ of $\mathcal{L}$. The construction of p still proceeds in ω steps. In each step we make decisions about the nth formula and *all* parameters from M. For each $\varphi(v, w)$ this is done by formal induction on parameters. This result is known as the MacDowell–Specker Theorem. The full proof is given later in this chapter.

2.1.1 *Minimal extensions*

Recall that if $M \prec N$, then N is a minimal extension if for all K, if $M \preccurlyeq K \preccurlyeq N$, then either $K = M$ or $K = N$. By Gaifman's Splitting Theorem, every minimal extension is either a cofinal or end extension. Also, notice that every minimal extension is simple (Do it!).

All constructions of minimal extensions are based on the next lemma and its variations.

Lemma 2.1.3 *Suppose that X is a definable unbounded subset of a model M and $t(v)$ is a Skolem term. Then there is an unbounded definable Y such that $Y \subseteq X$ and t is either constant or one-to-one on Y.*

Proof If there is $c \in M$ such that $t^{-1}(c) \cap X$ is unbounded, then let $Y = t^{-1}(c) \cap X$ for such c. Otherwise we define $Y = \{a_i : i \in M\}$ inductively in M. We let $a_0 = \min X$, and for $i \in M$, we let $a_{i+1} = \min\{x \in X : \forall j \leq i(t(x) \neq t(a_j))\}$. □

Corollary 2.1.4 *Every countable model has a minimal elementary and extension.*

Proof Using Lemma 2.1.3 define a descending sequence $X_0, X_1, \ldots$ of definable unbounded subsets of M such that, for each Skolem term t of $\mathcal{L}(M)$, there is $n < \omega$ such that t is either constant or one-to-one on X_n. If p is the type determined by $X_0, X_1, \ldots$, then the p-extension of M is a minimal elementary end extension of M. □

In the proof above "large" means unbounded. In the proof that every countable model has a minimal cofinal extension, "large" just means "of nonstandard size."

The following is a variant of Lemma 2.1.3. The proof is left to the reader.

Lemma 2.1.5 *Suppose that X is a definable bounded subset of a model M and $\mathrm{card}^M(X) = a$, and let $t(v)$ be a Skolem term. Then there is a bounded definable Y such that $Y \subseteq X$, $(\mathrm{card}^M(Y))^2 \geq a$, and t is either constant or one-to-one on Y.* □

The next corollary can be obtained from Lemma 2.1.5 in the same way as Corollary 2.1.4 is obtained from Lemma 2.1.3 (Do it!).

Corollary 2.1.6 *Every countable nonstandard model has a minimal cofinal extension.* □

Let N be an extension of a model M. Then N is a *conservative* extension of M if $\mathrm{Cod}(N/M) \subseteq \mathrm{Def}(M)$. In Chapter 6 we show that every countable model has undefinable inductive subsets. If M is countable and $X \subseteq M$ is inductive, then, as we have shown in Proposition 2.1.2, (M, X) has an elementary end extension. If $(M, X) \prec_{\mathsf{end}} (N, Y)$, then X is coded in N. From this we obtain the following result.

Corollary 2.1.7 *Every countable model has an elementary end extension which is not conservative.* □

As we show in Subsection 2.2.2, Corollary 2.1.7 does not generalize to the uncountable case.

For a direct example of an undefinable class coded in an elementary end extension we can apply the following argument. Let M be a countable model, and let $X = \{a_n : n < \omega\}$, where $\langle a_n : n < \omega\rangle$ is an increasing sequence cofinal in M. Clearly, X is a class of M. By the Compactness Theorem, there is a model N such that $M \prec N$ and there is $a \in N$ such that $(a)_n = a_n$, for all $n < \omega$. The extension splits into $M \preccurlyeq_{\mathsf{cof}} K \preccurlyeq_{\mathsf{end}} N$, so X is an undefinable class of K coded in N.

Theorem 2.1.8 *Let M be a countable model and suppose $X \subseteq M$ is coded in some elementary end extension of M. Then X is coded in some minimal elementary end extension of M.*

Proof Let the subset X of M be coded in an elementary end extension N. We will construct another extension of M by defining a descending sequence $\langle T_n : n < \omega\rangle$ of unbounded subsets of $2^{<M}$ (= the set of M-finite 0–1 sequences). For $a \in M$, let $\chi_a \in M$ be the code of the characteristic function of $X \cap a_M$, that is for $i < a$, $\chi_a(i) = 1$ iff $i \in X$.

We say that $T \subseteq 2^{<M}$ is large if

$$\forall a \in M \exists \sigma \in T\ (\chi_a \subseteq \sigma).$$

Claim. If a definable T is large and, for some $e \in M$, $f : T \longrightarrow e_M$ is a definable function, then there is $i < e$ such that $f^{-1}(i)$ is large.

To prove the claim, suppose to the contrary that

$$(M, X) \models \forall i < e \exists x \forall \sigma \in T\ (\chi_x \subseteq \sigma \longrightarrow f(\sigma) \neq i).$$

Then in (M, X) we have

$$\forall i < e \exists x, \tau [\forall j < x(\tau(j) = 1 \wedge \ell(\tau) = x \longleftrightarrow j \in X) \wedge \forall \sigma \in T\ (\tau \subseteq \sigma \longrightarrow f(\sigma) \neq i)].$$

Now, let $a \in N$ be such that $X = M \cap a_N$. Then in N, for each $d > M$,

$$\forall i < e \exists x, \tau < d [\forall j < x(\tau(j) = 1 \wedge \ell(\tau) = x \longleftrightarrow j \in a) \wedge \forall \sigma \in T(\tau \subseteq \sigma \longrightarrow f(\sigma) \neq i)].$$

By underspill, the same statement must be true in N for some $d \in M$, and this contradicts our assumption that T is large.

Equipped with the claim, now we construct $\langle T_n : n < \omega \rangle$, starting with $T_0 = 2^{<M}$, so that, for a given enumeration of all Skolem terms $\langle t_n : n < \omega \rangle$ of $\mathcal{L}(M)$, all sets T_n are large and t_n is either one-to-one or constant on T_{n+1}. Suppose we have a large T_n. For $t = t_n$ define the sequence σ_i by induction in M. Let $\sigma_0 = \min(T_n)$ and

$$\sigma_{i+1} = \begin{cases} \min\{\sigma \in T_n : \chi_i \subseteq \sigma \wedge \forall j < i\ t(\delta) \neq t(\delta_j)\} & \text{if such } \sigma \text{ exists,} \\ \sigma_0 & \text{otherwise.} \end{cases}$$

If $T = \{\delta_i : i \in M\}$ is unbounded, then it is large and $t \restriction T$ is one-to-one. In this case we set $T_{n+1} = T$. Otherwise, there is an $i_0 \in M$ such that

$$\forall \sigma \in T \big(\chi_{i_0} \subseteq \sigma \longrightarrow t(\sigma) \in \{t(\sigma_j) : j < i_0\} \big).$$

Since $T' = \{\sigma \in T : \chi_{i_0} \subseteq \sigma\}$ is large, by the claim, there is $j_0 < i_0$ such that $\{\sigma \in T' : t(\sigma) = t(\sigma_{j_0})\}$ is large, and we declare this set to be T_{n+1}.

If p is the type determined by $\langle T_n : n < \omega \rangle$, then $M(p)$ is a minimal elementary extension of M and the element realizing p in M codes X. □

Corollary 2.1.9 *Every countable model has a minimal elementary end extension which is not conservative.*

Proof Combine Corollary 2.1.7 with Theorem 2.1.8. □

2.1.2 *Superminimal extensions*

We say that the extension $M \prec N$ is *superminimal* if for each $a \in N \setminus M$, $N = \mathrm{Scl}(a)$. The next lemma is the key to constructing superminimal extensions.

Lemma 2.1.10 *Let X be an unbounded definable subset of a model M, and let $a \in M$. Then there are a Skolem term $t(x)$ and $Y \subseteq X$ which is unbounded and definable such that $M \models \forall x[x \in Y \longrightarrow t(x) = a]$.* □

Proof Note that, while the definitions of X and Y might have parameters from M, the Skolem term $t(x)$ is required to be parameter-free.

Let $\varphi(u,x)$ be a formula and $d \in M$ be such that $\varphi(d,x)$ defines X in M. We proceed rather informally.

For any u, let X_u be the set defined by $\varphi(u,x)$. In particular, $X = X_d$. We define two Skolem terms $t_1(y)$ and $t_2(y)$. Let $t_1(y)$ be the yth element of the set of pairs $\{u : X_u \text{ is unbounded}\} \times M$. (This is well-defined since, in particular, X_d is unbounded.) We next define $t_2(y)$ using formal induction. Letting $t_1(y) = \langle u, z\rangle$, we let $t_2(y)$ be the least x such that $x \in X_u$ and $x > t_2(y')$ for all $y' < y$.

Now let X'_u be the set defined by the formula

$$\exists y \exists z [t_1(y) = \langle u, z\rangle \wedge x = t_2(y)] \ .$$

The following facts are easy to check:

(1) if $X'_u \neq \emptyset$, then X'_u is unbounded;

(2) $X'_u \subseteq X_u$;

(3) if X_u is unbounded, then $X'_u \neq \emptyset$;

(4) if $u \neq v$, then $X'_u \cap X'_v = \emptyset$.

It follows from (1) and (3) that X'_d is unbounded. It follows from (4) that there is a Skolem term $t_3(x)$ which denotes the unique u (if such a u exists) such that $x \in X'_u$.

Now we define Y to be the set of all those $y \in X'_d$ such that y is the $\langle a, z\rangle$th element of X'_d for some z. Then $Y \subseteq X$ by (2). We now define the Skolem term $t(x)$ so that $t(x)$ is the unique w (if such a w exists) such that for some z, x is the $\langle w, z\rangle$th element of the set $X'_{t_3(x)}$. The reader can verify now, that Y and $t(x)$ have the required properties (DO IT!). □

Before we apply the lemma to superminimal extensions, let us note a corollary which is of independent interest.

Corollary 2.1.11 *If X is a definable unbounded subset of a model M, then* $\mathrm{Scl}(X) = M$. □

Theorem 2.1.12 *Every countable model has a superminimal elementary end extension.*

Proof Let M be a countable model. We construct a descending sequence $X_0, X_1, \ldots$ of definable unbounded subsets of M as in the construction of a minimal end extension of M. However, now we begin with an enumeration $\langle (t_n, a_n) : n < \omega\rangle$ of all pairs of the form (t, a), where t is a unary Skolem term and $a \in M$. For a given X_n, first we find an unbounded definable $Z \subseteq X_n$ on which t_n is either one-to-one or constant. If t_n is constant on Z, then, we use Lemma 2.1.10 to get an unbounded definable $Y \subseteq Z$ and a parameter-free Skolem term $t(x)$ such that, for all $x \in Y$, $t(x) = a_n$, and we set $X_{n+1} = Y$. Otherwise, we let Y be an unbounded subset of $t_n(Z)$ such that for some Skolem term t, given by Lemma 2.1.10, $t(x) = a_n$ for all $x \in Y$. Then we let $X_{n+1} = t_n^{-1}(Y) \cap X_n$.

Let p be the type determined by $X_0, X_1, \dots$, and let $M(a)$ be the p-extension of M with a realizing p. Notice that the construction guarantees that $M \subseteq \mathrm{Scl}(a)$ and that $M(a)$ is a minimal extension of M. We also made sure that, for all $b \in M(a) \setminus M$, $M \subseteq \mathrm{Scl}(b)$. Now let b be an element of $M(a) \setminus M$. Since the extension is minimal $a = t(b, c)$, for some $c \in M$. Since $c \in \mathrm{Scl}(b)$, this implies that $a \in \mathrm{Scl}(b)$. Hence $N = \mathrm{Scl}(b)$ as required. □

This theorem reappears in Chapter 4 as Lemma 4.3.1 with some additional frills.

A first-order structure $\mathfrak{A}$ is a *Jónsson* model if $\mathfrak{A}$ is infinite and has no proper elementary submodels of cardinality $|\mathfrak{A}|$.

Recall that, for a cardinal number κ, we say that a linearly ordered structure $(\mathfrak{A}, <, \dots)$ is *κ-like* if $|\mathfrak{A}| = \kappa$ and every proper initial segment of $(\mathfrak{A}, <)$ has cardinality smaller than κ.

Corollary 2.1.13 *Every countable model of* PA^* *has an ω_1-like elementary end extension which is Jónsson.*

Proof Let M_0 be a countable model. For each $\alpha < \omega_1$, let $M_{\alpha+1}$ be a superminimal elementary end extension of M_α, and let $M_\lambda = \bigcup_{\alpha<\lambda} M_\alpha$, for each limit ordinal $\lambda < \omega_1$. Let $N = \bigcup_{\alpha<\omega_1} M_\alpha$. Then N is the Skolem hull of each of its unbounded subsets (DO IT!). If $K \prec N$ and $|K| = |N| = \aleph_1$, then K is unbounded, so the result follows. □

2.1.3 *Greatest common initial segments*

If M is a submodel of N, then their *greatest common initial segment*, $\mathrm{GCIS}(M, N)$, is the set

$$\{x \in N : \forall y\ (y \le x \longrightarrow y \in M)\}.$$

If $M \prec N$, then N is an end extension of M iff $\mathrm{GCIS}(M, N) = M$. What can be said about $\mathrm{GCIS}(M, N)$ in other cases? This is settled by the following two results.

Proposition 2.1.14 *If $M \prec N$ then* $\mathrm{GCIS}(M, N)$ *is a cut which is closed under multiplication.*

Proof Clearly, $I = \mathrm{GCIS}(M, N)$ is a cut. To show that it is closed under multiplication, suppose $a^2 \in N \backslash I$. Then there is $b \le a^2$ such that $b \notin M$. There are $d, r \in N$ such that $b = ad + r$, $d \le a$, and $r < a$. Since $b \notin M$, one of a, d, r is not in M, implying that $a \notin I$. □

If $I \subseteq_{\mathsf{end}} M \prec N$, then we say that N is an *I-extension* of M if $I \subseteq \mathrm{GCIS}(M, N)$.

Lemma 2.1.15 *Suppose that $I \subseteq_{\mathsf{end}} M$ and a type p is determined by a decreasing sequence of M-finite sets $X_0, X_1, \ldots$ such that*

(1) *for each $n < \omega$, $\mathrm{card}^M(X_n) > I$;*
(2) *for each Skolem term $t(x)$ of $\mathcal{L}(M)$ there is $n < \omega$ such that $t(x)$ is either constant or one-to-one on X_n.*

Then $M(p)$ is a minimal I-extension of M.

Proof Let a realize p in $M(p)$. Suppose the contrary that for some $b \in I$ and some $c \in M(p) \setminus M$, $c < b$. Then there is $n < \omega$ and some term $t(x)$ such that $t(a) = c$ and $t(x) < b$ for all $x \in X_n$. Then, by (2), $t(x)$ is one-to-one on some $X_m \subseteq X_n$, which implies that $\mathrm{card}^M(X_m) \in I$, a contradiction. □

Theorem 2.1.16 *Suppose that M is countable and $I \subseteq_{\mathsf{end}} M$ is a cut closed under multiplication. Then M has a minimal elementary extension N such that $I = \mathrm{GCIS}(M, N)$.* □

Proof We will define a descending sequence $X_0, X_1, \ldots$ of of M-finite sets dovetailing two constructions. The first is the one used to construct minimal extensions. The second makes sure that the new elements in the extension occur arbitrarily low above I. The conditions to satisfy are:

(1) each term of $\mathcal{L}(M)$ is either constant or one-to-one on some X_n;
(2) $\inf\{\mathrm{card}^M(X_n) : n < \omega\} = I$.

Let $\langle a_n : n < \omega \rangle$ be a decreasing sequence of elements of M whose infimum is I. Use Lemma 2.1.5 and the fact that if $a^2 > I$, then $a > I$ to define $X_0, X_1, \ldots$ so that (1) holds and, for each $n < \omega$, $I < \mathrm{card}^M(X_n) < a_n$.

Let p be the complete type determined by $X_0, X_1, \ldots$, and let $N = M(a)$ be the p-extension of M.

If $b \in M \backslash I$, then there is $c \in N \backslash M$ such that $c < b$. Indeed, let $X = X_n$ be such that $\mathrm{card}^M(X) < b$. Let $t(x) = \mathrm{card}\{y \in X : y < x\}$. Let $c = t(a)$. Then $c < \mathrm{card}^M(X) < b$. Also, $c \notin M$ since $t(x)$ is one-to-one on X. This shows that $\mathrm{GCIS}(M, N) \subseteq I$.

By Lemma 2.1.15, $I \subseteq \mathrm{GCIS}(M, N)$, and the result follows. □

Corollary 2.1.17 *Let M be a countable model, and let I be a cut of M which is closed under multiplication. Then M has an elementary extension N such that $\mathrm{GCIS}(M, N) = I$, and, for each $a > I$, $|a_N| > \aleph_0$.*

Proof Iterate Theorem 2.1.16 $\aleph_1$ times. □

Two stronger results concerning greatest common initial segments, due to Paris & Mills [148], are discussed in Chapter 3.

2.2 The MacDowell–Specker Theorem

Rapid development of the model theory of Peano Arithmetic in the 1960's started with the theorem of MacDowell and Specker. This theorem says that every model of PA has an elementary end extension. For MacDowell and Specker this theorem was a lemma in the study of additive groups of nonstandard models. What is the additive group of a model M? This group is obtained from M by adding the negative elements $\{-n : n \in M\}$ and defining addition in the extended structure in the natural way. Let G_M be the additive group of the model M. Let a be an element of G_M. For each $1 < n < \omega$ there is a unique residue $0 \leq r_n < n$ such that $a = qn + r_n$, for some q. The elements whose all residues $r_n = 0$ form a subgroup of G_M which is often called the divisible part of G_M. The group G_M can be represented uniquely as $D \oplus R$, where D is the divisible part of G_M and R is the reduced part. MacDowell and Specker were interested in the relation between cardinalities of R and D. If M is a model of PA, the cardinality of R is the cardinality of the standard system of M. Every end extension of M has the same standard system as M, and by the MacDowell–Specker theorem every model of PA has an elementary end extension of every cardinality greater than or equal to $|M|$. Hence, for every infinite cardinal $\kappa \leq 2^{\aleph_0}$ and every cardinal $\lambda \geq \kappa$, there is a model of PA such that $|R| = \kappa$ and $|D| = \lambda$.

For us the theorem of MacDowell and Specker is a point of departure for further analysis of possible end extensions of models of PA. We begin with an important definition.

Definition 2.2.1 Let T be a completion of PA* in a possibly uncountable language $\mathcal{L}$. A 1-type $p(v)$ of T is *definable* if for each $\mathcal{L}$-formula $\varphi(u, v)$, there is an $\mathcal{L}$-formula $\sigma_\varphi(u)$ such that for any constant Skolem term t,

$$\varphi(t, v) \in p(v) \Longleftrightarrow T \vdash \sigma_\varphi(t).$$

We call the map $\varphi \mapsto \sigma_\varphi$ a *defining scheme* of $p(v)$.

Gaifman introduced the notion of a definable type over a model of a first-order theory in [45]. In our terminology we can say that $p \in S_1(M)$ is definable over M if p is a definable type of $\mathrm{Th}((M, a)_{a \in M})$. In particular, if T is a completion of PA*, then a type $p(v)$ of T is definable iff it is definable over the prime model M_T.

The concept of definability extends naturally to n-types. In the context of PA it suffices to consider 1-types, as the type of any tuple $v_0, \ldots, v_{n-1}$ can be replaced the type of the code $\langle v_0, \ldots, v_{n-1} \rangle$. For the same reason, we only consider single variable u in $\varphi(u, v)$.

Proposition 2.2.2 *Let T be a complete theory extending* PA*, *and let $p(v) \in S_1(T)$ be a definable type. Then for any model M of T, $p(v)$ has an extension to a definable type $p'(v) \in S_1(M)$.*

Proof Let $\varphi \mapsto \sigma_\varphi$ be a defining scheme of $p(v)$. Let $p'(v)$ be the collection of all $\varphi(b, v)$, with $b \in M$ such that $M \models \sigma_\varphi(b)$. We will show that $p'(v) \in S_1(M)$. Clearly, $p'(v)$ is complete. To show that it is finitely realizable, suppose $\varphi_0(b_0, v), \varphi_1(b_1, v), \ldots, \varphi_{n-1}(b_{n-1}, v)$ are in $p'(v)$. Since $p(v)$ is finitely realizable, we have

$$T \vdash \forall u_0, \ldots, u_{n-1} \big[\bigwedge_{i<n} \sigma(u_i) \longrightarrow \exists v \bigwedge_{i<n} \varphi_i(u_i, v) \big].$$

Since $M \models \bigwedge_{i<n} \sigma(b_i)$ and $M \models T$, there is $b \in M$ such that $M \models \bigwedge_{i<n} \varphi_i(u_i, b)$; hence $p'(v)$ is finitely realizable. □

Since every definable $p(x) \in S_1(T)$ has a canonical extension to a definable type in $S_1(M)$, for every model M of T, we will let $p_M(x)$ be this canonical extension. Then instead of referring to $M(a)$ as a $p_M(x)$-extension of M, we can unambiguously refer to it as a $p(x)$-extension of M.

Theorem 2.2.3 *Every completion of* PA^* *has nonprincipal definable types.* □

Before we give formal details, let us outline the idea of the proof. The construction is flexible, and it can be refined in many ways. Many different kinds of definable types are discussed in detail in Chapter 3.

Let T be a completion of $\mathsf{PA}^*(\mathcal{L})$. To prove the theorem, we work in the prime model M of T. Let $\langle \varphi_n(u, v) : n < \omega \rangle$ be an enumeration of all formulas of the language of $\mathcal{L}$ in the variables shown. We construct a sequence of definable sets $X_0 \supseteq X_1 \supseteq \cdots$ such that, for each n, X_n is eventually a subset of or is disjoint from every subset of M defined by $\varphi_n(b, v)$. Then for each $n < \omega$, the formula $\sigma_n(u)$ decides for which parameters b the set X_n is contained in the set defined by $\varphi_n(b, v)$ and for which it is disjoint. Then the definable type $p(v)$ is defined as

$$\{\varphi_n(b, v) : b \in M \text{ and } M \models \sigma_n(b)\}.$$

One can begin the construction with the additional requirement that a particular formula defining an unbounded subset of M is in $p(v)$. This proves that the set of definable types is dense, in fact comeager, in the set of nonprincipal types of $S_1(T)$.

The construction is based on the following main lemma. For this lemma we do not need to assume that the language is countable.

Lemma 2.2.4 *Let D be an unbounded definable subset of M, and let $\varphi(u, v)$ be a formula of the language of M in the variables shown. Then there is an* *unbounded definable $E \subseteq D$ such that in M*

$$\forall u\big[\exists w \forall v > w(v \in E \longrightarrow \varphi(u,v)) \vee \exists w \forall v > w(v \in E \longrightarrow \neg\varphi(u,v))\big]. \qquad \square$$

Proof To simplify notation, let us assume that $D = M$, the general case being derivable from the special case. Let $\varphi(u,v)$ be given. For $i \in M$, let $X_i = \{x : M \models \varphi(i,x)\}$, and let $Y_i = M \setminus X_i$. By formal induction, define a sequence of sets of indices $\langle I_i : i \in M\rangle$, with $I_i \subseteq \{0,\ldots,i\}$ and the set $E = \{e_i : i \in M\}$ as follows.

$$I_0 = \begin{cases} \{0\} & \text{if } X_0 \text{ is unbounded,} \\ \emptyset & \text{otherwise.} \end{cases}$$

$$e_0 = \begin{cases} \min(X_0) & \text{if } 0 \in I_0, \\ \min(Y_0) & \text{otherwise.} \end{cases}$$

Then for $i > 0$,

$$I_i = \begin{cases} I_{i-1} \cup \{i\} & \text{if } \bigcap\{X_j : j \in I_{i-1}\} \cap X_i \text{ is unbounded,} \\ I_{i-1} & \text{otherwise.} \end{cases}$$

$$e_i = \begin{cases} \min(\bigcap\{X_j : j \in I_{i-1}\} \cap X_i) \setminus \{e_0,\ldots,e_{i-1}\}) & \text{if } i \in I_i, \\ \min(\bigcap\{X_j : j \in I_{i-1}\} \cap Y_i) \setminus \{e_0,\ldots,e_{i-1}\}) & \text{otherwise.} \end{cases}$$

Notice that for all i and all sufficiently large $a \in E$

$$M \models \varphi(i,a) \longleftrightarrow i \in I_i,$$

which implies that E has the required property. $\square$

Now we can give a proof of Theorem 2.2.3.

Proof Let $\langle \varphi_n(u,v) : n < \omega\rangle$ be an enumeration of all $\mathcal{L}$-formulas in the variables shown. Let M be the prime model of T. Let X_0 be the set E in Lemma 2.2.4 for $D = M$ and $\varphi = \varphi_0$. Suppose that for $n \geq 0$ an unbounded definable set X_n has been defined. Then X_{n+1} is the E of the lemma for $D = X_n$ and $\varphi = \varphi_{n+1}$. For each $n < \omega$ we can define

$$\sigma_n(u) = \exists w \forall v > w(v \in X_n \longrightarrow \varphi_n(u,v)).$$

The definable type determined by $\langle \sigma_n(u) : n < \omega \rangle$ is

$$\{\varphi_n(b, v) : b \in M \wedge M \models \sigma_n(b)\},$$

the defining scheme being $\varphi_n \mapsto \sigma_n$. □

Corollary 2.2.5 *There are nonprincipal definable types over every model M of* PA^*.

Proof Apply Theorem 2.2.3 to $T = \mathrm{Th}(M)$, and use Proposition 2.2.2. □

For later applications, let us note another immediate corollary.

Corollary 2.2.6 *Let T be a completion of* PA^* *in a finite language. There are nonprincipal definable types of T which are recursive in T.* □

We now show how Theorem 2.2.3 implies the MacDowell–Specker Theorem. The conclusion we obtain is slightly stronger.

Recall that N is a conservative extension of M if, for every $X \in \mathrm{Def}(N)$, $X \cap M \in \mathrm{Def}(M)$. Hence, N is a conservative extension of M iff, for every $a \in N$, the type of a over M is a definable type of $\mathrm{Th}((M, x)_{x \in M})$ (DO IT!).

Proposition 2.2.7 *Every conservative extension is an end extension.*

Proof Let c be an element of $N \setminus M$. Since N is a conservative extension, the set $X = \{x \in M : N \models x < c\}$ is a definable cut of M; hence $X = M$. □

Theorem 2.2.8 (MacDowell–Specker Theorem) *Every model has a conservative elementary end extension.*

Proof The theorem follows immediately from Theorem 2.2.3 and Proposition 2.2.7, since every extension generated by a definable type is conservative (DO IT!). □

Corollary 2.2.9 *For every model M and every cardinal $\kappa > |M|$, M has a κ-like elementary end extension.* □

The MacDowell–Specker Theorem applies in a natural way to countable models of ACA_0.

Corollary 2.2.10 *Let $\mathfrak{X}$ be a countable family of subsets of M and suppose that $(M, \mathfrak{X})$ be a model of* ACA_0. *Then there is N such that $M \prec_{\mathsf{end}} N$ and $\mathfrak{X} = \mathrm{Cod}(N/M)$.*

Proof Let $M(\mathfrak{X})$ be the expansion of M obtained by adding all sets $X \in \mathfrak{X}$ as new relations. Then $M(\mathfrak{X})$ is a model of PA^* in a countable language. If N is the reduct to the language of M of a conservative elementary end extension of $M(\mathfrak{X})$, then $\mathfrak{X}$ is the family of subsets of M coded in N. □

The next proposition has a simple direct proof (DO IT!).

Proposition 2.2.11 *If a realizes a definable type over a model M and b realizes a definable type over $M(a)$, then $\langle a, b\rangle$ realizes a definable type over M.* □

In the next subsections we give three applications of the MacDowell–Specker Theorem.

2.2.1 *Superminimal conservative extensions*

The inductive construction of the definable type $p(v)$ in the proof of Theorem 2.2.3 can be dovetailed with other constructions of unbounded types. In particular, by combining the proofs of Theorems 2.1.12 and 2.2.3 we get the following corollary (DO IT!):

Corollary 2.2.12 *Every countable model has a superminimal conservative extension.* □

This result reappears in Chapter 4 with some additional frills as Lemma 4.3.1.

In Chapter 6 we prove that every countable model has inductive undefinable subsets. We also give a proof of a theorem of Simpson saying that every countable model M has an expansion (M, X) to a prime model of PA^*.

Theorem 2.2.13 *If M is a cofinal extension of a prime model, then M has a countable elementary end extension N such that, for every undefinable class X of N, (N, X) is prime.*

Proof Let N be a superminimal conservative extension of M, and let X be an undefinable class of N. Then $X \cap M$ is definable in M. Suppose that $X \cap M = \varphi(M, b)$ for some $\mathcal{L}(M)$ formula $\varphi(v, b)$. Let $c \in \mathrm{Scl}(0)$ be such that $b < c$. If $b_1 \in M$ is such that $X \cap M = \varphi(M, b_1)$, then $\varphi(N, b) = \varphi(N, b_1)$. Hence the set

$$\{z \in N : (N, X) \models \exists y < c \forall x < z[x \in X \longleftrightarrow \varphi(x, y)]\}$$

is bounded in M, because X is undefinable, and it has a maximum in N, because X is a class. Then $a > M$, and, by superminimality, $N = \mathrm{Scl}(a)$. Since a is definable in (N, X), this finishes the proof. □

2.2.2 *Rather classless models*

Although not mentioned explicitly, the stronger version of the MacDowell–Specker Theorem, the one involving conservative extensions, is implicit in the original proof of MacDowell and Specker. In fact, it would be hard to prove that every model of PA has an elementary end extension without proving that every model has a conservative elementary end extension. The reason is that there are models all of whose end extensions to models of PA are conservative. This is a property of the rather classless models: a model M is *rather classless* if every class of M is definable.

We show now that every model has a rather classless elementary end extension. This follows from the MacDowell–Specker Theorem and the next result.

A sequence of models $\langle M_\nu : \nu < \alpha \rangle$, where α is an ordinal, is a *continuous elementary end chain*, if $M_\nu \prec_{\mathsf{end}} M_\mu$, for $\nu < \mu < \alpha$, and $M_\lambda = \bigcup_{\nu<\lambda} M_\nu$, for limit ordinals $\lambda < \alpha$.

Theorem 2.2.14 *Suppose α is a limit ordinal, $\mathrm{cf}(\alpha) > \aleph_0$, and $\langle M_\nu : \nu < \alpha \rangle$ is a continuous elementary chain such that, for some stationary set $I \subseteq \alpha$, and for all $\nu \in I$, $M_{\nu+1}$ is a conservative elementary end extension of M_ν. Then $N = \bigcup_{\nu<\alpha} M_\nu$ is rather classless.*

Proof Since $\langle M_\nu : \nu < \alpha \rangle$ can be replaced with a continuous subchain indexed by $\nu < \mathrm{cf}(\alpha)$, without loss of generality, we can assume that α is an uncountable regular cardinal and that I consists of limit ordinals. Let X be a class of N. We will show that X is definable in M. For $\nu < \alpha$, let $X_\nu = X \cap M_\nu$. For each $\nu \in I$, X_ν is a definable subset of M_ν. Let $f : I \longrightarrow \alpha$ be defined by

$$f(\nu) = \min\{\beta : X_\nu \text{ is definable in } M_\nu \text{ with parameters from } M_\beta\}.$$

Since all ν in I are limit ordinals, for all $\nu \in I$, $f(\nu) < \nu$. Hence, by Fodor's Lemma, there exists $\beta < \alpha$ such that $f^{-1}(\beta)$ is a stationary subset of I. Let β be such an ordinal, and let ν_0 be the smallest ordinal in $f^{-1}(\beta)$.

Let $\varphi_0(v)$ be a formula with parameters from M_β defining X_{ν_0} in M_{ν_0}. Let $\nu > \nu_0$ be an element of $f^{-1}(\beta)$, and let $\varphi(v)$ be a formula with parameters from M_β defining X_ν in M_ν. Since $\varphi_0(v)$ and $\varphi(v)$ define the same subset of M_{ν_0}, we have $M_{\nu_0} \models \forall v(\varphi_0(v) \longleftrightarrow \varphi(v))$. Since $M_{\nu_0} \prec M_\nu$, the same equivalence holds in M_ν, which shows that $\varphi_0(v)$ defines X_ν in M_ν, and thus, from the fact that $f^{-1}(\beta)$ is unbounded in α, it follows that $\varphi_0(v)$ defines X in N, and the result follows. □

Corollary 2.2.15 *If $\kappa > |M|$ and $\mathrm{cf}(\kappa) > \aleph_0$, then M has a κ-like, rather classless, elementary end extension.* □

2.2.3 *Ramsey's Theorem in* ACA_0

If X is an infinite subset of ω, $a < \omega$, and $F : [X]^n \longrightarrow [0, a]$ is a coloring of the set of increasing n-tuples from X using $a + 1$ colors, then there is an infinite $Y \subseteq X$ which is homogeneous for F, which means that F assigns the same color to all tuples from $[Y]^n$. This is the statement of Ramsey's Theorem. The formalized Ramsey's Theorem asserts that if X as above is definable, then we can find an arithmetically definable homogeneous Y. Not all proofs of Ramsey's adapt in a straightforward way to give the formalized version. We give a direct proof of a stronger statement, of which the formalized Ramsey's Theorem is a special case.

Theorem 2.2.16 *Ramsey's Theorem is provable in* ACA_0.

Proof Let $(M, \mathfrak{X})$ be a model of ACA_0. The proof is by induction on $n < \omega$. The case of $n = 1$ is straightforward. So, let us assume that the theorem is true for n-tuples, and let $a \in M$, and $X, F \in \mathfrak{X}$, where X is unbounded and $F : [X]^{n+1} \longrightarrow [0, a]$ is a coloring. Then $(M, X, F) \models \mathsf{PA}^*$. Let (M', X', F') be a conservative elementary end extension of (M, X, F). Pick $b \in X' \setminus M$. In (M', X', F') we define a sequence σ of elements of M' by formal induction: let $\sigma_0 = \min(X)$ and for $k > 0$ let σ_k be the least $x > \sigma_{k-1}$ such that $x \in X'$ and

$$\forall (i_0, \ldots, i_{n-1}) \in [k]^n \; F'(\sigma_{i_0}, \ldots, \sigma_{i_{n-1}}, x) = F'(\sigma_{i_0}, \ldots, \sigma_{i_{n-1}}, b).$$

Let D' be the range of σ, and let $D = D' \cap M$. We claim that D is unbounded in M. Indeed, if not, then D would have a maximum element σ_k in M, but then, since $(M, X, F) \prec (N, X', F')$, one can show that $\sigma_{k+1} \in M$ (Do it!). Since D is coded in N, it is definable in (M, X, F). Now let us consider a coloring of $G' : [D']^n \longrightarrow [0, a]$ defined (in N) by $G'(x_0, \ldots, x_{n-1}) = F'(x_0, \ldots, x_{n-1}, b)$. Let G be the restriction of G' to M. Then G is a coloring of $[D]^n$ which is definable in (M, X, F). By the inductive assumption, there is $Z \in \mathfrak{X}$ which is homogeneous for G. One can easily check that Z is homogeneous for F as well. $\square$

The formalized Ramsey's Theorem is referred to later as the formalized *Infinite Ramsey Theorem*, or IRT. The formalized version of the well-known *Finite Ramsey Theorem*, FRT, is provable in PA. FRT is

$$\forall l, m, n \exists k \forall f : [k]^n \longrightarrow l \exists X \subseteq k \; (|X| = m \wedge |f([X]^n)| = 1).$$

FRT can be proved in PA by formalizing one of its direct proofs.

2.3 Amalgamations

First, we define what is meant by an amalgamation. Let M_0, M_1, M_2 be three models, and let $e_i : M_0 \longrightarrow M_i$ $(i = 1, 2)$ be elementary embeddings. Then an *amalgamation* (of this set-up) consists of a model M_3 and elementary embeddings $f_i : M_i \longrightarrow M_3$ $\;(i = 1, 2)$ such that $f_1 e_1 = f_2 e_2$, M_3 is generated by $f_1(M_1) \cup f_2(M_2)$ and $f_1 e_1(M_0) = f_1(M_1) \cap f_2(M_2)$. If $f_i' : M_i \longrightarrow M_3'$ $\;(i = 1, 2)$ is another amalgamation, then say that the two amalgamations are *isomorphic* if there is an isomorphism $h : M_3 \longrightarrow M_3'$ such that $f_1' = h f_1$ and $f_2' = h f_2$. In practice, when considering amalgamations, some or all of the embeddings e_1, e_2, f_1, f_2 are likely to be just identity maps. If they all are, then it must be that $M_0 = M_1 \cap M_2$. Unless stated otherwise, the convention when considering amalgamations is that all the embeddings are identity maps; thus, $M_1, M_2 \prec M_3$ and $M_0 = M_1 \cap M_2$. We then say that M_3 is an amalgamation of M_1 and M_2 over

M_0. Up to isomorphic amalgamations, there is no loss in generality in adopting this convention.

Models of PA do not, in general, amalgamate well.

Theorem 2.3.1 *Let M_0 be a countable recursively saturated model, and let $X \subseteq \omega$. Then M_0 has elementary end extensions M_1 and M_2 such that $M_0 \cong M_1 \cong M_2$, and whenever M_3 is an amalgamation of M_1 and M_2, then $X \in \mathrm{SSy}(M_3)$.*

Proof The proof uses minimal types. (See Section 10.2 for background material.) Let $p(x)$ be a minimal type realized in M_0, and let $\langle a_n : n < \omega \rangle$ be an increasing cofinal sequence of elements realizing $p(x)$. Then there are a countable recursively saturated model $M_1 \succ_{\mathsf{end}} M_0$ and an element $a \in M_1$ coding the sequence $\langle a_n : n < \omega \rangle$ and such that $M_1 \models \forall x, y[x < y < \ell(a) \longrightarrow (a)_x < (a)_y]$. Thus, $(a)_n = a_n$ for each $n < \omega$. Let x_i be the ith element of X, and let $\langle b_i : i < \omega \rangle$ be the subsequence where $b_i = a_{x_i}$. Similarly, there are a countable recursively saturated model $M_2 \succ_{\mathsf{end}} M_0$ and an element $b \in M_2$ coding the sequence $\langle b_i : i < \omega \rangle$ such that $M_2 \models \forall x, y[x < y < \ell(b) \longrightarrow (b)_x < (b)_y]$. Clearly (see Section 1.8), $M_0 \cong M_1 \cong M_2$. (It is even possible to arrange that $(M_1, a) \cong (M_2, b)$.)

Now let M_3 be an amalgamation of M_1 and M_2 over M_0. Then (Do it!)

$$X = \{n < \omega : M_3 \models \exists i[(a)_n = (b)_i]\} \in \mathrm{SSy}(M_3),$$

as required. $\square$

Neither of the extensions $M_0 \prec M_1$ and $M_0 \prec M_2$ constructed in the previous proof are conservative. In Theorems 2.3.2 and 2.3.3 we see that amalgamations can be better behaved when one of the extensions is conservative. Theorem 2.3.4 then characterizes conservative extensions in terms of amalgamations.

Theorem 2.3.2 *Suppose $M_0 \prec M_1$, $M_0 \prec M_2$ and M_2, is a conservative extension of M_0. Then there is an amalgamation M_3 of M_1 and M_2 over M_0 such that M_3 is a conservative extension of M_1. Furthermore, if $M_3' \succ_{\mathsf{end}} M_1$ is any amalgamation of M_1, M_2 over M_0, then M_3 and M_3' are isomorphic amalgamations.*

Proof Since the extension $M_0 \prec M_2$ is conservative, for each $\mathcal{L}(M_2)$-formula $\varphi(x)$, there is an $\mathcal{L}(M_0)$-formula $\sigma_\varphi(x)$ such that for each $a \in M_0$,

$$M_2 \models \varphi(a) \quad \text{iff } M_0 \models \sigma_\varphi(a).$$

Let T be the $\mathcal{L}(M_1 \cup M_2)$-theory consisting of all sentences $\varphi(b)$, where $\varphi(x)$ is an $\mathcal{L}(M_2)$-formula and $b \in M_1$ are such that $M_1 \models \sigma_\varphi(b)$. There are three important examples of sentences in T.

If φ is an $\mathcal{L}(M_2)$-sentence for which $M_2 \models \varphi$, then $\varphi \in T$, since $\sigma_\varphi(x)$ is such that $M_0 \models \forall x \sigma_\varphi(x)$ (DO IT!). If $\varphi(x)$ is an $\mathcal{L}(M_0)$-formula and $b \in M_1$ are such that $M_1 \models \varphi(b)$, then $\varphi(b) \in T$. For, since $M_0 \models \forall x[\sigma_\varphi(x) \leftrightarrow \varphi(x)]$, it follows that $M_1 \models \sigma_\varphi(b)$. The third example is that the sentence $b < c$ is in T whenever $b \in M_1$ and $c \in M_2 \backslash M_0$ (DO IT!).

We claim that T is a consistent and complete $\mathcal{L}(M_1 \cup M_2)$-theory.

For consistency, we consider the prototypical case of a single sentence $\varphi(b)$ in T. Then $M_1 \models \sigma_\varphi(b)$, so that $M_0 \models \exists x \sigma_\varphi(x)$. Let $a \in M_0$ be such that $M_0 \models \sigma_\varphi(a)$. Then $M_2 \models \sigma_\varphi(a)$, so that also $M_2 \models \varphi(a)$, thereby proving the consistency of $\varphi(b)$.

For completeness, consider an $\mathcal{L}(M_2)$-formula $\varphi(x)$ and an element $b \in M_1$. If $M_2 \models \sigma_\varphi(b)$, then $\varphi(b)$ is in T. If $M_2 \models \neg\sigma_\varphi(b)$, then since $M_0 \models \forall x[\sigma_{\neg\varphi}(x) \longleftrightarrow \neg\sigma_\varphi(x)]$, we get that $\neg\varphi(b)$ is in T.

Now let M_3 be the prime model of T. From the three examples of sentences in T, we get that $M_1 \prec M_3$, $M_2 \prec M_3$, and $M_1 = M_2 \cap M_3$. To complete the proof of the existence part of the theorem, we still need to show that M_3 is a conservative extension of M_1. Consider $D \in \mathrm{Def}(M_3)$, and suppose that $\varphi(x, b, c)$ is a formula defining D, where $\varphi(x, y, z)$ is an $\mathcal{L}$-formula and $b \in M_1$ and $c \in M_2$. Let $\sigma_\varphi(x, y)$ be an $\mathcal{L}(M_0)$-formula such that whenever $a, a' \in M_0$, then

$$M_2 \models \varphi(a', a, c) \quad \text{iff } M_0 \models \sigma(a', a).$$

We show that $\sigma_\varphi(x, b)$ defines $D \cap M_1$ in M_1. If $b' \in M_1$, then $b' \in D \Longleftrightarrow M_2 \models \varphi(b', b, c) \Longleftrightarrow \varphi(b', b, c) \in T \Longleftrightarrow M_1 \models \sigma_\varphi(b', b)$.

We next prove the uniqueness part of the theorem. Let M_3' be as in the theorem. It is enough to show that if $M_3' \models \varphi(b)$, where $\varphi(x)$ is an $\mathcal{L}(M_2)$-formula and $b \in M_1$, then $\varphi(b) \in T$. Let $\sigma_\varphi(x)$ be as before. In M_3', define d to be the least such that $M_3' \models \varphi(d) \leftrightarrow \sigma_\varphi(d)$. (Of course, there may not be such a d at all, but then things are even simpler.) Clearly, $d \in M_2$ since it is definable from parameters in M_2, and $d > M_0$ since $M_0 \prec_{\mathsf{end}} M_2$. But then, $d > M_1$ since $M_1 \prec_{\mathsf{end}} M_3'$ (DO IT!). Therefore, $M_1 \models \sigma_\varphi(b)$ and then $\varphi(b) \in T$. □

Theorem 2.3.3 *Suppose $M_0 \prec M_1$, $M_0 \prec M_2$, and M_1 is a conservative proper extension of M_0. Suppose there is an amalgamation M_3 of M_1 and M_2 over M_0 such that M_3 is an elementary end extension of M_1. Then M_3 is a conservative extension of M_0.*

Proof Suppose that $D \in \mathrm{Def}(M_3)$, intending to show that $D \cap M_0 \in \mathrm{Def}(M_0)$. Let $a \in M_1 \backslash M_0$, and then let $D_0 = D \cap a_{M_3}$. Then $D_0 \in \mathrm{Def}(M_1)$ since $M_1 \prec_{\mathsf{end}} M_3$, and then $D_0 \cap M_0 \in \mathrm{Def}(M_0)$ since M_1 is a conservative extension of M_0. Clearly, $D \cap M_0 = D_0 \cap M_0$, so $D \cap M_0 \in \mathrm{Def}(M_0)$. □

Theorems 2.3.2 and 2.3.3 together yield the following consequence characterizing conservative extensions. The MacDowell–Specker is also needed in the proof to get a conservative extension of M_1.

Theorem 2.3.4 *Suppose $M_0 \prec M_2$. Then M_2 is a conservative extension of M_0 iff whenever $M_0 \prec M_1$, there is an amalgamation M_3 of M_1 and M_2 over M_3 such that $M_1 \prec_{\mathsf{end}} M_3$.* □

The proof of the next theorem uses an argument similar to one in the proof of Theorem 2.3.2.

Theorem 2.3.5 *Suppose $M_0 \prec_{\mathsf{cof}} M_1$ and $M_0 \prec_{\mathsf{end}} M_2$. Then there is an amalgamation M_3 of M_1 and M_2 over M_0 such that $M_1 \prec_{\mathsf{end}} M_3$ and $M_2 \prec_{\mathsf{cof}} M_3$.*

Proof Let T be the $\mathcal{L}(M_1 \cup M_2)$-theory consisting of all sentences $\varphi(b)$, where $\varphi(x)$ is an $\mathcal{L}(M_2)$-formula and $b \in M_1$ are such that there are $a < d \in M_0$ such that:

(1) for all $b' \in a_{M_0}$, $M_2 \models \varphi(b')$ iff $M_0 \models b' \in d$;
(2) $M_1 \models b < a \wedge b \in d$.

There are three important examples of sentences in T.

If φ is an $\mathcal{L}(M_2)$-sentence for which $M_2 \models \varphi$, then $\varphi \in T$. (Do it!). If $\varphi(x)$ is an $\mathcal{L}(M_0)$-formula and $b \in M_1$ are such that $M_1 \models \varphi(b)$, then $\varphi(b) \in T$ (Do it!). The third example is that the sentence $b < c$ is in T whenever $b \in M_1$ and $c \in M_2 \backslash M_0$ (Do it!).

We claim that T is a consistent and complete $\mathcal{L}(M_1 \cup M_2)$-theory.

For consistency, as in the proof of Theorem 2.3.2, we consider the case of a single sentence $\varphi(b)$ in T. Let $a, d \in M_0$ be as in (1) and (2). From (2), it follows that $M_1 \models \exists x[x < a \wedge x \in d]$, so the same sentence is true in M_0. Let $b' \in M_0$ be a witness for this sentence. Then from (1) it follows that $M_2 \models \varphi(b')$, thereby proving the consistency of $\varphi(b)$.

For completeness, consider an $\mathcal{L}(M_2)$-formula $\varphi(x)$ and an element $b \in M_1$. Let $a, d \in M_0$ be such that $b < a$ and (1) holds. If (2) also holds, then $\varphi(b) \in T$. If (2) does not hold, then let $d' \in M_0$ be such that $M_0 \models \forall x < a[x \in d \leftrightarrow x \notin d']$. Then a, d' demonstrate that $\neg\varphi(b)$ is in T, proving completeness.

Now let M_3 be the prime model of T. From the three examples of sentences in T, we get that $M_1 \prec M_3$, $M_2 \prec M_3$, and $M_1 = M_2 \cap M_3$. In fact, we get that $M_1 \prec_{\mathsf{end}} M_3$ and $M_2 \prec_{\mathsf{cof}} M_3$ (Do it!). □

The similarity of the constructions of the amalgamations in Theorems 2.3.2 and 2.3.5 suggests that there should be a common generalization.

Definition 2.3.6 Suppose $M_0 \prec M_1$ and $M_0 \prec M_2$. Then we say that M_2 is a *conservative extension of M_0 relative to the extension M_1* if, whenever $\varphi(x)$ is an $\mathcal{L}(M_2)$-formula and $b \in M_1$, then there is an $\mathcal{L}(M_0)$-formula $\sigma(x)$ such that

$M_1 \models \sigma(b)$ and either for every $a \in M_0$, $M_0 \models \sigma(a) \Longrightarrow M_2 \models \varphi(a)$ or for every $a \in M_0$, $M_0 \models \sigma(a) \Longrightarrow M_2 \models \neg\varphi(a)$. □

Two of the important instances of this notion occur in Theorems 2.3.2 and 2.3.5 , and nothing more is needed to get the amalgamations therein. That is, in each of the proofs of these theorems, we define T to be the $\mathcal{L}(M_1 \cup M_2)$-theory consisting of all sentences $\varphi(b)$, where $\varphi(x)$ is an $\mathcal{L}(M_2)$-formula and $b \in M_1$, such that there is an $\mathcal{L}(M_0)$-formula $\sigma(x)$ for which $M_1 \models \sigma(b)$ and, for every $a \in M_0$, $M_0 \models \sigma(a) \Longrightarrow M_2 \models \varphi(a)$. This construction then yields the *principal* amalgamation.

Definition 2.3.7 Suppose $M_0 \prec M_1$, $M_0 \prec M_2$, and M_2 is a conservative extension of M_0 relative to M_1. Then we let $M_1 \star M_2$ be the principal amalgamation of M_1 and M_2 over M_0.

In this definition, the reference to M_0 is suppressed, so some context is necessary to interpret $M_1 \star M_2$ unambiguously.

Exercise 2.3.8 Suppose M_1, M_2, M_3 are elementary extensions of M_0, with M_2 and M_3 being conservative extensions. Then $(M_1 \star M_2) \star M_3 \cong M_1 \star (M_2 \star M_3)$.

Exercise 2.5.11 describes another situation in which there are principal amalgamations.

2.4 Nonelementary extensions

If an end extension of a model is not an elementary extension, how nonelementary can it be? Every end extension is Σ_0-elementary. We prove that every model has an end extension which is not Σ_1-elementary. We can require, in addition, that the model and its extension are elementarily equivalent.

Let us begin with a theorem characterizing the complete theories of end extensions of a given model. If $M \subseteq_{\mathsf{end}} N$, then $\mathrm{Th}_{\Pi_1}(N) \subseteq \mathrm{Th}(M)$, and any set represented in $\mathrm{Th}(N)$ must be in $\mathrm{SSy}(M)$ (because it is in $\mathrm{SSy}(N)$). It turns out that these are the only two restrictions. In the proof of this fact we will use the following lemma.

Lemma 2.4.1 *Let M be a model. Let $n < \omega$ and $T \in \mathrm{SSy}(M)$ be a theory such that for some model K, $M \prec_{\Sigma_n} K \models T$. Then there is a model K' such that $M \prec_{\mathsf{end},\Sigma_n} K' \models T$.*

Proof If M is standard, then let $K' = K$. Thus, we assume that M is nonstandard. Suppose that $M \prec_{\Sigma_n} K \models T$. Since $\mathsf{PA} \in \mathrm{SSy}(M)$, without loss of generality, we can assume that $\mathsf{PA} \subseteq T$. Let σ be the conjunction of all sentences in some finite fragment of T. We will show that

$$M \models \forall x[\mathsf{Tr}_{\Pi_n}(x) \longrightarrow \mathrm{Con}(\sigma \wedge x)].$$

Suppose not and let $\varphi \in M$ be a counterexample, that is,

$$M \models \mathsf{Tr}_{\Pi_n}(\varphi) \wedge \neg\operatorname{Con}(\sigma \wedge \varphi).$$

Since $M \prec_{\Sigma_n} K$ and $n > 0$, the same is true in K. Since $K \models \sigma \wedge \mathsf{Tr}_{\Pi_n}(\varphi)$, by Mostowski's Reflection Principle (see p. 19), $K \models \operatorname{Con}(\sigma + \varphi)$, which is a contradiction. Since $T \in \operatorname{SSy}(M)$, there is an M-finite $S \supseteq T$ such that $M \models \operatorname{Con}(S + \mathsf{Tr}_{\Pi_n}(M))$. Then $K' = \operatorname{ACT}(M, S + \mathsf{Tr}_{\Pi_n}(M))$ has the required properties. □

Theorem 2.4.2 *Let $M \models \mathsf{PA}^*(\mathcal{L})$ be a nonstandard model. Suppose that $\mathcal{L}$ is coded in M and let $T \supseteq \mathsf{PA}^*(\mathcal{L})$ be a complete theory. Then M has an end extension N such that $N \models T$ iff the following two conditions hold:*

(1) $\operatorname{Rep}(T) \subseteq \operatorname{SSy}(M)$;
(2) $T \cap \Pi_1 \subseteq \operatorname{Th}(M)$.

Proof Suppose $M \subseteq_{\mathsf{end}} N$ and $N \models T$. Then $\operatorname{Rep}(T) \subseteq \operatorname{SSy}(N) = \operatorname{SSy}(M)$, so (1) holds. Since Σ_0 formulas are absolute with respect to end extensions, (2) holds as well.

Now suppose that (1) and (2) hold.

We define models N_i, inductively so that for each $i < \omega$,

$$N_i \models T \cap \Pi_{i+1} \quad \text{and} \quad N_i \prec_{\mathsf{end},\Sigma_i} N_{i+1}.$$

Let $N_0 = M$ and suppose N_i has been defined. We will obtain N_{i+1} using Lemma 2.4.1, where K is a model of

$$T_0 = T \cup \big(\operatorname{Th}((N_i, a)_{a \in N_i}) \cap \Pi_i\big).$$

To show that T_0 is consistent we will use the familiar elementary diagram argument. Suppose T_0 is inconsistent. Then there are some $\varphi(x) \in \Pi_i$ and $a \in N_i$ such that

$$T \vdash \neg\varphi(a) \text{ and } N_i \models \varphi(a).$$

Since a does not occur in T, we get $T \vdash \forall x \neg\varphi(x)$. Since T is complete, $\forall x \neg\varphi(x) \in T$, and since $\forall x \neg\varphi(x) \in \Pi_{i+1}$, by the inductive hypothesis, $N_i \models \forall x \neg\varphi(x)$. In particular, $N_i \models \neg\varphi(a)$, which is a contradiction.

Let K be a model of T_0. Then $N_i \prec_{\Sigma_i} K \models T \cap \Pi_{i+2}$ and since, by (1), $T \cap \Pi_{i+2} \in \operatorname{SSy}(M) = \operatorname{SSy}(N_i)$ it follows from Lemma 2.4.1 that there is K' such that $N_i \prec_{\mathsf{end},\Sigma_i} K' \models T \cap \Pi_{i+2}$. Let $N_{i+1} = K'$. Then, $N = \bigcup_{i<\omega} N_i$ has the required properties. □

Theorem 2.4.3 *For each nonstandard model M there is N such that $N \equiv M$, $M \subseteq_{\text{end}} N$, but N is not a Σ_1-elementary extension of M.*

Proof It is enough to find a model N such that $M \subseteq_{\text{end}} N$, N is not an Σ_1-elementary extension of M and $\text{Th}(M) \cap \Pi_1 \subseteq \text{Th}(N)$, replacing $N \equiv M$, because then Theorem 2.4.2 can be applied to get the desired model.

Let $\varphi(x)$ be a Σ_1 formula defining a simple set as in Theorem 8.II of [158]. For this formula we have:

$$M \models \forall x \exists y (x \leq y \leq 2x \wedge \neg\varphi(y)), \tag{1}$$

and for any Σ_1 formula $\psi(x)$ either

$$M \models \exists x (\varphi(x) \wedge \psi(x)), \tag{2}$$

or else there is $n < \omega$ such that

$$M \models \forall x (\psi(x) \longrightarrow x \leq n). \tag{3}$$

By (1), there is a nonstandard a such that $M \models \neg\varphi(a)$. Let

$$T_a = \mathsf{PA} + \text{Th}(M) \cap \Pi_1 + \text{Th}(M, a) \cap \Sigma_1 + \varphi(a).$$

By (2) and (3), T_a is consistent. Then, as in the proof of Theorem 2.4.2, by the elementary diagram argument, we see that (M, a) embeds in a model $(K, a) \models T_a$. Hence, $(M, a) \prec_{\Sigma_0} (K, a) \models \text{Th}(M) \cap \Pi_1 + \varphi(a)$. Then, by Lemma 2.4.1, there is a also an end extension (N, a) of (M, a) with the same properties. □

2.5 Exercises

♦**2.5.1** Every countable model which is not prime has a minimal elementary end extension which is not superminimal. (This follows from the results discussed in Chapter 3. The reader is encouraged to look for a direct proof.)

♦**2.5.2** Every countable nonstandard model which is generated by a bounded set of generators has a superminimal cofinal extension.

♦**2.5.3** Let M be a countable model of PA. Find a counterexample to the following "dual splitting theorem": For every extension $M \prec N$ there is K such that $M \preccurlyeq_{\text{end}} K \preccurlyeq_{\text{cof}} N$.

♣**2.5.4** Every Jónsson model of cardinality $\aleph_1$ is either ω_1-like or short.

♦**2.5.5** There are short Jónsson models M of cardinality $\aleph_1$.

♣**2.5.6** There are countable nonstandard models M such that no cofinal extension of M is a simple extension of the prime elementary submodel of M.

♦**2.5.7** Every ω_1-like model has a minimal cofinal extension.

♦**2.5.8** Suppose that $\mathrm{SSy}(M) = \mathcal{P}(\mathbb{N})$. Assuming the continuum hypothesis, show that M has a minimal cofinal extension.

♦**2.5.9** If $|M| = \kappa$, then $\mathrm{Th}((M, a)_{a\in M})$ has $\kappa^{\aleph_0}$ definable types.

♣**2.5.10** If $\kappa \geq |M|+\aleph_1$, then M has a rather classless, elementary end extension of cardinality κ.

♣**2.5.11** Suppose the cut $I \subseteq_{\mathsf{end}} M_0$ is closed under exponentiation and that $M_0 \prec M_1$ and $M_0 \prec M_2$. Suppose further that $I \subseteq \mathrm{GCIS}(M_2, M_0)$ and that M_1 is generated by $M_0 \cup \{x \in M_1 : \text{there is } a \in I \text{ such that } M_1 \models x < a\}$. Then M_2 is a conservative extension of M_0 relative to M_1.

A model M is *short recursively saturated* if M is short and it realizes every bounded recursive type with a finite number of parameters which is finitely realizable in M.

♦**2.5.12** Every short recursively saturated model has a recursively saturated elementary end extension. (HINT: first do it for countable models.)

The *recursively saturated part* of a model M is the set of those $a \in M$ for which there is a recursively saturated model $K \prec M$ such that $a \in K$.

♣**2.5.13** For every model M, the recursively saturated part of M is either empty or is equal to M or is a recursively saturated elementary cut of M.

♥**2.5.14** For every countable recursively saturated model M there is a model N such that the recursively saturated part is M and there is an $a \in N$ coding an increasing sequence of nonstandard length such that $M = \sup\{(a)_n : n < \omega\}$.

♦**2.5.15** Adapt the proof of Theorem 2.4.2 to prove MacDowell–Specker theorem. Sketch: given a model M construct a sequence $N_0, N_1, N_2, \ldots$ such that, for each i, $N_i = ACT(M, T_i)$, for some T_i, and $N_i \prec_{\Sigma_{i+1}} N_{i+1}$ and $M \prec_{\mathsf{end},\Sigma_{i+2}} N_i$.

♦**2.5.16** Let M be any nonstandard model of PA, let T be a completion of PA, and let $n < \omega$. Then M has a Σ_n-elementary end extension N such that $N \models T$ iff the following two conditions hold:

(1) $\mathrm{Rep}(T) \subseteq \mathrm{SSy}(M)$;

(2) $T \cap \Pi_{n+1} \subseteq \mathrm{Th}(M)$.

(HINT: consider the model $(M, \mathsf{Tr}_{\Sigma_n}(M))$.)

♥**2.5.17** For each nonstandard model M and each $n < \omega$, there is N such that $N \equiv M$, $M \prec_{\mathsf{end},\Sigma_n} N$, but N is not a Σ_{n+1}-elementary extension of M.

Stuart Smith [190] defined a subset X of a model M to be *extendible* if, for every elementary extension N of M, there is $Y \subseteq N$ such that $(M, X) \prec (N, Y)$. He showed that the extendible subsets of countable models of PA^* are exactly the definable sets. The proof is outlined in the following four exercises. As indicated by ♥ two of these exercises are not easy. For complete proofs see [190].

A set $X \subseteq M$ is *end extendible* if it is extendible with respect to elementary end extensions, and it is *cofinally extendible* if it is extendible with respect to cofinal extensions.

♣**2.5.18** If X is a class of M, then X is end extendible iff X is definable in M.

♥**2.5.19** $\mathbb{N}$ is not a cofinally extendible subset of any countable nonstandard model.

♥**2.5.20** There are no cofinally extendible cuts in any nonstandard countable models.

♣**2.5.21** The only cofinally extendible subsets of a countable nonstandard model are the inductive sets.

♦**2.5.22** If M and N are nonstandard elementarily equivalent models with the same standard system, then $(M, \mathbb{N}) \equiv (N, \mathbb{N})$. (Hint: show that for each n, the existential player has a winning strategy in the Ehrenfeucht-Fraïssé game of length n involving $(M, \mathbb{N})$ and $(N, \mathbb{N})$. Notice that it follows that $\mathbb{N}$ is end extendible in all nonstandard models.)

♥**2.5.23** Every bounded subset of a countable model is end extendible.

2.6 Remarks & References

The theorem of MacDowell & Specker is from their article [123]. As we commented earlier, MacDowell and Specker were interested in elementary end extensions and their application to models of Presburger arithmetic, but in fact they proved the stronger Theorem 2.2.8 on conservative extensions, even though they did not isolate this notion. Conservative extensions were first recognized by Phillips [150]. Theorem 2.1.1, Corollary 2.1.6, and many other results on conservative extensions, nonconservative extensions and amalgamations were obtained by Blass in several papers [13–15, 17]. Blass' results were formulated mostly for models of *full arithmetic* (i.e. models of $\mathrm{Th}(\mathbb{N}, X)_{X \in \mathcal{P}(\omega)}$.) Nonconservative minimal extensions of the standard model were constructed by Philips [149] and, using another construction, by Potthoff [152]. The ultimate, in a sense, result on nonconservative extensions of countable models is Theorem 2.1.8, proved in [105].

Ali Enayat has been successfully pursuing a program of comparative, PA versus ZF, model theory. In particular, he proved several analogues of the MacDowell-Specker Theorem for models of set theory. See [30], [32], and [33].

The nonamalgamation Theorem 2.3.1 is due to Knight & Nadel [85]. The proof given here is ours. The connections between amalgamation and conservative extensions that appear in Theorems 2.3.2, 2.3.3, and 2.3.2 were studied by Blass [17] but only in the case of full arithmetic.

The existence of Jónsson model was proved independently by Gaifman [45], Knight [87], and Paris [144].

The study of κ-like models of PA^* and their classes begins with Schmerl [161]. Theorem 2.2.14 is from [164].

Rabin [156] proved that every model M of PA has an elementarily equivalent extension which solves a Diophantine equation having coefficients in M but having no solutions in M. Rabin's result predates Matiyasevich's solution to Hilbert's 10th problem from which it followed that an extension is Σ_1 iff it does not solve new Diophantine equations. Gaifman [44] asked if Rabin's theorem could be improved by requiring the extension to be an end extension. Partial answers were given by Manevitz [127] and Wikie [209]. The full answer, given by Theorem 2.4.3 and Theorem 2.4.2, characterizing possible theories of extensions of models of PA, are due to Wilkie [211]. Our proof is from [168] and is essentially the same as the unpublished proof by Lessan [121]. Lemma 2.4.1 is due to McAloon [128].

Theorem 2.2.13 improves upon a theorem of Enayat [31] that every prime model M has an elementary end extension whose every expansion obtained by adding an undefinable class is prime. Recursively saturated parts of models were introduced and studied by Kotlarski [112]. Exercise 2.5.14 is one of the results from [112].

Exercise 2.5.12 is from Smoryński [193] for the countable case and from Kossak [91] for the uncountable case.

3

MINIMAL AND OTHER TYPES

Minimal types, in the context of Peano Arithmetic, were introduced and developed in the 1960s by Gaifman. This chapter discusses these kinds of types and also some other related kinds of types, such as indiscernible, rare, end-extensional and selective types. *Throughout this chapter, we consider completions of* PA*, *which are referred to as* T, *and all types are* 1-*types relative to this theory* T.

3.1 Types related to indiscernibility

Indiscernibles are very important in model theory. In general, one distinguishes between indiscernible *sets* and indiscernible *sequences*. Since all models of PA are linearly ordered, we have no need to make such a distinction. For a model M of PA (or even for any structure M linearly ordered by $<$), a subset $I \subseteq M$ is a *set of indiscernibles* if, whenever $\varphi(x_0, x_1, \ldots, x_{n-1})$ is an n-ary formula and $a_0 < a_1 < \cdots < a_{n-1}$ and $b_0 < b_1 < \cdots < b_{n-1}$ are increasing n-tuples of elements of I, then $M \models \varphi(\bar{a}) \longleftrightarrow \varphi(\bar{b})$. Alternatively, we could avoid the mention of formulas in this definition of indiscernibility by saying instead that the two n-tuples $\bar{a}$ and $\bar{b}$ realize the same n-types. In a general model-theoretic setting, models having indiscernibles are obtained by a combination of the Compactness Theorem and Ramsey's Theorem. We also make use of Ramsey's Theorem as a formal theorem or scheme of theorems of PA*.

3.1.1 *Indiscernible types*

The following definition identifies a certain kind of type which is useful for producing indiscernibles. In this section we take a close look at indiscernible types. Along the way we will be introduced to some other kinds of types such as n-indiscernible, end-extensional, and rare types. The next section considers minimal types, which are precisely the unbounded indiscernible types.

It is a fact from model theory that if $p(x)$ is a nonprincipal type, then there are structures with large indiscernible sets consisting of elements which realize $p(x)$. On the other hand, it could well be that there is a set, each of which realizes the same type $p(x)$, which is not an indiscernible set. This is not possible for indiscernible types.

Definition 3.1.1 The type $p(x)$ is an *indiscernible* type if it is a nonprincipal type such that for any model M, if $I \subseteq M$ is a set of elements each realizing $p(x)$, then I is a set of indiscernibles.

There is a bit of shorthand that is useful when discussing indiscernibles. If $\varphi(x)$ and $\psi(x_0, x_1, \ldots, x_{n-1})$ are formulas, we say that $\varphi(x)$ *forces* $\psi(x_0, x_1, \ldots, x_{n-1})$ if the sentence

$$\forall x_0\, x_1, \ldots, x_{n-1}[x_0 < x_1 < \cdots < x_{n-1}$$
$$\wedge \varphi(x_0) \wedge \varphi(x_1) \wedge \cdots \wedge \varphi(x_{n-1}) \longrightarrow \psi(x_0, x_1, \ldots, x_{n-1})]$$

is a consequence of T. The type $p(x)$ *forces* $\psi(x_0, x_1, \ldots, x_{n-1})$ if some formula $\varphi(x)$ in $p(x)$ does. Thus a nonprincipal type $p(x)$ is indiscernible iff for each formula, $p(x)$ forces it or its negation.

Indiscernible types are complete (Do it!). The first order of business is to show that indiscernible types exist. There are two very similar constructions for doing this, one using Infinite Ramsey's Theorem (IRT) and the other Finite Ramsey's Theorem (FRT). Recall that we say that a type $p(x)$ is *bounded* if there is some constant Skolem term c such that the formula $x \leq c$ is in $p(x)$ and that it is *unbounded* if it fails to be bounded. The first theorem, whose proof makes use of a formalized version of IRT, shows the existence of unbounded indiscernible types, and a later result, Corollary 3.1.6, whose proof uses a formalized version of FRT, shows the existence of bounded indiscernible types.

Theorem 3.1.2 *Suppose $\varphi(x)$ is a formula which defines an unbounded set in some (or, equivalently, every) model of T. Then there is an unbounded indiscernible type $p(x)$ which contains the formula $\varphi(x)$.*

Proof The type $p(x)$ will be constructed inductively. That is, we inductively define a sequence $\langle \varphi_i(x) : i < \omega \rangle$ of formulas each one of which defines an unbounded set, and then let X_i be the set defined in the prime model by the formula $\varphi_i(x)$. (Thus each X_i is a "large" set, which in this proof means that X_i is unbounded.) The sequence $\langle X_i : i < \omega \rangle$ will be a decreasing sequence. We let $\varphi_0(x) = \varphi(x)$. In order to define the rest of the sequence, we need a sequence $\langle \chi_i(x_0, x_1, \ldots, x_{n_i}) : i < \omega \rangle$ consisting of all formulas having free variables as indicated.

Now suppose that we already have $\varphi_i(x)$ and that we wish to define $\varphi_{i+1}(x)$. By making use of a formalization of IRT in PA*, we can find a formula $\varphi_{i+1}(x)$ such that each of the following holds:

(1) $T \vdash \forall w \exists x[x > w \wedge \varphi_{i+1}(x)]$;

(2) $T \vdash \forall x[\varphi_{i+1}(x) \longrightarrow \varphi_i(x)]$;

(3) $\varphi_{i+1}(x)$ forces $\chi_i(x_0, x_1, \ldots, x_{n_i})$ or forces its negation.

This last sentence expresses the essence of IRT: the set X_{i+1} is homogeneous with regard to the partition of $[X_i]^{n_i+1}$ that χ_i defines.

Now let $p(x)$ consist of all those formulas $\theta(x)$ for which there is $i \in \omega$ such that the sentence $\forall x(\varphi_i(x) \longrightarrow \theta(x))$ is a consequence of T. We show that $p(x)$ is an unbounded indiscernible type to which $\varphi(x)$ belongs. Clearly, $\varphi(x)$ is in

$p(x)$ since $\varphi(x) = \varphi_0(x)$. Since $X_0 \supseteq X_1 \supseteq X_2 \supseteq \cdots$ and each of the X_i is nonempty, the set $p(x)$ is consistent with T. Each X_i is even unbounded, so $p(x)$ is an unbounded type.

We show that $p(x)$ is an indiscernible type. For some model $M \models T$, let $I \subseteq M$ be a set of elements each realizing $p(x)$. To show that I is a set of indiscernibles, consider elements $a_0 < a_1 < \cdots < a_n$ and $b_0 < b_1 < \cdots < b_n$ of I and some $(n+1)$-ary formula, say $\chi_i(x_0, x_1, \ldots, x_{n_i})$. Since each of the a_j and b_j satisfy the formula $\varphi_{i+1}(x)$, which is in $p(x)$, it follows from (3) that $M \models \chi_i(a_0, a_1, \ldots, a_{n_i})$ iff $M \models \chi_i(b_0, b_1, \ldots, b_{n_i})$. This proves that I is a set of indiscernibles and, therefore, that $p(x)$ is an indiscernible type. □

REMARK The construction of the type $p(x)$ in the previous theorem can be done effectively in the theory T. This is more easily appreciated when the language $\mathcal{L}$ is finite. There is a uniform way to do this construction. Thus, given a finite $\mathcal{L}$ and formula $\varphi(x)$, we can get a recursive set $\Gamma(x)$ of formulas such that for any completion T of $\mathsf{PA}^*(\mathcal{L})$, there is a unique complete type $p(x) \supseteq \Gamma(x) \cup T$, and this type $p(x)$ has the properties required by Theorem 3.1.2 whenever $\varphi(x)$ defines an unboounded set in a model of T.

The following exercise, an application of indiscernible types, should be contrasted with Ehrenfeucht's Lemma.

Exercise 3.1.3 There exist an unbounded type $p(x)$ and a Skolem term $t(x, y)$ such that whenever M is a model and $a, b \in M$ are distinct elements realizing $p(x)$, then $t(a, b)$ also realizes $p(x)$ and $t(a, b) \notin \{a, b\}$. (HINT: *let $q(x)$ be an indiscernible type, and then utilize the unique* 2*-type $p_2(u, v)$ containing $q(u) \cup q(v) \cup \{u < v\}$.*)

Many constructions are like getting olives out of a jar: the first takes some effort, and the rest just pour out. Indiscernible types are like that.

Theorem 3.1.4 *There are $2^{\aleph_0}$ unbounded indiscernible types.*

Proof For each subset $X \subseteq \omega$, we will construct an unbounded indiscernible type, with different X's resulting in different types. Using the notation from the proof of Theorem 3.1.2, let $\varphi(x)$ be $x = x$. Having $\varphi_i(x)$, let $\varphi'_i(x)$ be the same as $\varphi_{i+1}(x)$ in that proof, and then let $\delta(x)$ be the formula $\varphi'_i(x) \wedge$ "there are an even number of y's such that $y < x \wedge \varphi'_i(y)$." Then let $\varphi_{i+1}(x) = \delta(x)$ if $i \in X$, and let $\varphi_{i+1}(x) = \varphi'_i(x) \wedge \neg\delta(x)$ if $i \notin I$. □

The previous theorem can be given a topological interpretation, a quick statement of which is presented for the cognoscenti. The set $S_1(T)$ of complete 1-types is a topological space. Let $S_1^\infty(T)$ be the closed subset of $S_1(T)$ consisting of the unbounded types. With the subspace topology, $S_1^\infty(T)$ is homeomorphic to Cantor Space, so the Baire Category Theorem applies, implying that each comeager subset of $S_1^\infty(T)$ has cardinality $2^{\aleph_0}$. The topological version of Theorem 3.1.4 is: *the set of unbounded indiscernible types is a comegaer subset of $S_1^\infty(T)$.*

In the next theorem, FRT is used to get indiscernible types that are bounded.

Theorem 3.1.5 *Suppose $\varphi(x)$ is a formula that defines an infinite set in some (or, equivalently, every) model of T. Then there is an indiscernible type $p(x)$ that contains the formula $\varphi(x)$.*

Proof If the formula $\varphi(x)$ defines an unbounded set, then Theorem 3.1.2 can be used. So it suffices to assume that this is not the case and that the formula $x < c$, where c is some constant Skolem term, is implied by $\varphi(x)$.

The proof to follow is just like the proof of Theorem 3.1.2 with IRT being replaced by FRT. The type $p(x)$ will be constructed inductively. That is, we inductively define a sequence $\langle \varphi_i(x) : i < \omega \rangle$ of formulas each one of which defines an infinite (but bounded) set in the prime model; let X_i be the set defined by the formula $\varphi_i(x)$. (In this proof, each X_i is "large" in the sense of being infinite.) The sequence $\langle X_i : i < \omega \rangle$ will be a decreasing sequence. We let $\varphi_0(x) = \varphi(x)$. In order to define the rest of the sequence, we need a sequence $\langle \chi_i(x_0, x_1, \ldots, x_{n_i}) : i < \omega \rangle$ consisting of all formulas having free variables as indicated.

Now suppose that we already have $\varphi_i(x)$ and that we wish to define $\varphi_{i+1}(x)$. By making use of the formalization of FRT in Peano Arithmetic, we can find a formula $\varphi_{i+1}(x)$ such that (2) and (3) in the proof hold. Instead of (1), we require only that X_{i+1} be infinte. This can be accomplished by letting X_{i+1} be the largest among sets satisfying (2) and (3).

Now let $p(x)$ consist of all those formulas $\theta(x)$ for which there is $i \in \omega$ such that the sentence $\forall x(\varphi_i(x) \longrightarrow \theta(x))$ is a consequence of T. Just as in the proof of Theorem 3.1.2, we can see that $p(x)$ is an indiscernible type to which $\varphi(x)$ belongs. □

If T has a standard model, then all bounded types are principal and, consequently, there are no bounded indiscernible types. If T does not have a standard model, then just the opposite is true.

Corollary 3.1.6 *Suppose that T does not have a standard model. If $\varphi(x)$ is a formula which defines an infinite set in some (or, equivalently, every) model of T, then there is a bounded indiscernible type $p(x)$ containing the formula $\varphi(x)$.*

Proof Let c be a constant Skolem term representing a nonstandard element, such that the formula $\varphi(x) \wedge x < c$ defines an infinite set. Apply the previous theorem to this formula to obtain an indiscernible type, which is necessarily bounded. □

Remark The remark following Theorem 3.1.2 about the effectiveness of that theorem applies just as well to Theorem 3.1.5 and Corollary 3.1.6.

The types in Corollary 3.1.6 are also like olives: the conclusion can be strengthened to assert the existence of $2^{\aleph_0}$ bounded indiscernible type containing $\varphi(x)$ (Do IT!).

3.1.2 *n-indiscernible types*

There is a refinement of the notion of an indiscernible type. If $1 \leq n \in \omega$, then we say that the nonprincipal type $p(x)$ is *n-indiscernible* if, whenever $M \models T$ and $I \subseteq M$ is a set of at most $n+1$ elements each realizing $p(x)$, then I is a set of indiscernibles. Trivially, every nonprincipal complete type is 1-indiscernible. It is clear that a type is indiscernible iff it is n-indiscernible for each $n \geq 1$. Also, if $1 \leq m < n$ and $p(x)$ is n-indiscernible, then it is m-indiscernible (Do IT!).

The following two easy exercises give equivalent definitions.

Exercise 3.1.7 The nonprincipal type $p(x)$ is n-indiscernible iff whenever M is a model of T and $a_0 < a_1 < \cdots < a_{n-1}$ and $b_0 < b_1 < \cdots < b_{n-1}$ are elements of M which realize $p(x)$, then the n-tuples $\bar{a}$ and $\bar{b}$ realize the same n-type.

Exercise 3.1.8 The type $p(x)$ is n-indiscernible iff each n-ary formula or its negation is forced by $p(x)$.

In general, we find much more diversity among bounded types than among unbounded ones. This is the case with indiscernible types. For each $n \geq 1$, there are n-indiscernible types which are not $(n+1)$-indiscernible (see Theorem 3.1.10). If $n \geq 2$, these types are necessarily bounded. The main remaining goal of this section is Theorem 3.1.20, which asserts that every unbounded 2-indiscernible type is indiscernible. In the course of proving this we will see that all unbounded indiscernible types are definable. The notions of rare and end-extensional types will be defined, and we will see that all unbounded indiscernible types are rare and end-extensional.

The next theorem states that unbounded 2-indiscernible types are definable. See Exercise 3.6.5, in which there is the easier result that every unbounded 3-indiscernible type is definable.

Theorem 3.1.9 *Every unbounded* 2*-indiscernible type is definable.*

Proof Let $p(x)$ be an unbounded 2-indiscernible type. To show that it is definable, we consider a formula $\varphi(u,x)$ with the intent of showing that there is a *defining* formula $\sigma(u)$; that is, $\sigma(u)$ should have the property that whenever c is a constant Skolem term, then $\varphi(c,x) \in p(x)$ iff $\sigma(c) \in T$. We define the three formulas $\theta_0(x,y), \theta_1(x,y), \theta_2(x,y)$ which are as follows, respectively:

$$\begin{aligned}
&\forall z < x[\varphi(z,x) \longleftrightarrow \varphi(z,y)],\\
&\exists z < x[\forall w < z(\varphi(w,x) \longleftrightarrow \varphi(w,y)) \wedge \varphi(z,x) \wedge \neg\varphi(z,y)],\\
&\exists z < x[\forall w < z(\varphi(w,x) \longleftrightarrow \varphi(w,y)) \wedge \neg\varphi(z,x) \wedge \varphi(z,y)].
\end{aligned}$$

These three formulas induce partitions in the following way: for any model $M \models T$ and any $a, b \in M$ there is exactly one $i \in \{0,1,2\}$ such that $M \models \theta_i(a,b)$. We are interested in this partition only when restricted to the set of pairs (a, b) in which $a < b$.

We make some informal remarks about these formulas. Let us say, if $c < a < b$, that a and b *look the same over* c if $\varphi(c, a) \longleftrightarrow \varphi(c, b)$. Then $\theta_0(a, b)$ is true iff a and b look the same over all $c < a$. If they do not look the same over all $c < a$, then there is a least $c < a$ over which a and b do not look. Fixing this c, we have that $\neg(\varphi(c, a) \longleftrightarrow \varphi(c, b))$. Now, whether $\theta_1(a, b)$ or $\theta_2(a, b)$ holds depends on how a and b do not look the same over c: if $\varphi(c, a)$, then we have that $\theta_1(a, b)$, and if $\neg\varphi(c, a)$, then we have $\theta_2(a, b)$.

Returning to the proof, we see that since $p(x)$ is 2-indiscernible, then Exercise 3.1.8 implies there is $i \in \{0, 1, 2\}$ such that $p(x)$ forces $\theta_i(x, y)$. Thus, that exercise shows the existence of a formula $\psi(x) \in p(x)$ such that the sentence

$$\forall x \forall y[\psi(x) \wedge \psi(y) \wedge x < y \longrightarrow \theta_i(x, y)]$$

is in T.

We are now prepared to define $\sigma(u)$ to be the formula

$$\exists w \forall x > w[\psi(x) \longrightarrow \varphi(u, x)] .$$

There are three cases to consider, depending on what i is, in proving that $\sigma(u)$ is a defining formula for $\varphi(u, x)$. Suppose M is a model of T, a is an element realizing $p(x)$, and c is a constant Skolem term.

For the first case, let $i = 0$. Suppose $\sigma(c) \in T$. Then there is a constant Skolem term d such that the sentence $\forall x > d[\psi(x) \longrightarrow \varphi(c, x)]$ is in T. Thus, $M \models \psi(a)$ and, since $p(x)$ is unbounded, $M \models a > d$, from which it follows that $M \models \varphi(c, a)$. Therefore, $\varphi(c, x) \in p(x)$.

Conversely, suppose $\sigma(c) \notin T$. In particular, $\exists x > c[\psi(x) \wedge \neg\varphi(c, x)]$ is in T. Therefore, there is a constant Skolem term d such that the sentence $c < d \wedge \psi(d) \wedge \neg\varphi(c, d)$ is in T. Then $M \models \theta_0(d, a)$ so that $M \models \neg\varphi(c, a)$ and, therefore, $\varphi(c, x) \notin p(x)$.

Next let $i = 1$, and let $\delta(u, x)$ be the formula

$$\forall v \forall w \forall z < u[\psi(v) \wedge \psi(w) \wedge x < v < w \longrightarrow \big(\varphi(z, v) \longleftrightarrow \varphi(z, w)\big)] .$$

It is to be shown by formal induction that $T \vdash \forall u \exists x \delta(u, x)$. The basis step is that $T \vdash \exists x \delta(0, x)$, which is trivial since even $T \vdash \forall x \delta(0, x)$. For the inductive step, suppose that $a, b \in M \models T$ and $M \models \delta(a, b)$. If $M \models \delta(a + 1, b)$, then we are done, so suppose otherwise. Then there are $a_2 > a_1 > b$ such that $M \models \psi(a_1) \wedge \psi(a_2) \wedge (\varphi(a, a_1) \longleftrightarrow \neg\varphi(a, a_2))$. We claim that $M \models \delta(a + 1, a_2)$. If the claim were false, then there would be $b_2 > b_1 > a_2$ such that $M \models \psi(b_1) \wedge \psi(b_2) \wedge \neg(\varphi(a, b_1) \longleftrightarrow \varphi(a, b_2))$. Then, since $M \models \theta_1(b_1, b_2)$, it must be that $M \models \varphi(a, b_1) \wedge \neg\varphi(a, b_2)$, so that $M \models \neg\theta_1(a_2, b_1)$, which is a contradiction.

To show that $\varphi(c, x) \in p(x)$ iff $\sigma(c) \in T$, first choose some constant Skolem term d such that $\delta(c, d) \in T$, and then argue as in the case $i = 0$.

Finally, when $i = 2$, just observe that this is just the case of $i = 1$ when the formula $\neg\varphi(u, x)$ is considered in place of $\varphi(u, x)$. □

The following theorem is in contrast to Theorem 3.1.20 which implies that types as in the following theorem are necessarily bounded. Notice that Corollary 3.1.6 shows the existence of bounded types that are n-indiscernible for all n.

Theorem 3.1.10 *Suppose T has no standard model and $1 \leq n < \omega$. Then there is an n-indiscernible type $p(x)$ that is not $(n+1)$-indiscernible.* □

The proof is like the proof of Theorem 3.1.5 but with the another combinatorial theorem replacing FRT. For $n < \omega$, let $[V]^n$ be the set of all n-element subsets of V. Then $H = (V, E)$ is an *n-uniform hypergarph* if $E \subseteq [V]^n$. An n-uniform hypergraph (V, E) is *sparse* if there is no $K \in [V]^{n+1}$ such that $[K]^n \subseteq E$. If $H = (V, E)$, $H' = (V', E')$ are n-uniform hypergraphs, then H is a subhypergraph of H' iff $V \subseteq V'$ and $E = E' \cap [V]^n$.

> NEŠETŘIL–RÖDL THEOREM: *Suppose $1 \leq n < \omega$ and $H = (V, E)$ is a finite, sparse $(n+1)$-uniform hypergraph. Then there is a finite, sparse $(n+1)$-uniform hypergraph $H' = (V', E')$ such that whenever $[V']^{n+1} = F_1 \cup F_2$, then (V', E') has a subhypergraph (W, F) which is isomorphic to (V, E) and such that either $F \subseteq F_1$ or $F \subseteq F_2$.*

The Nešetřil–Rödl Theorem is both formalizable and provable in PA.

Exercise 3.1.11 Prove Theorem 3.1.10 using the Nešetřil–Rödl Theorem. (HINT: work in the prime model M. In M, let (V, E) be a sparse $(n+1)$-uniform hypergraph such that any standard sparse $(n+1)$-uniform hypergraph is isomorphic to a subhypergraph of (V, E). Let a definable subset $X \subseteq V$ be "large" if every standard sparse $(n+1)$-uniform hypergraph is isomorphic to a subhypergraph of $(X, E \cap [X]^{n+1})$.)

3.1.3 *End-extensional types*

It was shown in Theorem 3.1.9 that all unbounded 2-indiscernible types are definable. In fact, they have a property even stronger than definability. For a definable type $p(x)$, every formula $\varphi(u, x)$ has a defining formula $\sigma(u)$, by which it is meant that for any constant Skolem term c the formula

$$\sigma(c) \longleftrightarrow \varphi(c, x)$$

is in $p(x)$. Moreover, the formula

$$\forall u \leq c[\sigma(u) \longleftrightarrow \varphi(u, x)]$$

is in $p(x)$ (Do IT!). The following definition imposes some even stronger conditions on the defining formula.

Definition 3.1.12 A type $p(x)$ is *end-extensional* if it is unbounded and for any formula $\varphi(u,x)$, there are a Skolem term $t(u)$ and a formula $\sigma(u)$ such that the formula

$$\forall u[t(u) < x \longrightarrow (\sigma(u) \longleftrightarrow \varphi(u,x))]$$

is in $p(x)$.

It follows right from the definition that every end-extensional type is definable (Do IT!).

Lemma 3.1.13 *Every unbounded* 2*-indiscernible type is end-extensional.*

Proof Let $p(x)$ be unbounded, 2-indiscernible and, hence, definable by Theorem 3.1.9. Consider some $\varphi(u,x)$ and its defining formula $\sigma(u)$. We show that there is a Skolem term $t(u)$ which, together with $\sigma(u)$, has the stronger property of Definition 3.1.12. Let M be the prime model of T, and let $M_0 = M(a_0)$ be a $p(x)$-extension of M and $M_1 = M_0(a_2)$ a $p(x)$-extension of M_0. Let $\theta(w,x)$ be the formula

$$\forall u \leq w[\sigma(u) \longleftrightarrow \varphi(u,x)].$$

Clearly, $M_1 \models \theta(a_0, a_2)$. [This is not a typo: the a_1 will appear later.]

Let $s(x)$ be the Skolem term which denotes the least $u \leq x$ such that $\neg[\sigma(u) \longleftrightarrow \varphi(u,x)]$. Now suppose that there is no such Skolem term $t(u)$. Thus, if $t(u)$ is a Skolem term, then the formula $t(s(x)) < x$ is in $p(x)$. It follows that $\mathrm{gap}(s(a_2)) < a_2$, so by compactness, in some model $M_2 \succ M_1$, there is a_1 such that $s(a_2) < \mathrm{gap}(a_1) < a_2$ and a_1 realizes $p(x)$. But then we have that $M_2 \models a_0 < s(a_2) < a_1$, contradicting the 2-indiscernibility of $p(x)$. □

3.1.4 *Rare types*

The following contains a definition of a rare type. Other possible definitions come from Lemma 3.1.15 and from Corollary 3.1.17, which is perhaps more reminiscent of the origins of rare types in ultrafilter theory.

Definition 3.1.14 A type $p(x)$ is *rare* if it is a nonprincipal complete type and for any Skolem term $t(x)$, there is a formula $\varphi(x) \in p(x)$ such that the sentence

$$\forall x \forall y[\varphi(x) \wedge \varphi(y) \wedge x < y \longrightarrow t(x) < y]$$

is a consequence of T.

In other words, the nonprincipal type $p(x)$ is rare iff it forces each formula $t(x) < y$. Every rare type is unbounded (Do IT!).

The next lemma gives a more model-theoretic definition.

Lemma 3.1.15 *A nonprincipal type $p(x)$ is rare iff in any model M of T, no two distinct elements in the same gap realize $p(x)$.*

Proof Let $p(x)$ be a nonprincipal type. For one direction, assume that $p(x)$ is rare and that $a \in M$ realizes $p(x)$ and $a < b \in \mathrm{gap}(a)$. There is a Skolem term $t(x)$ such that $M \models b < t(a)$, and then, according to Definition 3.1.14, there corresponds a formula $\varphi(x) \in p(x)$. But then $M \models \neg\varphi(b)$, so b does not realize $p(x)$.

Conversely, assume that $a, b \in M$ both realize $p(x)$ where $a < b \in \mathrm{gap}(a)$. Let $t(x)$ be a Skolem term such that $M \models b < t(a)$. But now, for any $\varphi(x) \in p(x)$, $M \models a < b < t(a) \wedge \varphi(a) \wedge \varphi(b)$, verifying that $p(x)$ is not rare. □

Theorem 3.1.16 *Let $p(x)$ be an unbounded complete type. Then $p(x)$ is rare iff whenever $a \in M$ realizes $p(x)$ and $b \in \mathrm{gap}(a)$, then $a \in \mathrm{Scl}(b)$.*

Proof Suppose $p(x)$ is rare. Let $a \in M$ realizes $p(x)$, and let $b \in \mathrm{gap}(a)$. First, suppose $b \leq a$. Then let $t(x)$ be a Skolem term such that $M \models b \leq a < t(b)$ and also such that $t(x)$ defines an increasing function in M. Let $\varphi(x) \in p(x)$ be as in Definition 3.1.14, and the let $s(x)$ be a Skolem term where $s(x) = \min\{y : x \leq y \wedge \varphi(y)\}$. Clearly, $M \models s(b) = a$, so $a \in \mathrm{Scl}(b)$.

Next, suppose $a < b$, and then let $t(x)$ be a Skolem term such that $M \models a < b < t(a)$ and also such that $t(x)$ defines an increasing function in M. Let $\varphi(x) \in p(x)$ be as in Definition 3.1.14, and the let $s(x)$ be a Skolem term where $s(x) = \max\{y : y < x \wedge \varphi(y)\}$. Clearly, $M \models s(b) = a$, so $a \in \mathrm{Scl}(b)$.

Conversely, suppose $p(x)$ is not rare. Then there are a model M and $a, b \in M$ such that $a \neq b \in gap(a)$ and both a, b realize $p(x)$. By Ehrenfeucht's Lemma, $a \notin \mathrm{Scl}(b)$. □

Corollary 3.1.17 *Let $p(x)$ be a complete type, and let N be a model generated by an element a realizing $p(x)$. Then $p(x)$ is rare iff there is $M \prec_{\mathsf{end}} N$ such that N is a minimal extension of M.*

Proof Suppose $p(x)$ is rare, and let $M = \{x \in N : \mathrm{gap}(x) < \mathrm{gap}(a)\}$. Clearly, $M \prec N$. If $b \in N \backslash M$, then $\mathrm{gap}(b) = \mathrm{gap}(a)$, so, by Theorem 3.1.15, $a \in \mathrm{Scl}(b)$, proving that N is a minimal extension of M.

Conversely, suppose N is a minimal elementary end extension of M. Then, if $b \in \mathrm{gap}(a)$, then $b \in N \backslash M$, so by the minimality of the extension, $b \in \mathrm{Scl}(a)$. Then, by Theorem 3.1.15, $p(x)$ is a rare type. □

Lemma 3.1.18 *Every unbounded 2-indiscernible type is rare.*

Proof Suppose $p(x)$ is an unbounded 2-indiscernible type. Consider a model M in which the distinct elements a, b realize the type $p(x)$. Assume that $a < b$. Let $N \succ M$ be such that there is $c \in N$ realizing $p(x)$ and $b < \mathrm{gap}(c)$. Thus, $N \models t(b) < c$ for any Skolem term $t(x)$, so by the 2-indiscernibility of $p(x)$, it is also the case that $M \models t(a) < b$ for every Skolem term $t(x)$. Therefore, a and b are in different gaps, proving that $p(x)$ is rare. □

Lemma 3.1.19 *Every end-extensional rare type is indiscernible.*

Proof Suppose that $p(x)$ is an end-extensional rare type. We will show by induction on n that $p(x)$ is n-indiscernible for each $n \geq 1$. Trivially, $p(x)$ is 1-indiscernible. Suppose that $p(x)$ is n-indiscernible. Let M be a model of T with elements $a_0 < a_1 < \cdots < a_n$ and $b_0 < b_1 < \cdots < b_n$, where each of these elements realizes $p(x)$. Assume that $M \models \varphi(a_0, a_1, \ldots, a_n)$ with the aim of showing that $M \models \varphi(b_0, b_1, \ldots, b_n)$. Then $a_{n-1} < \text{gap}(a_n)$ and $b_{n-1} < \text{gap}(b_n)$ since $p(x)$, is rare. By the end-extensionality of $p(x)$, there is a formula $\sigma(u_0, u_1, \ldots, u_{n-1})$ which not only is a defining formula for $\varphi(u_0, u_1, \ldots, u_{n-1}, x)$ but also has the stronger property required by end-extensionality. Thus, $M \models \sigma(a_0, a_1, \ldots, a_{n-1})$. The n-indiscernibility of $p(x)$ implies that $M \models \sigma(b_0, b_1, \ldots, b_{n-1})$, so that $M \models \varphi(b_0, b_1, \ldots, b_n)$, completing the proof that $p(x)$ is $(n+1)$-indiscernible. □

Putting together the previous lemmas yields the following theorem.

Theorem 3.1.20 *Every unbounded 2-indiscernible type is indiscernible.*

Proof Follows from Lemmas 3.1.13, 3.1.18, and 3.1.19. □

Unbounded indiscernible types might properly be called *Ramsey types.* The following exercise, the first of three giving some stronger properties of unbounded indiscernible types, suggests that they might even be called *uniformly* Ramsey types. These three exercises are not routine, but not difficult, either. They define properties of indiscernible types that are used elsewhere.

Exercise 3.1.21 Suppose that $p(x)$ is an unbounded indiscernible type and $\theta(u, x_0, x_1, \ldots, x_n)$ is an $(n+2)$-ary formula. Then there is a formula $\varphi(x) \in p(x)$ such that the sentence

$$\begin{aligned}\forall u\, \exists w [\forall \bar{x}\, \forall \bar{y} (w < x_0 < x_1 < \cdots < x_n \ \wedge\ w < y_0 < y_1 < \cdots < y_n \\ \wedge\, \varphi(x_0) \wedge \varphi(x_1) \wedge \cdots \wedge \varphi(x_n) \wedge \varphi(y_0) \wedge \varphi(y_1) \wedge \cdots \wedge \varphi(y_n) \\ \longrightarrow (\theta(u, \bar{x}) \longleftrightarrow \theta(u, \bar{y}))]\end{aligned}$$

is a consequence of T.

The property that Exercise 3.1.21 claims that $p(x)$ has could be informally rephrased as: for each $(n+2)$-ary formula $\theta(u, \bar{x})$ and each u, $p(x)$ eventually forces $\theta(u, \bar{x})$ or eventually forces $\neg\theta(u, \bar{x})$.

Another stronger property that unbounded indiscernible types possess suggests that they might also be called *canonical* Ramsey types. If $t^*(x_0, x_1, \ldots, x_{m-1})$ is a Skolem term, then we say that $\varphi(x)$ *forces that*

$t^*(x_0, x_1, \ldots, x_{m-1})$ *is one-to-one* if, whenever $i_0 < i_1 < \cdots < i_{m-1} < \omega$ and $j_0 < j_1 < \cdots < j_{m-1} < \omega$ are not exactly the same m-tuples, then $\varphi(x)$ forces the formula

$$t^*(x_{i_0}, x_{i_1}, \ldots, x_{i_{m-1}}) \neq t^*(x_{j_0}, x_{j_1}, \ldots, x_{j_{m-1}}).$$

Given a Skolem term $t(x_0, x_1, \ldots, x_n)$ and a formula $\varphi(x)$, we say that $t(x_0, x_1, \ldots, x_n)$ is *canonical* on $\varphi(x)$ if there are $m \leq n+1$ and $0 \leq i_0 < i_1 < \cdots < i_{m-1} \leq n$ and an m-ary Skolem term $t^*(y_0, y_1, \ldots, y_{m-1})$ such that $\varphi(x)$ forces that $t^*(\bar{x})$ is one-to-one and $\varphi(x)$ forces the formula

$$t(x_0, x_1, \ldots, x_n) = t^*(x_{i_0}, x_{i_1}, \ldots, x_{i_{m-1}}).$$

Exercise 3.1.22 Suppose that $p(x)$ is an $(n+2)$-indiscernible type and $t(x_0, x_1, \ldots, x_n)$ is an $(n+1)$-ary Skolem term. Then there is a formula $\varphi(x) \in p(x)$ such that the Skolem term $t(x_0, x_1, \ldots, x_n)$ is canonical on $\varphi(x)$.

Exercise 3.1.23 (*Uniformly Canonical Ramsey types*) Formulate and prove a statement asserting that unbounded indiscernible types have a property which encompasses both the uniform Ramsey and canonical Ramsey properties.

3.2 Minimal types

The notion of minimal types is due to Gaifman, who exposed most of their fundamental properties. This section contains various characterizations of minimal types, perhaps the most notable being that the minimal types are precisely the unbounded indiscernible types. This and some other characterizations are presented in Theorem 3.2.10.

Minimal extensions were considered in Chapter 2. Recall from that chapter that whenever $M \prec N$, then we say that N is a *minimal* extension of M if there is no K such that $M \prec K \prec N$. It was proved in Theorem 2.1.4 that every countable model has a minimal elementary end extension. Minimal types were originally introduced in order to generalize this to models of arbitrary cardinality as Theorem 3.3.1 shows.

The following is Gaifman's definition.

Definition 3.2.1 A type $p(x)$ is *minimal* if it is an unbounded complete type and whenever $M \prec M(a)$, where a realizes $p(x)$ and $a > M$, then $M(a)$ is a minimal extension of M.

3.2.1 *Selective types*

It follows from Corollary 3.1.17 that not only are all minimal types rare, but also so are all selective types, the definition of which follows.

Definition 3.2.2 A type $p(x)$ is *selective* if it is a nonprincipal complete type and the $p(x)$-extension of the prime model M of T is a minimal extension of M.

Every minimal type is selective. The following exercise presents a more syntactical way to characterize selective types.

Exercise 3.2.3 If $p(x) \in S_1(T)$ is a nonprincipal type, then $p(x)$ is selective iff for every Skolem term $t(x)$ either $p(x)$ forces $t(x_0) = t(x_1)$ or $p(x)$ forces $t(x_0) \neq t(x_1)$.

The particular case of $n = 0$ of Exercise 3.1.22 on canonical Ramsey types says that all 2-indiscernible types are selective.

Lemma 3.2.4 *Every unbounded selective type is rare.*

Proof Let $p(x)$ be an unbounded selective type. We will use Theorem 3.1.16 to show that $p(x)$ is rare. Assume that $p(x)$ is not rare. Then there are a model M and $a, b \in M$ such that a realizes $p(x)$ and $a \in \operatorname{gap}(b) \setminus \operatorname{gap}(b)$. Let $M_1, M_2 \preccurlyeq M$ be generated by a, b respectively. Let $M_0 = M_1 \cap M_2$. Then $a \notin M_0$, since $a \notin M_2$. By Theorem 2.1.1, $M_0 \prec_{\mathsf{cof}} M_1$, so M_0 is not the prime model. Therefore, M_1 is not a minimal extension of its prime submodel, contradicting the selectivitity of $p(x)$. □

In particular, every minimal type is rare.

Lemma 3.2.5 *Every minimal type is definable.*

Proof Suppose that $p(x)$ is not definable, and let $\theta(u, x)$ be a formula for which there is no defining formula. Let $\Gamma(x, y, z)$ be the following set of formulas:

$$\begin{aligned} p(y) \cup p(z) \cup \{&t_0 < x : t_0 \text{ is a constant Skolem term}\} \\ &\cup\{t_1(x) < y < z : t_1(x) \text{ is a Skolem term}\} \\ &\cup\{\forall u < x[\theta(u, y) \longleftrightarrow \theta(u, z)]\} \cup \{\theta(x, y) \longleftrightarrow \neg\theta(x, z)\}. \end{aligned}$$

This set is consistent with T (Do it!), so let M be a model of T having elements a, b, c satisfying $\Gamma(x, y, z)$, and let M_0 be the prime substructure of M. Clearly $M_0 < a < \operatorname{gap}(b)$, so that $a \notin M(b)$, and also $b < \operatorname{gap}(c)$ since $p(x)$ is rare, so that $c \notin M(a, b)$. Therefore, $M(b) \prec M(a, b) \prec M(a, b, c) = M(b, c)$, so $M(b, c)$ is not a minimal extension of $M(b)$. Therefore, $p(x)$ is not a minimal type. □

Lemma 3.2.6 *Every definable selective type is end-extensional.*

Proof Let $p(x)$ be a definable selective type. Consider the formula $\varphi(u, x)$, and let $\sigma(u)$ be a defining formula for it. Let $s(x)$ be the Skolem term which, informally, is the largest w such that $w \leq x \wedge \forall u[u < w \longrightarrow (\sigma(u) \longleftrightarrow \varphi(u, x))]$. Let $M(a)$ be a $p(x)$-extension of the prime model M, and let $c \in M$. Then

$M \models c < s(a)$. Since $p(x)$ is selective, there is a Skolem term $s'(y)$ such that $M \models s'(s(a)) = a$. Let $t(x)$ be the Skolem term for which the sentence

$$\forall x \forall y[x < y \longrightarrow s(x) < t(x) < t(y)]$$

is a consequence of T. Then, the formula

$$\forall u[t(u) < x \longrightarrow (\sigma(u) \longleftrightarrow \varphi(u, x))]$$

is in $p(x)$, verifying that $p(x)$ is end-extensional. □

Lemma 3.2.7 *Every rare end-extensional type is minimal*

Proof Let $p(x)$ be a rare end-extensional type. Since $p(x)$ is definable, we can let $M(a)$ be a $p(x)$-extension of M. Consider some $b \in M(a)\backslash M$, and let the Skolem term $t(u, x)$ and element $c \in M$ be such that $M(a) \models b = t(c, a)$. Since $p(x)$ is end-extensional, $\text{gap}(b) = \text{gap}(a)$ (Do IT!), so by Corollary 3.1.17 there is a Skolem term $s(y)$ such that $M \models s(b) = a$, thereby proving that $p(x)$ is minimal. □

Definition 3.2.8 If $I \subseteq M$ is a set of indiscernibles in a model M, then I is a set of *strong indiscernibles* if whenever $c_0, c_1, \dots, c_k \leq a \in I$, then $\{x \in I : x > a\}$ is a set of indiscernibles in the structure $(M, c_0, c_1, \dots, c_k)$. The type $p(x)$ is a *strongly indiscernible* type if it is a nonprincipal type such that for any model M, if $I \subseteq M$ is a set of elements each realizing $p(x)$, then I is a set of strong indiscernibles.

Lemma 3.2.9 *Every definable indiscernible type is strongly indiscernible.*

Proof Let $p(x)$ be a definable indiscernible type, and let I be a set of elements in a model M realizing $p(x)$. Let $a < a_0 < a_1 < \cdots < a_{n-1}$ and $a < b_0 < b_1 < \cdots < b_{n-1}$ be elements of I. Let $c_0, c_1, \dots, c_k \leq a$, and consider a formula $\varphi(\bar{u}, x_0, x_1, \dots, x_{n-1})$, with the goal of showing that $M \models \varphi(\bar{c}, \bar{a}) \longleftrightarrow \varphi(\bar{c}, \bar{b})$. Because $p(x)$ is definable, it is unbounded, so there is $M_0 \prec_{\text{end}} M$ such that $a \in M_0 < a_0, b_0$. The n-type $q(x_0, x_0, \dots, x_{n-1})$ of $\langle a_0, a_1, \dots, a_{n-1} \rangle$ (which is the same as the type of $\langle b_0, b_1, \dots, b_{n-1} \rangle$) is definable (Do IT!), so there is a defining formula $\sigma(\bar{u})$ for $\varphi(\bar{u}, \bar{x})$. Then $M \models \varphi(\bar{c}, \bar{a}) \leftrightarrow \sigma(\bar{c}) \leftrightarrow \varphi(\bar{c}, \bar{b})$. □

3.2.2 *Characterizing minimal types*

The are many different ways to characterize minimal types.

Theorem 3.2.10 *Let $p(x)$ be a type. Then $p(x)$ is minimal iff $p(x)$ is any one of the following:*

(1) *indiscernible and unbounded;*

(2) *rare and end-extensional;*

(3) *selective and definable;*

(4) *2-indiscernible and unbounded;*

(5) *strongly indiscernible and definable.*

Proof This is just a compilation of Lemmas 3.2.5, 3.2.6, 3.2.7, 3.1.13, 3.1.18, 3.1.19, 3.2.4, and 3.2.9. □

It is natural to ask, upon viewing the list in Theorem 3.2.10, whether (2) or (3) can be replaced by weaker conditions. How about replacing "end-extensional" with "definable" in (2) or replacing "definable" with "unbounded" in (3)? The purpose of Exercises 3.6.14 and 3.6.17 is to show that in each of these cases there are counterexamples showing this cannot be done.

In the previous section we saw in Theorem 3.1.4 that every completion of PA^* has $2^{\aleph_0}$ unbounded, indiscernible types. Since these types are precisely the minimal types, we get the existence of $2^{\aleph_0}$ minimal types.

Unbounded rare types which are not definable are even more plentiful, as the following theorem demonstrates.

Theorem 3.2.11 *Suppose that M is a model and that $d \in M$ and formula $\varphi(u, x)$ are such that $\varphi(d, x)$ defines an unbounded subset of M. Then M has a conservative extension N in which some $a > M$ realizes a rare type and is such that $N \models \varphi(d, a)$.*

Proof In the language of the structure (M, d), let $q(d, x)$ be a minimal type containing the formula $\varphi(d, x)$, and let (N, d) be a $q(d, x)$-extension of M generated by the element a realizing $q(d, x)$. Then (N, d) is a conservative extension of (M, d), so N is a conservative extension of M. We show that the type $p(x)$ realized by a is rare by showing that Definition 3.1.14 holds.

Let $t(x)$ be a Skolem term. Since $q(d, x)$ is minimal, hence rare, there is a formula $\theta(d, x) \in q(d, x)$ such that $M \models \forall xy[\theta(d, x) \wedge \theta(d, y) \wedge x < y \longrightarrow t(x) < y]$. Let $\sigma(d)$ be this sentence, and then let $\psi(x)$ be the formula $\exists u[\sigma(u) \wedge \varphi(u, x)]$. Clearly, $\psi(x) \in p(x)$ and $M \models \forall xy[\psi(x) \wedge \psi(y) \wedge x < y \longrightarrow t(x) < y]$, proving that $p(x)$ is rare. □

REMARK The rare type constructed in the previous proof can be made recursive in $\mathrm{tp}(d)$. See the remark following the proof of Theorem 3.1.2.

Corollary 3.2.12 *There is a rare type which is not definable.*

Proof Let $\varphi(u, x)$ be the formula $\exists v[x = \langle u, v\rangle]$, and then let M be a model with $d \in M$ not being definable. Apply Theorem 3.2.11. □

More syntactical characterizations of minimal types can be given.

Theorem 3.2.13 *Let $p(x)$ be an unbounded type. Then $p(x)$ is minimal iff for every Skolem term $t(u,x)$ there are a formula $\varphi(x) \in p(x)$ and Skolem terms $t_1(u)$ and $t_2(y)$ such that the sentence*

$$\forall u \exists w \big[\forall x > w\big(\varphi(x) \longrightarrow t(u,x) = t_1(u)\big) \vee \forall x > w\big(\varphi(x) \longrightarrow t_2(t(u,x)) = x\big)\big]$$

is a consequence of T.

Proof There is another condition that a type $p(x)$ can have that appears weaker than the one in the theorem: *for every Skolem term $t(u,x)$ there are Skolem terms $t_1(u)$ and $t_2(u,y)$ such that the formula*

$$\forall u \exists w[x > w \longrightarrow \big(t(u,x) = t_1(u) \ \vee \ t_2(u, t(u,x)) = x\big)]$$

is in $p(x)$.

Notice that the main difference between the formal sentence in the theorem and the above formula is the arity of the Skolem term t_2. For the one direction of the theorem, we assume that $p(x)$ satisfies this weaker condition. Consider a model M of T and an elementary extension $M(a)$, where $a > M$ and a realizes $p(x)$. Suppose that $M \prec N \preccurlyeq M(a)$. Then, there is a Skolem term $t(u,x)$ and there is an element $b \in M$ such that $t(b,a)$ denotes an element $c \in N \backslash M$. Then there are the claimed Skolem terms $t_1(u)$ and $t_2(u,y)$, and either

$$M(a) \models t(b,a) = t_1(b)$$

or

$$M(a) \models t_2(b, t(b,a)) = a\,.$$

The first alternative is impossible since $c \in M$, and the second alternative implies that $N = M(a)$, thereby proving that $M(a)$ is a minimal extension of M.

Conversely, suppose that $p(x)$ is a minimal type. Then $p(x)$ is indiscernible, so let $\varphi(x) \in p(x)$ be the formula in Exercise 3.1.21 corresponding to the formula $\theta(u,x,y) \equiv t(u,x) = t(u,y)$. [Roughly speaking, for each u, the function defined by $t(u,x)$ is either eventually constant or eventually one-to-one on $\varphi(x)$.] Let $t_1(u)$ be defined so that if $t(u,x)$ is eventually constant on $\varphi(x)$, then the value of $t_1(u)$ is that constant. Let $t_2(u,x)$ be defined so that if $t(u,x)$ is eventually one-to-one on $\varphi(x)$, then $t_2(u,x)$ is its inverse. Clearly $t_1(u)$ and $t_2(u,x)$ satisfy the weaker condition.

To complete the proof of the theorem, we must show that the parameter u in the Skolem term $t_2(u,x)$ can be dispensed with. Let $\theta(u)$ be the formula

$$\forall w \exists x \exists y[\varphi(x) \wedge \varphi(y) \wedge w < x < y \wedge t(u,x) \neq t(u,y)]\,.$$

In words, $\theta(u)$ says that the function defined by $t(u,x)$ is eventually one-to-one on $\varphi(x)$. We next consider a Skolem term $f(z)$ which is defined by formal induction but which we describe more informally. Let $f(0)=0$. Inductively, $f(z+1)$ is the least y such that the formula

$$\forall u \leq z \forall v \leq z \forall w \leq f(z) \forall x \geq y$$
$$[\varphi(x) \wedge \varphi(w) \wedge \theta(u) \wedge \theta(v) \longrightarrow t(u,x) > t(v,w)]$$

holds. Since $p(x)$ is rare (by Lemma 3.2.4), there is a formula $\psi(x)$ in $p(x)$ such that the sentences $\forall x[\psi(x) \longrightarrow \varphi(x)]$ and $\forall x \forall y[\psi(x) \wedge x < y \leq f(x) \longrightarrow \neg\psi(y)]$ are in T. The significance of the formula $\psi(x)$ is that if $t(u_1,x)$ and $t(u_2,x)$ define functions which are eventually one-to-one on $\varphi(x)$, then they eventually have disjoint images on $\psi(x)$. Moreover, this is true not just for two such functions but for any bounded set of functions. Thus, we can get $t_2(y)$ which defines a one-to-one function such that the sentence

$$\forall x \forall y \forall u[\psi(x) \wedge t(u,x) = y \wedge \neg\exists z \exists v < u(\psi(z) \wedge t(v,z) = y) \longrightarrow t_2(y) = x]$$

is in T. Then $t_1(u)$ and $t_2(y)$ satisfy the requirement of the theorem. □

Lemma 3.2.14 *Let $p(x)$ be an unbounded type. Then $p(x)$ is minimal iff for every Skolem term $t(u,x)$ there is a formula $\varphi(x) \in p(x)$ such that the sentence*

$$\forall u \exists w\big[\forall x \forall y\big(\varphi(x) \wedge \varphi(y) \wedge w \leq x < y \longrightarrow t(u,x) \neq t(u,y)\big)$$
$$\vee \forall x \forall y\big(\varphi(x) \wedge \varphi(y) \wedge w \leq x < y \longrightarrow t(u,x) = t(u,y)\big)\big]$$

is a consequence of T.

In words, the formal sentence in the theorem asserts that each of the functions defined by $t(u,x)$ is either eventually one-to-one or eventually constant on the set defined by $\varphi(x)$.

Proof For one direction, suppose that $p(x)$ is a minimal type and, therefore, an indiscernible type. Let $\theta(u,x,y)$ be the formula $t(u,x) = t(u,y)$, and apply Exercise 3.1.21 to get $\varphi(x)$.

For the converse, let $p(x)$ satisfies the condition in the lemma with the aim of showing that it is minimal. Let $t_1(u)$ be a Skolem term such that the sentence

$$\forall u \forall w\Big[\forall x \forall y\big(\varphi(x) \wedge \varphi(y) \wedge w \leq x < y \longrightarrow t(u,x) = t(u,y)\big)$$
$$\longrightarrow \forall x\big(\varphi(x) \wedge w \leq x \longrightarrow t(u,x) = t_1(u)\big)\Big]$$

is in T. Similarly, let $t_2(u, y)$ be such that

$$\forall u \forall w \Big[\forall x \forall y \big(\varphi(x) \wedge \varphi(y) \wedge w \leq x < y \longrightarrow t(u,x) \neq t(u,y)\big) \\ \longrightarrow \forall x \big(\varphi(x) \wedge w \leq x \longrightarrow t_2(u, t(u,x)) = x\big)\Big]$$

is in T. Then apply the previous theorem (or even the weaker version of that theorem stated in its proof). □

3.2.3 *An example*

It is evident that every 2-indiscernible type is selective (DO IT!). This subsection contains an example of a bounded selective type that is not 2-indiscernible. This example (or a variation of it) will reappear in Theorem 8.9.4.

The crucial combinatorial fact that is used in the proof of Theorem 3.2.15 is the canonical version of a Ramsey-style theorem due to Erdős and Rado. Let I be a finite index set, and for each $i \in I$, let A_i be a finite set. Let $X = \prod_{i \in I} A_i$. Then a function f is said to be *canonical* on X if there is $K \subseteq I$ such that whenever $x, y \in X$, then $f(x) = f(y)$ iff $x_j = y_j$ for each $j \in K$. If we wish to indicate the set K, we say that f is *K-canonical* on X. Notice that if f is I-canonical, then f is one-to-one on X, and if f is $\emptyset$-canonical, then f is constant on X.

> ERDŐS–RADO THEOREM: *For any $n, k < \omega$, there is $m < \omega$ such that whenever $|J| = k$, $|A_j| \geq m$ for each $j \in J$, and f is a function on $\prod_{j \in J} A_j$, then there are $B_j \subseteq A_j$ for each $j \in J$ such that each $|B_j| \geq n$ and f is canonical on $\prod_{j \in J} B_j$.*

We can think of this theorem as saying that if we start with a k-dimensional m-box, then it has a k-dimensional n-subbox with the required property. The ER Theorem has a consequence which is important in the proof of Theorem 3.2.15. Suppose that f is K-canonical on the k-dimensional box $Y = \prod_{j \in J} B_j$ as in the conclusion of the ER Theorem, and let z be an arbitrary element of the box. Then f is constant on the set

$$Z_1 = \{x \in Y : x_j = z_j \text{ for all } j \in K\}$$

and is one-to-one on

$$Z_2 = \{x \in Y : x_i = z_i \text{ for all } i \in J \backslash K\}.$$

Just (DO IT!). The sets Z_1 and Z_2, while technically not boxes, are naturally isomorphic to boxes. For example, the set Z_2 can be thought of as a $|K|$-dimensional n-box by considering the projection map $x \mapsto x{\restriction}K$. Thus, we refer to Z_1 as a $|J \backslash K|$-dimensional subbox of Y and Z_2 as a $|K|$-dimensional subbox of Y. Since $k = |K| + |J \backslash K|$, there is the following corollary.

COROLLARY OF THE ER THEOREM: *For any $n, k < \omega$, there is $m < \omega$ such that whenever Y is a k-dimensional m-box and f is a function on Y, then there is a $\lfloor \frac{1}{2}(k+1) \rfloor$-dimensional n-subbox $Z \subseteq Y$ on which f is either constant or one-to-one.*

The Erdős–Rado Theorem and its corollary are formalizable in PA and their proofs, which just involve some applications of FRT, are also formalizable in PA.

Theorem 3.2.15 *Assume T does not have a standard model. There is a bounded, selective type that is not 2-indiscernible.*

Proof Let M_0 be the prime model of T, and let $r \in M_0$ be nonstandard. Let $h(x)$ be a Skolem term which defines a function that grows sufficiently fast with respect to the Erdős–Rado Theorem and its corollary. What this means should become apparent in the proof. Let $\langle f_n : n < \omega \rangle$ be a list of all the definable functions $f : M_0 \longrightarrow M_0$.

Working in M_0, let $A = [0, h(r)-1]$, and then let $X = A^r$. In other words, let X be the set of (codes of) sequences of elements of A having length r. As usual, if $x \in X$ and $i < r$, then we denote the ith element of this sequence by $(x)_i$.

We construct a complete type $p(x)$ by defining a decreasing sequence of definable large subsets $X_0 \supseteq X_1 \supseteq X_2 \supseteq \cdots$ of M_0, and then letting

$$p(x) = \{\varphi(x) : M_0 \models \forall x[x \in X_n \longrightarrow \varphi(x)], \text{ for some } n < \omega\}.$$

A subset $Y \subseteq X$ is *large* if, for some $m < \omega$, Y is k-dimensional $h(a-m)$-subbox of X, where $k^m > a$.

Let $X_0 = X$, which clearly is large.

Now suppose that we have large X_n. In particular, X_n is a k-dimensional $h(a-m)$-box, where $k^m > a$. At this stage we consider a definable function $f_n : M_0 \longrightarrow M_0$. Applying the formalized ER Theorem and its corollary, we get $Y \subseteq X_n$ which is a $\lfloor \frac{1}{2}(k+1) \rfloor$-dimensional $h(a-m-1)$-box on which f_n is either one-to-one or constant.

Since the characteristic functions of definable subsets of M_0 appear in the list of functions, it is evident that this construction produces a complete type $p(x) \in S_1(T)$. Clearly, this type is selective using the criterion of Exercise 3.2.3.

Notice, for each $n < \omega$, that there are $a, b \in X_n$ such that $|\{i < r : (a)_i \neq (b)_i\}| = 1$. Therefore, there are $M \models T$ and $a, b \in M$ such that $|\{i < r : (a)_i \neq (b)_i\}| = 1$. And similarly, there are $M \models T$ and $a, b \in M$ such that $|\{i < r : (a)_i \neq (b)_i\}| = 2$. Thus there are at least two different complete 2-types $q(x, y)$ containing $p(x) \cup p(y)$. (EXERCISE: How many are there?) Hence, $p(x)$ is not 2-indiscernible. □

3.3 Canonical extensions

The fundamental MacDowell–Specker Theorem asserting that every model of PA* has an elementary end extension was proved in Chapter 2. The extensions constructed in that proof were conservative extensions generated by an arbitrary definable type. We saw in the previous sections that minimal types are definable, so minimal types could also be used in the proof of the MacDowell–Specker Theorem. Indeed, this is just what Gaifman did to obtain the following improvement of the MacDowell–Specker Theorem.

Theorem 3.3.1 *Every model of* PA^* *has a conservative, minimal end extension.*

Proof By Theorem 3.1.2, there are unbounded indiscernible types which, by Theorem 3.2.10, are minimal and definable. Such a type can be used to get a conservative extension which, following the definition of minimal type, is a minimal extension. □

The minimal elementary end extension of M constructed in the previous proof is just a $p(x)$-extension of $M(a)$, where $p(x)$ is a minimal type. Such a model is referred to as a *canonical* extension of M. More generally, if $(I, <)$ is a linearly ordered index set, then an I-*canonical* extension of M is a model $M(\langle a_i : i \in I\rangle)$ which is a $(p(x))^I$-extension of M for some minimal type $p(x)$, a notion defined in the next subsection.

3.3.1 *Products of types*

There is a natural way to iterate $p(x)$-extensions, where $p(x)$ is a definable type. Considerations of such iterations lead to the notion of the product of definable types. If $p(x), q(y) \in S_1(T)$ are definable types. (Actually, only the definability of $q(y)$ is crucial.) Then $p(x) \times q(y)$ is the 2-type $r(x, y) \in S_2(T)$ obtained as follows: Let $M(a)$ be a prime model of $p(x)$, and let $M(a)(b)$ be a $q(y)$-extension of $M(a)$. Then $r(x, y)$ is the type realized by the pair (a, b). It is also possible to give a syntactic definition of the product using defining schemes (DO IT!). Either way, the definition easily extends to the product of an n-type and a definable type, producing an $(n+1)$-type. Thus, given types $p_0(x_0), p_1(x_1), \ldots, p_n(x_n) \in S_1(T)$, where all (except possibly $p_0(x_0)$) are definable types, we can make the inductive definition

$$p_0(x_0) \times p_1(x_1) \times \cdots \times p_n(x_n) = \big(p_0(x_0) \times p_1(x_1) \times \cdots \times p_{n-1}(x_{n-1})\big) \times p_n(x_n)$$

resulting in a complete $(n + 1)$-type.

This product of types has all the nice properties you would expect. Two of these appear in the following two exercises.

Exercise 3.3.2 If $p_0(x_0), p_1(x_1), \ldots, p_n(x_n)$ are definable types, then the $(n+1)$-type $p_0(x_0) \times p_1(x_1) \times \cdots \times p_n(x_n)$ is also definable.]

Exercise 3.3.3 Suppose that $p_0(x_0), p_1(x_1), \ldots, p_n(x_n) \in S_1(T)$ are definable types and $0 \leq i_0 < i_1 < \cdots < i_k \leq n$. Then $p_{i_0}(x_{i_0}) \times p_{i_1}(x_{i_1}) \times \cdots \times p_{i_k}(x_{i_k}) \subseteq p_0(x_0) \times p_1(x_1) \times \cdots \times p_n(x_n)$.]

The previous exercise has a more model-theoretic rendition. Suppose we have the types as in Exercise 3.3.3 and that $M(a_0, a_1, \ldots, a_n)$ and $M(b_0, b_1, \ldots, b_k)$ are $p_0(x_0) \times p_1(x_1) \times \cdots \times p_n(x_n)$-extensions and $p_{i_0}(x_{i_0}) \times p_{i_1}(x_{i_1}) \times \cdots \times p_{i_k}(x_{i_k})$-extensions, respectively, of M. Then there is a unique elementary embedding of $M(b_0, b_1, \ldots, b_k)$ into $M(a_0, a_1, \ldots, a_n)$ that is the identity on M and takes each b_j to a_{i_j}. With all these elementary embeddings available, it is possible to define infinite products.

Let $(I, <)$ be a (possibly infinite) linearly ordered set. For each $i \in I$, let $p_i(x_i)$ be a definable type, where to each distinct i there corresponds a distinct variable x_i. We can then form a big elementary extension $M(\langle a_i : i \in I \rangle)$ of M by taking the union of all possible $M(a_{i_0}, a_{i_1}, \ldots, a_{i_k})$, where $i_0 < i_1 < \cdots < i_k$ are in I and $M(a_{i_0}, a_{i_1}, \ldots, a_{i_k})$ is a $p_{i_0}(x_{i_0}) \times p_{i_1}(x_{i_1}) \times \cdots \times p_{i_k}(x_{i_k})$-extension of M. Then $M(\langle a_i : i \in I \rangle)$ is a $\prod\{p_i(x_i) : i \in I\}$-*extension* of M.

Exercise 3.3.4 Let $(I, <)$ be a linearly ordered set, and let $p_i(x_i)$ be a definable type for each $i \in I$. Suppose that $N = M(\langle a_i : i \in I \rangle)$ is a $\prod\{p_i(x_i) : i \in I\}$-extension of M. Let g be an automorphism of M and h an automorphism of $(I, <)$ such that $p_i(x) = p_{f(i)}(x)$ for every $i \in I$. Then there is a unique automorphism h of N extending g such that $g(a_i) = a_{f(i)}$ for each $i \in I$. □

If $p(x)$ is a minimal type and each $p_i(x_i) = p(x_i)$ for $i \in I$, where $(I, <)$ is a linearly ordered set, then $(p(x))^I = \prod\{p_i(x_i) : i \in I\}$. We refer to a $(p(x))^I$-extension as an I-*canonical* extension.

It can be that when considering an I-canonical extension, we are not so much interested in the ordered set I as in its order type. In such a case, we refer to a τ-canonical extension, where τ is the order type of I. For example, τ may be an ordinal. If τ is the order type of $(I, <)$, then τ^* is the order type of its reversal $(I, >)$.

The I-canonical extensions of a model are more amenable to successful investigations than are arbitrary extensions. This applies, in particular, to their automorphism groups and their substructure lattices. Both of these topics, automorphism groups and substructure lattices, are considered in this section, but in later chapters they are studied more intensively.

However, our first example has to do with the order type of the set of gaps of a model. Clearly, every model has a first gap, namely gap(0). In general, there is nothing more that can be said.

Theorem 3.3.5 *Let M be a model whose set of gaps has order type α. Let τ be an order type, and let N be a τ-canonical extension of M. Then the set of gaps of N has order type $\alpha + \tau$.*

Proof Let I be an ordered set having order type τ, and let N be an I-canonical extension of M using the minimal type $p(x)$. (We can assume that I is chosen so that its elements are actually the elements of N which generate the extension of M.) Every element b of N has the form $t(a, i_0, i_1, \ldots, i_n)$, where $t(u, x_0, x_1, \ldots, x_n)$ is a Skolem term, $a \in M$, and $i_0 < i_1 < \cdots < i_n$ are elements of I. Consider such an element b. It suffices to show that either $b \in M$ or $\operatorname{gap}(b) = \operatorname{gap}(i)$ for some $i \in I$.

The reader had been asked in Exercise 3.1.23 to formulate and prove a statement asserting that indiscernible types (and so in particular $p(x)$) are uniformly canonical Ramsey types. The intent was for the reader to show that for each subset $J \subseteq \{0, 1, 2, \ldots, n\}$, there is a formula $\sigma_J(u)$ and there is a Skolem term $t^*_J(u, y_0, y_1, \ldots, y_{m-1})$, where $m = |J|$, such that the sentence

$$\forall u \bigvee_J \sigma_J(u)$$

is a consequence of T as are the sentences

$$\forall u \forall \bar{y} \forall \bar{z} [\sigma_J(u) \wedge t^*_J(u, \bar{y}) = t^*_J(u, \bar{z}) \longrightarrow \bar{y} = \bar{z}]$$

and

$$\forall u \exists w \forall \bar{x} [\sigma_J(u) \wedge w < x_0 < x_1 < \cdots < x_n \wedge \varphi(x_0) \wedge \varphi(x_1) \wedge \cdots \wedge \varphi(x_n)$$
$$\longrightarrow t(u, x_0, x_1, \ldots, x_n) = t^*_J(u, x_{i_0}, x_{i_1} \ldots, x_{i_{m-1}})],$$

where $J = \{i_0, i_1, \ldots, i_{m-1}\}$ and $i_0 < i_1 < \cdots < i_{m-1} \leq n$. Now let $J = \{i_0, i_1, \ldots, i_{m-1}\}$ be such that $M \models \sigma_J(a)$. If $m = 0$, then $b \in M$. If $m > 0$, then there is a term $s(u, y)$ such that $N \models s(a, b) = i_{m-1}$, from which it follows that $\operatorname{gap}(b) = \operatorname{gap}(i_{m-1})$. □

Corollary 3.3.6 *If τ is any order type, then there is a model N of T whose set of gaps has order type $1 + \tau$.*

Proof In Theorem 3.3.5, let M be the prime model of T. □

3.3.2 *The automorphism group*

We next look at automorphism groups. Recall that if G is a group acting on a set X and if $A \subseteq X$, then $G_{\{A\}}$ is the *setwise stabilizer* of A, which is the subgroup of G consisting of those elements of G which fix the set A, and $G_{(A)}$ is the *pointwise stabilizer* consisting of those elements of G which fix each point in A. In particular, if M is a model and $A \subseteq M$, then $\operatorname{Aut}(M)_{\{A\}} = \{f \in \operatorname{Aut}(M) : f(a) \in A \text{ iff } a \in A\}$ and $\operatorname{Aut}(M)_{(A)} = \{f \in \operatorname{Aut}(M) : f(a) = a \text{ for all } a \in A\}$. The following theorem gives us a description of those automorphisms of an I-canonical extension which pointwise fix the ground model. We see in the corollary that all such automorphisms are uniquely determined by an automorphism of I.

Theorem 3.3.7 *Let N be an I-canonical extension of M, where $N = M(\langle a_i : i \in I\rangle)$.*

(1) *If $f \in \mathrm{Aut}(N)_{\{M\}}$, then there is $h \in \mathrm{Aut}(I, <)$ such that $f(a_i) = a_{h(i)}$ for each $i \in I$.*
(2) *Conversely, if $h \in \mathrm{Aut}(I, <)$ and $g \in \mathrm{Aut}(M)$, then there is a unique $f \in \mathrm{Aut}(N)$ such that f extends g and $f(a_i) = a_{h(i)}$ for each $i \in I$.*

Proof Let $p(x)$ be the minimal type used in obtaining the I-canonical extension.

(1) As we saw in the proof of Theorem 3.3.5, every new gap contains an element realizing $p(x)$. Since $p(x)$ is a minimal (hence, rare) type, there are no other new elements realizing $p(x)$. Thus f just permutes the new gaps and, therefore, determines $h \in \mathrm{Aut}(I, <)$.
(2) This is a special case of Exercise 3.3.3. □

Corollary 3.3.8 *If $(I, <)$ is a linearly ordered set and N an I-canonical extension of M, then $\mathrm{Aut}(M)_{(M)} \cong \mathrm{Aut}(I, <)$.*

Corollary 3.3.9 *If $(I, <)$ is a linearly ordered set and M an I-canonical extension of the prime model, then $\mathrm{Aut}(M) \cong \mathrm{Aut}(I, <)$.*

Corollary 3.3.10 *For every cardinal $\kappa \geq \aleph_1$, every completion of* PA^* *has a rigid κ-like model.*

Proof Let I have order type κ. □

The question of which groups can appear as (being isomorphic to) the automorphism group of a model of Peano Arithmetic is answered in a later chapter. The following lemma is useful in answering this question. For now, the lemma is used to answer the question in some special cases.

Lemma 3.3.11 *If M is a model, then for some ordered set I of cardinality at most $|M|$, every I-canonical extension N_0 of M satisfies the following:*

(1) *each automorphism of M is extendible to exactly one automorphism of N_0;*
(2) *if $N \succcurlyeq_{\mathsf{end}} N_0$, then $\mathrm{Aut}(N) = \mathrm{Aut}(N)_{\{M\}} = \mathrm{Aut}(N)_{\{N_0\}}$.* □

Letting $N = N_0$ in (2), we see that $\mathrm{Aut}(N_0) = \mathrm{Aut}(M)$, essentially.

Proof We begin with two remarks about order types. The first one concerns ordinals. An ordinal τ is *indecomposable* if whenever $\tau = \alpha + \beta$, then either $\beta = 0$ or $\beta = \tau$. Then, an ordinal τ is indecomposable iff either $\tau = 0$ or $\tau = \omega^\gamma$ for some γ (DO IT!). Thus, *for any uncountable cardinal κ, there are arbitrarily large indecomposable ordinals $\tau < \kappa$.*

The second remark also concerns ordinals but also concerns an arbitrary order type τ. *If τ is the order type of a linearly ordered set $(J,<)$ having cardinality κ, then there is an ordinal $\alpha < \kappa^+$ such that whenever $\tau = \tau_0 + \beta + \tau_1$, where β is an ordinal, then $\beta < \alpha$.* For each $i \in J$, let

$$B_i = \{j \in J : i \leq j \text{ and } [i,j] \text{ is well-ordered by } <\}.$$

Since κ^+ is a regular cardinal, we can let $\alpha < \kappa^+$ be greater than the order type of each of the B_i's.

For an order type τ, we let τ^* be the *reverse* order type. Thus, if $(J,<)$ has order type τ, the $(J,>)$ has order type τ^*.

We return to the proof of the lemma per se, but we first prove a weaker version in which (2) is replaced by:

(2′) *if $N_1 \succcurlyeq_{\mathsf{end}} N_0$, then* $\mathrm{Aut}(N_1) = \mathrm{Aut}(N_1)_{\{M\}}$.

Let τ be the order type of the set of gaps of M, and let $\kappa = |M|$. Applying the second remark, there is an ordinal $\alpha < \kappa^+$ such that whenever $\tau^* = \tau_0 + \beta + \tau_1$, where β is an ordinal, then $\beta < \alpha$. The first remark implies that we can assume that α is indecomposable. Let N_0 be an α^*-canonical extension of M.

Clearly, (1) follows from Theorem 3.3.7 since α^*, considered as a linearly ordered set, is rigid. For (2′), let $N_0 = M(\langle a_i : i < \alpha\rangle)$, let N be an elementary end extension of N_0, and let $f \in \mathrm{Aut}(N)$. For a contradiction, assume that $f \notin \mathrm{Aut}(N)_{\{M\}}$. By considering f^{-1} instead of f if needed, we can assume that $f(M) \subseteq M$. Suppose $f(M) \subsetneq M$ and, therefore, $f(a_i) \in M$ for some $i < \alpha$. Then the order type of the set of those gaps $\mathrm{gap}(f(a_j))$, where $i \leq j < \alpha$, must be α^*, by the indecomposability of α. This contradicts the choice of α.

Finally, to get N_0 as in the lemma, let $M_0 \succ M$ be an extension of M as in the lemma with the weaker condition (2′). Obtain an elementary chain $\langle M_n : n < \omega\rangle$, where each M_{n+1} is an elementary extension of M_n as in the weaker version of the lemma. Finally, letting $N_0 = \bigcup_{n<\omega} M_n$, conditions (1) and (2) are easily checked (DO IT!), completing the proof. □

Corollary 3.3.12 *Let M be a model and $(I,<)$ a linearly ordered set. Then M has a conservative extension N such that $|N| = |M| + |I|$ and* $\mathrm{Aut}(N) \cong \mathrm{Aut}(M) \times \mathrm{Aut}(I,<)$.

Proof Let N_0 be as in the previous lemma, and then let N be a canonical I-extension of N_0. □

Strengthening the notion of a rigid model, we say that M is *very rigid* if it is rigid and whenever $N \prec_{\mathsf{end}} M$, then $N \not\cong M$. We see in Theorem 3.3.14 that every model has a very rigid elementary end extension. A preliminary lemma is needed.

Lemma 3.3.13 *For every M there is a $N \succ_{\mathsf{end}} M$ such that $|N| = |M|$ and whenever $N' \preccurlyeq_{\mathsf{end}} N$ and $f : N \longrightarrow N'$ is an isomorphism, then f is the identity on M.*

Proof This proof repeatedly applies Theorem 3.2.11, with $\varphi(u, x)$ being the formula $\exists y[x = \langle u, y\rangle]$. This particular formula was chosen because, first of all, $\mathsf{PA}^* \vdash \forall u \forall w \exists x[x > w \wedge \varphi(u, x)]$, and, second, there is a Skolem term $t(x)$ such that $\mathsf{PA}^* \vdash \forall x \forall u[\varphi(u, x) \rightarrow t(x) = u]$.

Let $\kappa = |M|$, and let $M = \{d_\alpha : \alpha < \kappa\}$. We construct a continuous elementary chain $\langle N_\alpha : \alpha < \kappa\rangle$, where $N_0 = M$ and $|N_\alpha| = \kappa$ for each $\alpha < \kappa$. We then let $N = \bigcup\{N_\alpha : \alpha < \kappa\}$.

Now suppose we have N_α and wish to get $N_{\alpha+1}$. First, use Theorem 3.2.11 to get $N'_\alpha = N_\alpha(a_\alpha)$ which is a conservative extension of N_α generated by the element a_α that realizes a rare type such that $N'_\alpha \models \varphi(d_\alpha, a_\alpha)$. Second, use Lemma 3.3.11 to get $N_{\alpha+1} \succ_{\mathsf{end}} N'_\alpha$ satisfying the requirements of that lemma.

Clearly $N \succ_{\mathsf{end}} M$ and $|N| = \kappa = |M|$.

To show that N is as required, consider $N' \preccurlyeq_{\mathsf{end}} N$ and isomorphism $f : N \longrightarrow N'$, intending to show that f is the identity on M. By Lemma 3.3.11(2), for each $\alpha < \kappa$, $f(N_{\alpha+1}) = N_{\alpha+1}$ and $f(N'_\alpha) = N'_\alpha$. Therefore, $N' = N$ so that $f \in \mathrm{Aut}(N)$. Since $f{\restriction}N'_\alpha \in \mathrm{Aut}(N'_\alpha)$ for each $\alpha < \kappa$ and a_α realizes a rare type in the last gap of N'_α, it follows that $f(a_\alpha) = a_\alpha$. Since $d_\alpha = t(a_\alpha)$, it follows that $f(d_\alpha) = d_\alpha$. Thus, f is the identity on M. □

Theorem 3.3.14 *For every M there is a very rigid $N \succ_{\mathsf{end}} M$ such that $|N| = |M|$.*

Proof Let $N = \bigcup\{N_n : n < \omega\}$, where $N_0 = M$ and each N_{n+1} is an elementary extension of N_n obtained by applying Lemma 3.3.13. □

3.3.3 *The substructure lattice*

Given any model M we define $\mathrm{Lt}(M)$, the set of elementary substructures of M considered as a lattice, to be the substructure lattice of M. Chapter 4 of this book is devoted to substructure lattices, where a more detailed definition can be found. The following proposition yields some corollaries and also serves as an introduction to substructure lattices which are more fully discussed in that chapter.

Proposition 3.3.15 *Let N be an I-canonical extension of M, where $N = M(\langle a_i : i \in I\rangle)$.*

(1) *If $N_0 \prec N$, then $N_0 = \mathrm{Scl}((M \cup \{a_i : i \in I\}) \cap N_0)$;*

(2) *If $N_1 \preccurlyeq M$, $J \subseteq I$, and $N_0 = \mathrm{Scl}(N_1 \cup \{a_i : i \in J\})$, then $N_1 = N_0 \cap M$, and $\{a_i : i \in J\} = \{a_i : a_i \in N_0\}$.*

Proof (1) This can be proved the way that (1) of Theorem 3.3.7 was (Do it!).

(2) Clearly, $N_1 \preccurlyeq N_0 \cap M$, so for the first equality, it suffices to show that $N_0 \cap M \preccurlyeq N_1$. Let $t(x)$ be a Skolem term, possibly involving some a_i for $i \in J$, and let $b \in N_1$ be such that $t(b) = c \in M$. Thus, c is a typical element of $N_0 \cap M$. It follows from Proposition 2.2.11 that there is a defining formula $\sigma(x, y)$ (with no additional parameters) such that whenever $b', c' \in M$, then $M \models \sigma(b', c')$ iff $N \models t(b') = c'$. Then c is the unique element in M such that $M \models \sigma(b, c)$, so c is definable from b. Therefore, $c \in N_1$.

The second equality can be proved like (2) of Theorem 3.3.7 (Do it!). □

Corollary 3.3.16 *For any set I each model M has an elementary end extension N such that* $\mathrm{Lt}(N) \cong \mathrm{Lt}(M) \times \mathcal{P}(I)$.

Corollary 3.3.17 *Let I be any set and T any completion of* PA^*. *Then T has a model M such that* $\mathrm{Lt}(M) \cong \mathcal{P}(I)$.

Exercise 3.3.18 There is M such that $|M| = 2^{\aleph_0}$, no two distinct elements of M realize the same type, and $\mathrm{Lt}(M) \cong \mathcal{P}(\mathbb{R})$.

The next exercise puts a restriction on possible generalizations of Theorem 2.1.1.

Exercise 3.3.19 Let M be a model and $(I, <)$ a linearly ordered set having no last element. Let N be an I-canonical extension of M. Then N has cofinal substructures N_1, N_2 such that $M = N_1 \cap N_2$.

3.4 Resolute types

Suppose that M_1 is a model, $p(x)$ is a definable type, and $M_1(b)$ is a $p(x)$-extension of M_1. Let M_0 be the prime submodel of M_1, and let $M_0(b)$ the elementary submodel of $M_1(b)$ generated by b. Then $M_0(b)$ is a $p(x)$-extension of M_0. Now consider a model N_1 where $M_1 \preccurlyeq N_1 \preccurlyeq M_1(b)$. Then $M_0 \preccurlyeq N_1 \cap M_0 \preccurlyeq M_0(b)$. Thus, there is a function $N_1 \mapsto M_1 \cap M_0(b)$ which maps an arbitrary model between M_1 and $M_1(b)$ to an elementary substructure of $M_0(b)$. In general, this function is onto; in fact, whenever $N \preccurlyeq M_0(b)$ and N_1 is the elementary substructure of $M_1(b)$ generated by $M_1 \cup N_0$, then $N = N_1 \cap M_0(b)$ (Do it!). Resolute types are those definable types for which this function is always one-to-one.

Definition 3.4.1 Suppose that $p(x)$ is a nonprincipal definable type. Then $p(x)$ is a *resolute* type if whenever $M_0 \prec M_1$ are models of T, $M_0(b), M_1(b)$ are $p(x)$-extensions of M_0, M_1 respectively, and $M_1 \preccurlyeq N_0, N_1 \preccurlyeq M_1(b)$ are distinct, then $N_0 \cap M_0(b)$ and $N_1 \cap M_1(b)$ are distinct.

In this section we take a look at resolute types, which lie strictly between minimal types and definable types. That is, all minimal types are resolute and all resolute types are definable, with examples showing that both inclusions are proper. The significance of resolute types becomes more apparent in Chapter 4.

The next proposition gives a more syntactic characterization of resolute types.

Proposition 3.4.2 *Suppose that $p(x)$ is a nonprincipal definable type. Then $p(x)$ is resolute iff whenever $M(b)$ is a $p(x)$-extension of M, $t(u,x)$ is a Skolem term, and $c \in M$, then there are Skolem terms $f'(u,y), f''(u,z)$, and $f(x)$ such that the sentence*

$$f'(c, f(b)) = t(c,b) \wedge f''(c, t(c,b)) = f(b)$$

is true in $M(b)$.

Proof For the one direction, assume that $p(x)$ satisfies the condition in this proposition. To see that Definition 3.4.1 is satisfied, consider models $M_0, M_1, M_0(b), M_1(b), N_0$ and N_1 in that definition. Without loss of generality, assume that $a \in N_1 \backslash N_0$. Let $t(u,x)$ be a Skolem term and $c \in M_1$ be such that $a = t(c,b)$. Then there are the Skolem terms $f'(u,y), f''(u,z), f(x)$ having the requisite property. Then, we claim that $f(b) \in (N_1 \cap M_0(b)) \backslash (N_0 \cap M_0(b))$.

Clearly, $f(b) \in M_0(b)$, also $f(b) = f''(c,a) \in N_1$. Finally, $f(b) \notin N_0$, as otherwise $a = f''(c, f(b)) \in N_0$, which is a contradiction. Thus, $p(x)$ is resolute.

For the other direction, assume that $p(x)$ is resolute and that $M(b)$ is a $p(x)$-extension of M. Let $t(u,x)$ be a Skolem term and $c \in M$. Then let $a = t(c,b)$, and let $N = M(a) \cap M(b)$. Due to the resoluteness, we can assume that M is generated by c.

We claim that N is finitely generated. For, if not, then there is an increasing chain $\langle N_i : i < \omega \rangle$ of elementary submodels of N whose union is N. Let M_i be the submodel of $M(b)$ generated by $M \cup N_i$. Then $\langle M_i : i < \omega \rangle$ is also an increasing elementary chain; let M_ω be its union. Then $M_\omega \cap M(b) = M(a) \cap M(b)$ (Do it!), so by resoluteness, $M_\omega = M(a)$. But M_ω is not a finitely generated extension of M and $M(a)$ is, which is a contradiction.

Hence, there is a Skolem term $f(x)$ such that $f(b) \in N$ generates N. Notice that by resoluteness, $a \in \mathrm{Scl}(c, f(b))$, thereby showing the existence of $f'(u,y)$. Similarly, $f(b) \in \mathrm{Scl}(c,a)$, thereby showing the existence of $f''(u,z)$. □

In the characterization of resolute types in the previous proposition, the choice of the Skolem terms depends upon the model M and the definable function $t(c,\cdot)$. In the case of a minimal type, the Skolem term $f(x)$ can always be the one defining the constant function 0 or the identity function. The choice depends upon whether $t(c,b) \in M$ or $t(c,b) \notin M$ (Do it!). If a resolute type is such that there is a supply of $n+1$ Skolem terms from which the choice of $f(x)$ can always be made, then we say that the type is n-resolute. The following is the formal definition:

Definition 3.4.3 Let $n < \omega$ and $p(x) \in S_1(T)$ be a type. Then $p(x)$ is an *n-resolute* type if it is unbounded and there are Skolem terms

$t_0(x), t_1(x), \dots, t_n(x)$ such that for every Skolem term $t(u,x)$ there is a formula $\varphi(x) \in p(x)$ such the sentence

$$\forall u \bigvee_{i \leq n} \exists w \forall x > w \forall y > w[\varphi(x) \wedge \varphi(y) \longrightarrow \big(t(u,x) = t(u,y) \longleftrightarrow t_i(x) = t_i(y)\big)]$$

is in T.

The 0-resolute types are just the principal types.

Proposition 3.4.4 *If $1 \leq n < \omega$, then every n-resolute type is resolute.*

Proof Let $p(x)$ be n-resolute, with n being minimal. Let $t_0(x), t_1(x), \dots, t_n(x)$ be Skolem terms as in Definition 3.4.3. Without loss of generality, we can assume that $T \vdash \forall x[t_0(x) = 0 \wedge t_1(x) = x]$.

We first show that $p(x)$ is definable. Consider a formula $\psi(u,x)$ with the intent of finding a defining formula $\sigma(u)$ for it. We might as well assume that $\psi(d,x) \in p(x)$ for some constant term d, as otherwise we could just let $\sigma(u)$ be the formula $u \neq u$. Let $t(u,x)$ be a Skolem term such that

$$T \vdash \forall u \forall x[\big(\psi(u,x) \longrightarrow t(u,x) = 0\big) \wedge \big(\neg\psi(u,x) \longrightarrow t(u,x) = x+1\big)] .$$

Let $\varphi(x) \in p(x)$ be as in Definition 3.4.3. Let $\sigma(u) = \exists w \forall x > w[\varphi(x) \longrightarrow t(u,x) = 0]$. We claim that $\sigma(u)$ is a defining formula for $\varphi(u,x)$. To prove this claim, consider a constant term c.

If $T \vdash \sigma(c)$, then $T \vdash \exists w \forall x > w[\varphi(x) \longrightarrow t(c,x) = 0]$, so that $T \vdash \exists w \forall x > w[\varphi(x) \longrightarrow \psi(c,x)]$. Therefore, $\psi(c,x) \in p(x)$.

Conversely, assume that $\psi(c,x) \in p(x)$. Then there is $i \leq n$ for which

$$\exists w \forall x > w \forall y > w[\big(\varphi(x) \wedge \varphi(y)\big) \longrightarrow \big(t(c,x) = t(c,y) \longleftrightarrow t_i(x) = t_i(y)\big)]$$

is a consequence of T. It cannot be that $i > 1$. For, notice that the function defined by $t(c,x)$ is constant on $\{x : t(d,x) = 0\}$ and is one-to-one on its complement. Therefore, there is a formula $\delta(x) \in p(x)$ on which $t_i(x)$ is either constant or one-to-one. Therefore, $t_i(x)$ is superfluous, contradicting the minimality of n. Thus, $i = 0$ or $i = 1$. But $i \neq 1$ since $\psi(c,x) \in p(x)$, so that $i = 0$. Therefore, $T \vdash \sigma(c)$, completing the proof that $p(x)$ is definable. □

It is now clear that the 1-resolute types are precisely the minimal types (Do it!).

Recall that if $p(x)$ is a definable type and $M_0(a)$ is a $p(x)$-extension of the prime model M_0, then every element of $M_0(a)$ realizes a definable type. Resolute types behave in a similar way: if $p(x)$ is a resolute type and $M_0(a)$ is a $p(x)$-extension of the prime model M_0, then every nondefinable element of $M_0(a)$ realizes a resolute type (Do it!). This suggests a definition. We say that a model M is *resolute* if each nondefinable element in M realizes a resolute type. The definition of a resolute model can be generalized to a resolute extension. If $M \prec N$, then N is a *resolute* extension of M if its expansion $(N, a)_{a \in M}$ is resolute. Thus, a model is resolute iff it is a resolute extension of its prime submodel.

Let M be any model and N a conservative extension of the prime model. We make use of the amalgamation $M \star N$ over the prime model as discussed in Theorem 2.3.2 and defined in Definition 2.3.7. Recall that $M \star N$ is a conservative extension M.

Corollary 3.4.5 *The model N is resolute iff N is a conservative extension of its prime submodel M_0 and the following holds: whenever $M_0 \preccurlyeq M \preccurlyeq N_0 \prec N_1 \preccurlyeq M \star N$, then $N_0 \cap N \prec N_1 \cap N$.* □

Corollary 3.4.6 *If M_1, M_2 are resolute, then so is $M_1 \star M_2$.*

Proof Corollary 3.4.5 is used. Let M_0 be the prime model, and consider

$$M_0 \preccurlyeq M \preccurlyeq N_0 \prec N_1 \preccurlyeq M \star (M_1 \star M_2),$$

intending to show that $N_0 \cap (M_1 \star M_2) \prec N_1 \cap (M_1 \star M_2)$. We consider two cases.

First case: assume that $(M \star M_1) \cap N_0 = (M \star M_1) \cap N_1$. Then $(M \star M_1) \preccurlyeq N_0 \prec N_1 \preccurlyeq (M \star M_1) \star M_2$. Since M_2 is resolute, then $N_0 \cap M_2 \prec N_1 \cap M_2$, so that $N_0 \cap (M_1 \star M_2) \prec N_1 \cap (M_1 \star M_2)$.

Second case: assume that $(M \star M_1) \cap N_0 \prec (M \star M_1) \cap N_1$. Let $N_i' = (M \star M_1) \cap N_i$. Since M_1 is resolute, then $N_0' \cap M_1 \prec N_1' \cap M_1$, so that $N_0 \cap M_1 \prec N_1 \cap M_1$. Therefore, $N_0 \cap (M_1 \star M_2) \prec N_1 \cap (M_1 \star M_2)$. □

The previous corollaries allow us to construct new resolute types from old ones. It has already been noted that minimal types are resolute, so that minimal types can be used. The following theorem shows that not just minimal types can be used.

Theorem 3.4.7 *If M is resolute and N is a minimal, conservative extension of M, then N is resolute.*

Proof Let M_0 be the prime model. Suppose that $M_0 \preccurlyeq M_1 \preccurlyeq N_0 \prec N_1 \preccurlyeq M_1 \star N$, with the intention of showing that $N_0 \cap N \prec N_1 \cap N$. (See Corollary 3.4.5.)

We can assume that $N_0 \cap (M_1 \star M) = N_1 \cap (M_1 \star M)$, as otherwise the resoluteness of M would imply that $N_0 \cap M \prec N_1 \cap M$.

Let $N_0' = \mathrm{Scl}(N_0 \cup M)$ and $N_1' = \mathrm{Scl}(N_1 \cup M)$. We claim that $N_0' \prec N_1'$. Assume that this is not so, and choose some $c \in N_1 \backslash N_0$. Then $c \in N_0'$, so there are elements $a \in N_0$ and $b \in M$ and a Skolem term $t(u, v)$ such that $c = t(a, b)$. Now let $d \in M_1 \star N$ be the least element such that $c = t(a, d)$. Since $d \leq b \in M_1 \star M \prec_{\mathsf{end}} M_1 \star N$, it follows that $d \in M_1 \star M$. Clearly, $d \in N_1$. Thus,

$d \in N_1 \cap (M_1{\star}M)$, and then, by the earlier assumption, $d \in N_0 \cap (M_1{\star}M)$. It follows then that $c \in N_0$, which is a contradiction, proving the claim that $N_0' \prec N_1'$.

Since N is a minimal, resolute extension of M, the previous claim implies that $N_0' = M_1{\star}M$ and $N_1' = M_1{\star}N$. But then $N_0 \preccurlyeq M_1{\star}M$, so that $N_0 = N_1$, which is a contradiction. □

There is a companion result to Theorem 3.4.7 for cofinal extensions, but it is not quite as neat. Suppose $I \subseteq_{\mathsf{end}} M$ and $p(x) \in S_1(M)$. Then say that the type $p(x)$ is *uniformly selective over* I if it is selective and, in addition, whenever $t(u,x)$ is a Skolem term in the language $\mathcal{L}(M)$ and $i \in I$, then there is a formula $\varphi(x)$ in $p(x)$ such that the sentence

$$\forall u \leq i[\forall x(\varphi(x) \longrightarrow t(u,x) > i) \vee \exists y \leq i \forall x(\varphi(x) = y)]$$

is in T.

Theorem 3.4.8 *Let M be a resolute model generated by a proper cut $I \subseteq_{\mathsf{end}} M$. Let $p(x) \in S_1(M)$ be a bounded type that is uniformly selective over I. If N is a $p(x)$-extension of M, then N is resolute.*

Proof First observe that N is a minimal extension of N since $p(x)$ is selective. Let M_0 be the prime model. Since M is resolute, it is a conservative, hence end, extension of M_0. Thus, we can assume that $M_0 < I$.

The proof now parallels the proof of Theorem 3.4.7.

Suppose that $M_0 \preccurlyeq M_1 \preccurlyeq N_0 \prec N_1 \preccurlyeq M_1{\star}N$, with the intention of showing that $N_0 \cap N \prec N_1 \cap N$. As in the proof of Theorem 3.4.7, we can assume that $N_0 \cap (M_1{\star}M) = N_1 \cap (M_1{\star}M)$.

We again let $N_0' = \mathrm{Scl}(N_0 \cup M)$ and $N_1' = \mathrm{Scl}(N_1 \cup M)$ and claim that $N_0' \prec N_1'$. The same proof works here except for one point: we do not have that $M_1{\star}M \prec_{\mathsf{end}} M_1{\star}N$. But we do have a replacement fact that suffices.

Since $I \subseteq_{\mathsf{end}} M \prec M_1 \star M$, we let $J = \sup(I) \subseteq_{\mathsf{end}} M_1 \star M$. Then the replacement fact is : $M_1 \star N$ is a J-extension of $M_1 \star M$. To see this, suppose that $q \in J$, and let $i \in I \backslash M_0$ be such that $q \leq i$. The goal is to show that $q \in M_1 \star M$.

There are a Skolem term $t(u,x)$ in the language $\mathcal{L}(M)$ and $c \in M_1$ such that $t(c,b) = q$. Let $\varphi(x) \in p(x)$ be as in the definition of a uniformly selective type; that is, $\varphi(x)$ is an $\mathcal{L}(M)$-formula such that $M_1 \star N \models \varphi(b)$ and the displayed sentence above is true in $M_1 \star N$. Since $M_0 < i \in I$, then $c < i$. Thus, either

$$\forall x[\varphi(x) \longrightarrow t(c,x) > i]$$

or

$$\exists y \leq i \forall x[\varphi(x) \longrightarrow t(c,x) = y]$$

is true in $M_1 \star M$. Since $t(c,b) \leq i$, it must be that the second sentence holds, and therefore, $t(c,b) \in \mathrm{Scl}(M \cup \{i,c\}) \subseteq M_1 \star M$. □

Corollary 3.4.9 *Let M be a nonstandard countable resolute model generated by a proper cut $I \subseteq_{\mathsf{end}} M$. Then M has a superminimal cofinal I-extension N that is resolute.*

Proof Recall (Exercise 2.5.2) that M has a superminimal cofinal I-extension N. When constructing the type $p(x) \in S_1(M)$ such that N is $p(x)$-extension of M, interlace into the construction the requirements to make $p(x)$ uniformly selective over I (Do it!). □

3.5 The Paris–Mills theorems

Indiscernible types are used in this section to prove the Second Paris–Mills Theorem. The proof of the First Paris–Mills Theorem, which is also presented in this section, uses types related to indiscernible types.

It was shown in Proposition 2.1.14 that if $M \succ N$, then $\mathrm{GCIS}(M,N)$ is a cut of M closed under multiplication, and a converse to this was proved in Theorem 2.1.16. An improvement to Theorem 2.1.16 will be given in Theorem 3.5.3. A definition is needed.

Definition 3.5.1 If M is a model of PA^* and $X \subseteq M$, then we define the *outer cardinality* of X to be $OC(X) = \min\{|D| : X \subseteq D \in \mathrm{Def}(M)\}$.

Notice that $OC(X)$ is a cardinal number which depends not only on the set X but also the model M which it is a subset of. Our interest in outer cardinality is restricted to just cuts. If $I \subseteq_{\mathsf{end}} M$ is a proper cut, then $OC(I) = \min\{|a_M| : I < a \in M\}$ (Do it!).

Corollary 3.5.2 *If $I \subseteq_{\mathsf{end}} M$ is such that $OC(I) > |I|$, then I is closed under multiplication.*

Proof Let $M_0 = \mathrm{Scl}(I)$. Then $|M_0| < OC(I)$, so $\mathrm{GCIS}(M_0, M) = I$. Hence, I is closed under multiplication by Proposition 2.1.14. □

The previous corollary has a converse, at least for countable M.

Theorem 3.5.3 (The First Paris–Mills Theorem) *Suppose that M is countable and $I \subseteq_{\mathsf{end}} M$ is closed under multiplication. Then M has an elementary extension N such that $\mathrm{GCIS}(M,N) = I$ and $OC^N(I) = 2^{\aleph_0}$.*

This theorem raises the question of whether we can increase the cardinal $2^{\aleph_0}$ occurring in it. The next proposition places a restriction on doing this.

Proposition 3.5.4 *If $I \subseteq_{\mathsf{end}} M$ is a cut such that $OC(I) > 2^{|I|}$, then I is closed under exponentiation.*

Proof Suppose that $a \in I$. Then, in M, there is a definable bijection f from $[0, 2^a - 1]$ to what M thinks is the powerset of $[0, a-1]$. But then, f is in actuality an injection from $[0, 2^a - 1]$ into $\mathcal{P}([0, a-1])$. Since $|[0, a-1]| \leq |I|$, we get that $|[0, 2^a - 1]| \leq 2^{|I|}$, so that $2^a \in I$. □

This proposition also has a converse, at least for countable M.

Theorem 3.5.5 (The Second Paris–Mills Theorem) *Suppose that M is countable and $I \subseteq_{\mathsf{end}} M$ is closed under exponentiation. Then for any infinite cardinal λ, M has an elementary extension N such that* $\mathrm{GCIS}(M, N) = I$ *and* $OC^N(I) = \lambda$.

The proofs of the two Paris–Mills theorems will now be presented. For the proof of the second, to be given first, the following lemma is quite useful.

Lemma 3.5.6 *Assume that $1 \leq n < \omega$. Let M be a countable model and let $p(x) \in S_1(M)$ be an indiscernible bounded type. Let $I = \inf\{\mathrm{card}^M(\varphi(x)) : \varphi(x) \in p(x)\}$. Let $M(\bar{a})$ be a $p_n(\bar{x})$-extension of M, where $p_n(\bar{x})$ is the unique complete n-type extending $\{x_0 < x_1 < \cdots < x_{n-1}\} \cup p(x_0) \cup p(x_1) \cup \cdots \cup p(x_{n-1})$. Then* $I = \mathrm{GCIS}(M, M(\bar{a}))$.

Proof First we show that $\mathrm{GCIS}(M, M(\bar{a})) \subseteq I$. It suffices to show this just for the case $n = 1$. Suppose $b \in M \backslash I$. Then there is a formula $\varphi(x) \in p(x)$ such that $\mathrm{card}^M(\varphi(x)) \leq b$. Proceeding informally, let $t(x)$ be the Skolem term such that: if $\varphi(x)$ and $t(x) = y$, then x is the yth element in the set defined by $\varphi(x)$. Clearly, $t(a) < \mathrm{card}^M(\varphi(x)) \leq b$, so $t(a) > I$. (Notice that this part of the proof made no use of the indiscernibility of $p(x)$.)

Next, we show the converse inclusion, that $I \subseteq \mathrm{GCIS}(M, M(\bar{a}))$, by induction on n. Consider some n, assuming the inclusion holds for all smaller values. Suppose $b \in I$ and $c \in M(\bar{a}) \backslash M$, aiming for a contradiction. Let $t(\bar{x})$ be an n-ary Skolem term such that $c = t(\bar{a})$. It follows from the inductive hypothesis (Do it!) that there is a formula $\varphi(x) \in p(x)$ which forces that $t(\bar{x})$ is a one-to-one function (as defined just before Exercise 3.1.22). Suppose $c < b$, and then by indiscernibility and Exercise 3.1.8, we can also assume that all values of $t(\bar{x})$ on $\varphi(x)$ are less than b. Thus, $\mathrm{card}^M(\varphi(x)) \leq b$, contradicting that $b \in I$. □

Proof of Theorem 3.5.5 The way the theorem is stated, it is possible that $I = M$. If that is the case, we make the preliminary move of taking a countable elementary end extension of M. Thus, without loss of generality, we can assume that I is bounded.

This proof not only makes use of FRT inside the model M but also makes use of numerical bounds for Ramsey numbers. For natural numbers h, n, c, let $R(h, n, c)$ be the least number r such that whenever $|X| \geq r$ and $P : [X]^n \longrightarrow c$, then there is a subset $H \subseteq X$ such that $|H| = h$ and P is constant on $[X]^n$. Ramsey's Theorem for $n = 1$ is just the Pigeon-hole Principle, and in this case the exact value of $R(h, 1, c)$ is easily calculated: $R(h, 1, c) = c(h-1)+1$. There are various proofs of FRT, some yielding no information about the size of $R(h, n, c)$. On the other hand, there are proofs that yield the following:

$$1 \leq h, n, c \Longrightarrow R(h, n+1, c) \leq R(h, n, c)^{nR(h,n,c)}.$$

Moreover, this is provable in PA. Thus we get the following lemma:

> LEMMA: *Suppose $1 \leq n < \omega$. Let $I \subseteq_{\mathsf{end}} M$ be a cut closed under exponentiation, and let $Y \subseteq M$ be definable and* $\operatorname{card}(Y) > I$, *and let $F : [Y]^n \longrightarrow [0, c]$, where f is definable and $c < I$. Then there is a definable $X \subseteq Y$ such that f is constant on $[X]^n$ and* $\operatorname{card}(X) > I$.

Now proceed with the construction of an indiscernible type $p(x)$ as in the proof of Theorem 3.1.5 but using the above lemma and some additional care to guarantee that $I = \inf\{\operatorname{card}^M(\varphi(x)) : \varphi(x) \in p(x)\}$.

Now let N be a model generated by a set of λ elements each realizing the indiscernible type $p(x)$. It easily follows from Lemma 3.5.6 that $\operatorname{GCIS}(M, N) = I$ (DO IT!). It also follows from Lemma 3.5.6 that $OC(I) = \lambda$. For, letting a be one of the indiscernibles and $b \in M$ such that $I < b$, there is a Skolem term $t(x)$ such that $I < t(a) < b$ and $t(a) \notin M$. Then there is a formula $\varphi(x) \in p(x)$ such that

$$M \models \forall x, y[\varphi(x) \wedge \varphi(y) \wedge x < y \longrightarrow t(x) \neq t(y) < b].$$

Thus, for each one of the λ indiscernibles a, there is a distinct $t(a)$ such that $I < t(a) < b$. This completes the proof of the Second Paris–Mills Theorem. □

Proof of Theorem 3.5.3 The proof of this theorem involves the construction of continuum many types which fit together in a very compatible way. A rather detailed account of the construction of this type is given here. Most of the details of showing that it does what it should do are left for the reader to verify. The key concept underlying this construction is an appropriate notion of a large set.

As we have noticed before when dealing with countable models, we can assume that M is the prime model of its theory. Also, as we saw in the proof of Theorem 3.5.5, we can assume that I is a proper cut.

For $1 \leq n < \omega$, we say that a subset $A \subseteq M^n$ is *large* if it is definable and there are $a > I$ and definable subsets $X_0, X_1, \ldots, X_{n-1} \subseteq M$ such that $\operatorname{card}(X_0) = \operatorname{card}(X_1) = \cdots = \operatorname{card}(X_{n-1}) = a$, $A \subseteq X_0 \times X_1 \times \cdots \times X_{n-1}$, and $a^n / \operatorname{card}(A) \in I$. This last phrase is to be interpreted as: there is $b \in I$ such that $b \cdot \operatorname{card}(A) \geq a^n$. There are three needed lemmas concerning large sets.

Lemma 3.5.7 *Suppose A is large and $g : A \longrightarrow M$ is a definable function. Then there is a large $B \subseteq A$ such that either g is constant on B or else $g(x) > I$ for each $x \in B$.*

Proof There is (unique) $c \in M$ such that $M \models \operatorname{card}(\{x \in A : g(x) < c\}) < \frac{1}{2}\operatorname{card}(A) \leq \operatorname{card}(\{x \in A : g(x) \leq c\})$. If $c > I$, then let $B = \{x \in A : g(x) \geq c\}$.

If $c \in I$, then $B = \{x \in A : g(x) = i\}$, where $i \leq c$ is chosen to make $\mathrm{card}(B)$ as big as possible. In the first case $\mathrm{card}(B) \geq \mathrm{card}(A)/2$, and in the second case $\mathrm{card}(B) \geq \mathrm{card}(A)/(2(c+1))$. Either way, B is large. □

Lemma 3.5.8 *Suppose A is large and $c > I$. Then there are bounded, definable $X_0, X_1, \ldots, X_{n-1} \subseteq M$ such that* $\mathrm{card}(X_0) = \mathrm{card}(X_1) = \cdots = \mathrm{card}(X_{n-1}) = c$ *and $A \cap (X_0 \times X_1 \times \cdots \times X_{n-1})$ is large.*

Proof This argument is intended to be formalized in M. Without loss of generality, let $A \subseteq X^n$, and let $b < I < a$ be such that $|X| = a$ and $\mathrm{card}(A) \geq a^n/b$. We can assume that $c < a$ as otherwise just let each $X_i \supseteq X$. [The ensuing argument is essentially a probabilistic one, showing that for random choices of $X_0, X_1, \ldots, X_{n-1} \subseteq X$, the expected value of $\mathrm{card}(A \cap (X_0 \times X_1 \times \cdots \times X_{n-1}))$ is at least c^n/b.] We use $C(i,j)$ for the binomial coefficient which is the number of j-element subsets of an i-element set. Then

$$\sum \mathrm{card}(A \cap (X_0 \times X_1 \times \cdots \times X_{n-1})) = \mathrm{card}(A) \cdot (C(a-1, c-1))^n,$$

where the sum is taken over all $X_0, X_1, \ldots, X_{n-1} \subseteq X$ with each $\mathrm{card}(X_i) = c$. Since there are $(C(a,c))^n$ choices for the X_i's, there is some choice for which

$$\begin{aligned}&\mathrm{card}(A \cap (X_0 \times X_1 \times \cdots \times X_{n-1}))\\ &\qquad \geq \mathrm{card}(A) \cdot (C(a-1, c-1))^n/(C(a,c))^n \geq c^n/b,\end{aligned}$$

and, therefore, for this choice $A \cap (X_0 \times X_1 \times \cdots \times X_{n-1})$ is large. □

Lemma 3.5.9 *Suppose $A \subseteq M^n$ is large. Then there is $c \in M$ such that*

$$\begin{aligned}B_c = \{&\langle x_0, x_1, \ldots, x_{n-1}, x_n\rangle \in M^{n+1} :\\ &\langle x_0, x_1, \ldots, x_{n-1}\rangle, \langle x_0, x_1, \ldots, x_{n-2}, x_n\rangle \in A \text{ and } x_{n-1} < c \leq x_n\}\end{aligned}$$

is large.

Proof This argument is intended to be formalized in M. Without loss of generality, let $b < I < a$ be such that $A \subseteq (a_M)^n$ and $\mathrm{card}(A) \geq a^n/b$. Our object is to obtain c so that B_c is large. [In fact, we show that for a randomly chosen $c \leq a$, the expected value of $\mathrm{card}(B_c)$ is at least $a^{n+1}/8b^3$.]

We use $\bar{x}$ exclusively to denote elements of $(a_M)^{n-1}$. If $\bar{x} = \langle x_0, x_1, \ldots, x_{n-2}\rangle$, then let $Y(\bar{x}) = \{y < a : \langle x_0, x_1, \ldots, x_{n-2}, y\rangle \in A\}$ and let $e(\bar{x}) = \mathrm{card}(Y(\bar{x}))$. Then,

$$|B_c| = \sum_{\bar{x}} |(Y(\bar{x}) \cap c_M)| \cdot |(Y(\bar{x}) \backslash c_M)|,$$

so that

$$\sum_{c=0}^{a} |B_c| = \sum_{\bar{x}} \sum_{c=0}^{a} |(Y(\bar{x}) \cap c_M)| \cdot |(Y(\bar{x}) \backslash c_M)|.$$

For each $\bar{x}$ we have that

$$\begin{aligned}\sum_{c=0}^{a} |(Y(\bar{x}) \cap c_M)| \cdot |(Y(\bar{x}) \backslash c_M)| &\geq \sum_{c=0}^{a} |(e(\bar{x})_M \cap c_M)| \cdot |(e(\bar{x})_M \backslash c_M)| \\ &= \frac{1}{6}(e(\bar{x}) - 1)e(\bar{x})(e(\bar{x}) + 1).\end{aligned}$$

Therefore,

$$\sum_{c=0}^{a} |B_c| \geq \frac{1}{6} \sum_{\bar{x}} (e(\bar{x}) - 1)e(\bar{x})(e(\bar{x}) + 1).$$

The right-hand side of the previous inequality is bounded below by the sum obtained by replacing each occurrence of $e(\bar{x})$ with $|A|/a^{n-1}$ since

$$|A| = \sum_{\bar{x}} e(\bar{x}),$$

resulting in

$$\sum_{c=0}^{a} |B_c| \geq \frac{a^{n-1}}{6} \left(\frac{|A|}{a^{n-1}} - 1 \right)^3 > \frac{a^{n-1}}{7} \left(\frac{|A|}{a^{n-1}} \right)^3 \geq \frac{a^{n+2}}{7b^3}.$$

Thus, there is some c such that

$$|B_c| > \frac{a^{n+2}}{(a+1)7b^3} > \frac{a^{n+1}}{8b^3},$$

so that this B_c is large. □

Let $c_0 > c_1 > c_2 > \cdots$ be a sequence for which $I = \inf\{c_i : i < \omega\}$. We also need a list of all Skolem functions $f : M^n \longrightarrow M$, with each Skolem function appearing infinitely often in the list. We now inductively construct a sequence $A_0, A_1, A_2, \ldots$ of large sets, where $A_i \subseteq M^{2^i}$ for each $i < \omega$. With each A_i we also have subsets demonstrating that A_i is large, with the additional property

$$I < \operatorname{card}(X_0^{(i)}) = \operatorname{card}(X_1^{(i)}) = \cdots = \operatorname{card}(X_{2^i-1}^{(i)}) \leq c_i \,.$$

Getting A_0 is easy enough: let $A_0 = X_0^{(0)} = [0, c_0 - 1]$.

Now suppose that we have $A_i \subseteq X_0^{(i)} \times X_1^{(i)} \times \cdots \times X_{2^i-1}^{(i)}$. We get A_{i+1} by applying the three lemmas. Suppose the ith Skolem function in our list is the n-ary function f. For each n-tuple $k_0 < k_1 < \cdots < k_{n-1} < 2^i$, there is the induced function $g : A \longrightarrow M$ where

$$g(\bar{x}) = f(x_{k_0}, x_{k_1}, \ldots, x_{k_{n-1}}).$$

Apply Lemma 3.5.7 repeatedly to get large $B \subseteq A_i$ such that each of the functions g which f induces is either constant on B or else $g(\bar{x}) > I$ for each $\bar{x} \in B$. Next, apply Lemma 3.5.8 and get $X'_j \subseteq X_j$ for $j < 2^i$ such that

$$I < \mathrm{card}(X'_0) = \mathrm{card}(X'_1) = \cdots = \mathrm{card}(X'_{2^i-1}) \leq c_{i+1}$$

and $B \cap (X'_0 \times X'_1 \times \cdots \times X'_{2^i-1})$ is large. Finally, apply Lemma 3.5.9 2^i times, getting c_j for $j < 2^i$, such that, letting

$$X_{2j}^{(i+1)} = \{x \in X'_j : x < c_j\}, \qquad X_{2j+1}^{(i+1)} = \{x \in X'_j : x \geq c_j\}$$

and letting A_{i+1} be the set of those $\bar{x} \in X_0^{(i+1)} \times X_1^{(i+1)} \times \cdots \times X_{2^{i+1}-1}^{(i+1)}$ such that whenever $\bar{y}$ is a 2^i-tuple which is a subsequence of $\bar{x}$, with $y_j \in \{x_{2j}, x_{2j+1}\}$, then $\bar{y} \in B$, we have that A_{i+1} is large.

The A_i's that we just constructed are used to construct a massive type Γ. We consider $\{0,1\}^\omega$, the set of ω-sequences of 0's and 1's. For each $k < \omega$ and each such sequence s, let $s|k$ be the initial subsequence of s of length k. We consider $\{0,1\}^\omega$ to be ordered lexicographically; thus $s < t$ iff there is $k < \omega$ such that $s|k = t|k$ and $s(k) < t(k)$. For each $s \in \{0,1\}^\omega$, we introduce a variable v_s. Then Γ is a type in all of these $2^{\aleph_0}$ variables.

Consider a formula $\varphi(v_{s_0}, v_{s_1}, \ldots, v_{s_{n-1}})$, where, without loss of generality, $s_0 < s_1 < \cdots < s_{n-1}$. This formula is in Γ iff there is $i < \omega$ such that the following holds:

Let $f : M^n \longrightarrow \{0,1\}$ be the Skolem function such that $f(\bar{a}) = 0$ iff $M \models \varphi(\bar{a})$. There is $i < \omega$ such that $s_0|i < s_1|i < \cdots < s_{n-1}|i$ and such that if $g : M^{2^i} \longrightarrow \{0,1\}$ is the induced function such that $g(\bar{x}) = f(\bar{y})$, where $\bar{x}$ is a 2^i-tuple, $\bar{y}$ is an n-tuple which is a subsequence of $\bar{x}$ for which $y_j = x_k$ if $s_j|i$ is the kth element of $\{0,1\}^i$, and g is constant 0 on A_i.

This type Γ is consistent with $\mathrm{Th}(M)$ (Do it!) and is complete in the sense that for any appropriate formula φ, either $\varphi \in \Gamma$ or $\neg\varphi \in \Gamma$. Then Γ can be used to get an elementary extension N of M by first obtaining some model in which Γ is realized (say by $\langle a_s : s \in \{0,1\}^\omega \rangle$), and then letting $N = \mathrm{Scl}(\{a_s : s \in \{0,1\}^\omega\})$. It can be checked (Do it!) that $I = \mathrm{GCIS}(M, N)$ and $OC^N(I) = 2^{\aleph_0}$. □

3.6 Exercises

Reminder: in this chapter, including in this Exercises section, we have a fixed completion T of PA^*.

♥**3.6.1** There is a model M of T which has a minimal elementary end extension $M(a)$ in which the type of a is not selective.

♦**3.6.2** There are end-extensional types which are not rare.

♦**3.6.3** Suppose $M \prec N$ and $a \in M$, $b \in N$ are such that a, b are in the same gap of N and b realizes a rare type. Then $b \in M$.

♦**3.6.4** If γ is a gap of M, then the number of elements of γ realizing rare types is either 0 or $\aleph_0$.

♣**3.6.5** Without Theorem 3.1.9, prove that every unbounded 3-indiscernible type is definable by showing that for any formula $\varphi(u, x)$, a defining formula $\sigma(u)$ can be obtained as the formula $\forall x \forall y[\theta(x) \wedge \theta(y) \wedge u \leq x < y < z \longrightarrow \varphi(u, y)]$, where $\theta(x)$ is a formula in the type which forces $\forall u[u \leq x \longrightarrow (\varphi(u, y) \longleftrightarrow \varphi(u, z))]$.

Suppose that $t(x)$ is a Skolem term and $p(x)$ is a type. We then define $t(p(x))$ to be the set of all formulas $\theta(x)$ such that for some $\varphi(x) \in p(x)$, the sentence $\forall x[\varphi(x) \longrightarrow \theta(t(x))]$ is in T. A set X of minimal types is *independent* if whenever $p(x) \in X$ and $t(x)$ is a Skolem term, then either $t(p(x)) = p(x)$ or $t(p(x)) \notin X$. The following four exercises concern these notions.

♣**3.6.6** If $p(x)$ is a complete type and $t(x)$ is a Skolem term, then $t(p(x))$ is a complete type. Furthermore, if $a \in M$ realizes $p(x)$, then $t(a)$ realizes $t(p(x))$.

♣**3.6.7** If $p(x)$ is a minimal type, then $t(p(x))$ is either a minimal type or a principal type.

♣**3.6.8** If $p(x)$ is an end-extensional type, then $t(p(x))$ is either an end-extensional type or a principal type.

♦**3.6.9** There is a set of $2^{\aleph_0}$ independent minimal types.

♦**3.6.10** Let $M \prec_{\mathsf{end}} N$ and $A \subseteq N \backslash M$ be such that $N = \mathrm{Scl}(M \cup A)$, where A is a set of elements realizing minimal types. Then there is $B \subseteq N \backslash M$ and an independent set X of minimal types such that each element of B realizes a type in X and $N = \mathrm{Scl}(M \cup B)$.

♦**3.6.11** There is M such that $|M| = 2^{\aleph_0}$, no two distinct elements of M realize the same type, and $\mathrm{Lt}(M) \cong \mathcal{P}(\mathbb{R})$.

The following two exercises give characterizations of end-extensional types which could be used as definitions.

♦**3.6.12** The type $p(x)$ is end-extensional iff whenever M is a model of T and $M(a)$ is an elementary extension of M generated by a, where $a > M$ and a realizes $p(x)$, then $M(a)$ is an end extension of M.

♣**3.6.13** The type $p(x)$ is end-extensional iff whenever M is a model of T and $M(a)$ is an elementary extension of M generated a, then $M(a)\backslash M$ is a gap of $M(a)$.

♦**3.6.14** There are unbounded rare types that are not end-extensional.

♣**3.6.15** For each completion of PA^*, there are $2^{\aleph_0}$ rare types which are not definable.

♦**3.6.16** Show that Theorem 3.2.11 becomes false when "rare" is replaced by "selective."

♦**3.6.17** There are unbounded selective types that are not definable. (HINT: *modify the proof of Theorem 1.2.12.*)

♦**3.6.18** If $p(x)$ is an n-resolute type and M is a $p(x)$-extension of the the prime model, then $|\,\mathrm{Lt}(M)| \leq n+1$.

♦**3.6.19** Let M be a nonstandard countable n-resolute model generated by a proper cut $I \subseteq_{\mathsf{end}} M$. Then M has a superminimal cofinal I-extension N that is $(n+1)$-resolute.

The definition of weakly Ramsey type is adapted from a definition originating in ultrafilter theory. Suppose $p(x)$ is a type and $k \in \mathbf{N}$. Then $p(x)$ is a *k-weakly Ramsey* if there are at most k 2-types containing $p(x) \cup p(y) \cup \{x < y\}$. Thus, the 0-weakly Ramsey types are just the principal types, and the nonprincipal 1-weakly Ramsey types are just the 2-indiscernible types. A type which is k-weakly Ramsey but is not m-weakly Ramsey for any $m < k$ is called *strictly* k-weakly Ramsey.

♣**3.6.20** Give an easy construction which shows: there are definable strictly 6-weakly Ramsey types. (HINT: *see Exercise 3.1.3.*)

♣**3.6.21** Make a small change to the type constructed in Exercise 3.6.20 to show: there are definable strictly 5-weakly Ramsey types.

♦**3.6.22** Make a small change to the type constructed in Exercise 3.6.21 to show: there are definable strictly 2-weakly Ramsey types.

♦**3.6.23** Every definable 3-weakly Ramsey type is end-extensional.

♥**3.6.24** If $p(x)$ is 4-weakly Ramsey and M is a $p(x)$-extension of the prime model, then $\mathrm{Lt}(M)$ is linearly ordered.

♦**3.6.25** are definable 4-weakly Ramsey types that are not end-extensional.

♥**3.6.26** There is an end-extensional 5-weakly Ramsey type $p(x)$ such that if M is a $p(x)$-extension of the prime model of T, then $\mathrm{Lt}(M)$ is not linearly ordered.

♦**3.6.27** Suppose that $p(x)$ is k-weakly Ramsey and that $M(a)$ is $p(x)$-extension of the prime model M. Then $M(a)$ has at most k proper elementary substructures.

♥**3.6.28** Suppose $p(x)$ is a definable k-weakly Ramsey type and $M(a)$ is a $p(x)$-extension of the prime model. If $M(a)$ has exactly k proper elementary substructures, then (a) $\mathrm{Lt}(M(a))$ is linearly ordered and (b) $p(x)$ is end-extensional.

♦**3.6.29** For each $k \in \omega$, there is an end-extensional strictly k-weakly Ramsey type.

♦**3.6.30** There is a selective type which is not k-weakly Ramsey for any $k \in \omega$.

Refer to Section 2.3 for definitions related to amalgamations. The next exercise concerns the number of inequivalent amalgamations that two isomorphic copies of a model M can have.

♦**3.6.31** Let $p(x)$ be a type and let $M(a)$ be a $p(x)$-extension of the prime model M. If $k \in \omega$, then the following are equivalent:

$p(x)$ is strictly k-weakly Ramsey;
$M(a)$ has exactly $2k+1$ inequivalent amalgamations with itself.

We define another kind of type. Suppose that $p(x)$ is a nonprincipal type and $k \in \omega$. Then $p(x)$ is a *k-arrow* type if for any formula $\theta(x, y)$, either

$$T \vdash \exists x_0, x_1, \ldots, x_{k-1}[x_0 < x_1 < \cdots < x_{k-1} \wedge \bigwedge_{i<j<k} \neg\theta(x_i, x_j)]$$

or there is a formula $\varphi(x) \in p(x)$ such that

$$T \vdash \forall x \forall y[\varphi(x) \wedge \varphi(y) \wedge x < y \longrightarrow \theta(x, y)] \,.$$

Every nonprincipal type is 2-arrow.

♣**3.6.32** Every 2-indiscernible type is a k-arrow type for each $k \in \omega$.

♥**3.6.33** If T does not have a standard model and $3 \leq k < \omega$, then there is a k-arrow type that is not $(k+1)$-arrow.

The definition of amalgamation (for two models of T) can easily be extended to k models, whenever $2 < k \in \omega$, and the notion of equivalent amalgamations can be generalized to this context. If $M_0, M_1, \ldots, M_{k-1}$ are models and

$e_i : M_i \longrightarrow N$ are elementary embeddings which form an amalgamation, then for any $i < j < k$, there is an *induced* amalgamation of M_i and M_j, namely $e_i : M_i \longrightarrow N_{ij}$ and $e_j : M_j \longrightarrow N_{ij}$, where N_{ij} is the elementary substructure of N generated by $e_i(M_i) \cup e_j(M_j)$.

♦**3.6.34** For $k \in \omega$, the type $p(x)$ is a k-arrow type iff whenever M is a $p(x)$-extension of the prime model and $e, f : M \longrightarrow N$ is an amalgamation of M with itself, then there is an amalgamation of k copies of M such that each induced amalgamation of two copies of M is equivalent to the amalgamation e, f.

♦**3.6.35** Every 3-weakly Ramsey type is a k-arrow type for each $k < \omega$.

♦**3.6.36** Let $p(x)$ be a nonprincipal type such that for any formula $\theta(x, y, z)$, either there is a formula $\varphi(x) \in p(x)$ such that

$$T \vdash \forall x, y, z[\varphi(x) \wedge \varphi(y) \wedge \varphi(z) \wedge x < y < z \longrightarrow \theta(x, y, z)]$$

or

$$T \vdash \exists w, x, y, z[w < x < y < z \\ \wedge \neg\theta(x, y, z) \wedge \neg\theta(w, y, z) \wedge \neg\theta(w, x, z) \wedge \neg\, \theta(w, x, y)].$$

Then $p(x)$ is 2-indiscernible.

Let I be a proper cut of M. A type $p(x) \in S_1(M)$ is *indiscernible over* I if for any $\mathcal{L}(M)$-formula $\theta(u, x_0, x_1, \ldots, x_n)$, there is a formula $\varphi(x) \in p(x)$ such that whenever $a \in I$, then $\varphi(x)$ forces $\theta(a, \bar{x})$. (See Exercise 3.1.8.)

♣**3.6.37** If M is a nonstandard model and I is a proper cut of M, then there is a bounded type $p(x) \in S_1(M)$ which is indiscernible over I.

♣**3.6.38** If $p(x)$ is indiscernible over I, then $p(x)$ is selective over I.

If I is a bounded cut of M and $M \prec_{\mathsf{end}} N$, then N *fills* I if there is $b \in N$ such that whenever $a < I < c$ are elements of M, then $a < b < c$. The next two exercises refine the two Paris–Mills theorems.

♦**3.6.39** To the conclusion of Theorem 3.5.5, the requirement that N does not fill I can be added.

♦**3.6.40** To the conclusion of Theorem 3.5.3, the requirement that N does not fill I can be added.

A type $p(x) \in S_1(T)$ is *ubiquitous* if, whenever M is recursively saturated and $a \in M$ realizes $p(x)$, then the set of elements in gap(a) realizing $p(x)$ is cofinal in gap(a). A type $p(x) \in S_1(T)$ is *locally ubiquitous* if, whenever M is recursively saturated, $a \in M$ realizes $p(x)$, and $c \in M$ is nonstandard, then there is $b \in [a+1, c]$ realizing $p(x)$.

♣**3.6.41** If $p(x)$ is ubiquitous, M is recursively saturated, $a \in M$ realizes $p(x)$, and $b \in \text{gap}(a)$, then there is c realizing $p(x)$ such that $b > c \in \text{gap}(a)$.

♦**3.6.42** Every recursively saturated model has an element realizing a ubiquitous type.

♥**3.6.43** Every arithmetically saturated model has an element realizing a locally ubiquitous type.

♠**3.6.44** Does every recursively saturated model have an element realizing a locally ubiquitous type?

3.7 Remarks & References

Much of the material in this section concerning definable, indiscernible, minimal, and end-extensional types had its origins in the seminal and highly influential work of Gaifman [43], where many of the results first appeared. Definable types, which originated in [43], evolving from the notion of a conservative extension, are also of central importance in stability theory. A complete theory T is stable iff whenever A is a subset of a model of T, then every type in $S_1(A)$ is definable.

Theorem 3.1.20 is from [103]. Theorem 3.1.9 is from [103]. Rare types are the analogues of rare ultrafilters (also called Q-points) over ω. Theorem 3.2.11 is from Schmerl [174] but is quite similar to a theorem in Kossak & Schmerl [107]. The proof presented here, due to Ermek Nurkhaidarov, is simpler than the one in [174]. Theorem 3.3.14 and its proof are taken directly from [174].

The concept of a resolute type appears here for the first time.

The material in Section 3.5 on the two Paris–Mills theorems is due to Jeff Paris and George Mills, and it is taken from their paper [148].

The notion of a k-arrow type appearing in the Exercises was influenced by Baumgartner & Taylor [7], where k-arrow ultrafilters were introduced. Blass in several papers, especially in [16], has stressed the close connection between types and ultrafilters. Many of the exercises, including 3.6.28 and 3.6.34, are suggested by this connection. In this regard, Exercise 3.6.33 is essentially in [7].

There are other kinds of types not mentioned in this chapter. For example, the weakly definable types introduced by Kirby [81] and further studied by him and Anand Pillay [84] and by Schmerl [170], while interesting, have not yet proved to be useful.

4

SUBSTRUCTURE LATTICES

The set of all elementary substructures of a model of Peano Arithmetic is a lattice. This is most easily seen by observing that if M is a model and $\mathcal{C}$ is an arbitrary collection of elementary substructures of M, then its intersection $\bigcap \mathcal{C}$ is also an elementary substructure of M. This lattice is called the *substructure lattice* of M and is denoted by $\mathrm{Lt}(M)$. In this chapter we take a detailed look at substructure lattices, with the unrealized goal of characterizing those lattices which can appear as substructure lattices. More generally, suppose that $M \preccurlyeq N$, then define the *interstructure lattice* $\mathrm{Lt}(N/M)$ to be the sublattice of $\mathrm{Lt}(N)$ consisting of those models M_0 for which $M \preccurlyeq M_0 \preccurlyeq N$. Given a model M, we also consider in this chapter the question of which lattices can appear as some interstructure lattice $\mathrm{Lt}(N/M)$.

These questions are a natural outgrowth of the existence of minimal extensions. For, if we let **2** be the 2-element lattice, then a consequence of Gaifman's Theorem 3.3.1 is that every model M has an elementary end extension N such that $\mathrm{Lt}(N/M) \cong \mathbf{2}$. Blass' Theorem 2.1.1 has a similar sort of consequence concerning cofinal extensions of countable nonstandard models. As to which lattices can appear as substructure and interstructure lattices, there are three main results in this chapter. Many finite lattices are shown in Section 4.5 to be substructure lattices. In Section 4.6 it is proved that the pentagon lattice always can appear as an interstructure lattice. Finally, those distributive lattices which can appear as substructure lattices are characterized in Section 4.7.

4.1 Lattices

Lattices can be thought of in two ways: as partially ordered sets or as algebras. This section contains definitions of a lattice and also definitions of an algebra and other related concepts. Examples of lattices that arise from algebras are also discussed. The reader having some familiarity with lattices may want to skip this section at first, referring to it as needed.

For the first of the two approaches to lattices, consider $(P, \leq)$ which is a partially ordered set (hereinafter referred to as a *poset*). Small finite posets are conveniently represented by their Hasse diagrams. Some examples are given in Figure 4.1.

Given any subset $X \subseteq P$ and element $b \in P$, we say that b is the *supremum* (also called the *least upper bound*) of X, denoted by $\sup(X)$ or $\bigvee X$, if b is the unique element of P such that for any $y \in P$, if $x \leq y$ for each element $x \in X$,

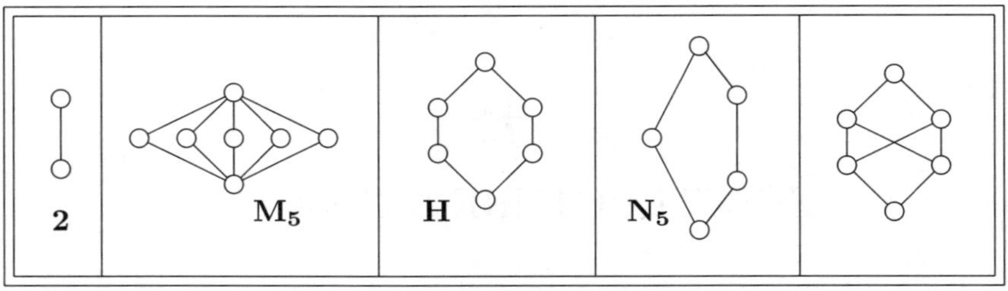

FIG. 4.1. The Hasse diagrams of several small posets.

then $b \leq y$. The subset X may or may not have a supremum, but if it does, then it is unique. (Of course, the uniqueness was built into the definition, but it was not essential to do so.) We similarly define the *infimum* of X, denoted by $\inf(X)$ or $\bigwedge X$, to be the unique $a \in P$ such that for any $y \in P$, if $y \leq x$ for each element $x \in X$, then $y \leq a$. If $X = \{x, y\}$ and $b = \sup(X)$, then we also write $b = x \vee y$, and if $a = \inf(X)$, then we write $a = x \wedge y$. The poset $(P, \leq)$ is a *lattice ordered set* if both $x \vee y$ and $x \wedge y$ exist whenever $x, y \in P$.

The first of the posets depicted in Figure 4.1 is a chain of length 2. For any natural number $n \geq 1$, the chain $\mathbf{n}$ of length n is a lattice ordered set. The second of the depicted posets is also one of an infinite family of lattice ordered sets: for each $n < \omega$, $\mathbf{M_n}$ is the lattice ordered set having $n + 2$ elements, n of which are pairwise incomparable. All but the last of the depicted posets is a lattice ordered set.

We next take a look at the other, more algebraic, approach. Consider a set L on which there are two binary operations $\vee$ and $\wedge$. Then $(L, \vee, \wedge)$ is an example of an algebra; this concept is given a precise definition later in this section. The algebra $(L, \vee, \wedge)$ is a *lattice* if it obeys the following four pairs of laws:

$$\text{commutative laws}: \quad x \vee y = y \vee x, \qquad x \wedge y = y \wedge x;$$

$$\text{associative laws}: \quad (x \vee y) \vee z = x \vee (y \vee z), \qquad (x \wedge y) \wedge z = x \wedge (y \wedge z);$$

$$\text{idempotency laws}: \quad x \vee x = x, \qquad x \wedge x = x;$$

$$\text{absorption laws}: \quad x \vee (x \wedge y) = x, \qquad x \wedge (x \vee y) = x.$$

The operation $\vee$ is the *meet* and $\wedge$ is the *join* of L. These two concepts, that of a lattice ordered set and a lattice, are essentially the same. Given a lattice ordered set $(P, \leq)$, we have already seen how to define the binary operations $\vee$ and $\wedge$, and with these operations, $(P, \vee, \wedge)$ is lattice. Conversely, if $(P, \vee, \wedge)$ is a lattice, then there is a binary relation $\leq$ which can be defined in either of two equivalent ways to get the poset $(P, \leq)$: let $x \leq y$ iff $x \vee y = y$, or let $x \leq y$

iff $x \wedge y = x$. These two procedures, going from $(P, \leq)$ to $(P, \vee, \wedge)$ and going from $(P, \vee, \wedge)$ to $(P, \leq)$, are inverses of one another. Because these two ways of defining a lattice are so similar, no effort is made to distinguish them: a lattice is viewed as a poset and/or an algebra. However, a little care must be used when considering sublattices. If $(L_1, \vee_1, \wedge_1)$ and $(L_2, \vee_2, \wedge_2)$ are lattices, then L_1 is a *sublattice* of L_2 iff $L_1 \subseteq L_2$ and whenever $x, y \in L_1$, then $x \vee_1 y = x \vee_2 y$ and $x \wedge_1 y = x \wedge_2 y$. If $(L_1, \leq_1)$ and $(L_2, \leq_2)$ are lattice posets and L_1 is a subposet of L_2, then it is possible that L_1 is *not* a sublattice of L_2.

If $(L, \leq)$ is a poset and if every subset $X \subseteq L$ has a supremum, then $(L, \leq)$ is a lattice. For, given $x, y \in X$, we see that $x \wedge y = \bigvee\{z \in L : z \leq x, y\}$. In fact, for every subset $X \subseteq L$, $\bigwedge X = \bigvee\{z \in P : z \leq x \text{ for all } x \in X\}$. If $(L, \leq)$ is a poset for which $\sup(X)$ and $\inf(X)$ exist whenever $X \subseteq L$, then $(L, \leq)$ is a *complete* lattice. In a complete lattice L we often let $0 = \bigvee \emptyset = \bigwedge L$ and $1 = \bigwedge \emptyset = \bigvee L$. Every finite lattice is complete.

For any set X, let $\mathcal{P}(X)$, the powerset of X, be the set of all subsets of X. Then $(\mathcal{P}(X), \subseteq)$ is a poset which is a complete lattice ordered set with join $\cup$ and meet $\cap$. If $X = n = \{0, 1, 2, \ldots, n-1\}$, then $\mathbf{B_n} = (\mathcal{P}(n), \cup, \cap)$ is the *Boolean* lattice having 2^n elements. Notice that $\mathbf{B_0} \cong \mathbf{1}$, $\mathbf{B_1} \cong \mathbf{2}$, and $\mathbf{B_2} \cong \mathbf{M_2}$.

For another example of a complete lattice, let $\mathrm{Eq}(X)$ be the set of equivalence relations on X. Each equivalence relation is a subset of $X \times X$, so $(\mathrm{Eq}(X), \subseteq)$ is a subposet of $\mathcal{P}(X \times X)$. It is not a sublattice of $(\mathcal{P}(X \times X), \cup, \cap)$ (unless $|X| \leq 2$) as the union of two equivalence relations need not be an equivalence relation. However, the intersection of two equivalence relations is an equivalence relation. Notice that if Θ_1 and Θ_2 are equivalence relations in $\mathrm{Eq}(X)$, then the equivalence classes of $\Theta_1 \cap \Theta_2$ are the nonempty sets of the form $X_1 \cap X_2$, where X_1, X_2 are equivalence classes of Θ_1, Θ_2 respectively. Moreover, the intersection of an arbitrary collection of equivalence relations is also an equivalence relation, so that $(\mathrm{Eq}(X), \subseteq)$ is a complete lattice. If $\Theta_1 \subseteq \Theta_2$ are equivalence relations on X, then Θ_1 is a *refinement* of Θ_2.

For any lattice $(L, \vee, \wedge)$, there is its *dual* lattice $(L, \wedge, \vee)$, and for any poset $(P, \leq)$ there is its *dual* poset $(P, \geq)$. We let $L^{\perp}$ and $P^{\perp}$ denote the duals of the lattice L and the poset P respectively. It makes no difference whether we view the lattice L as an algebra or a poset, its dual $L^{\perp}$ is well-defined. The dual of a complete lattice is also a complete lattice.

The two *extreme* equivalence relations in $\mathrm{Eq}(X)$ are $\mathbf{0}_X = \bigwedge \mathrm{Eq}(X)$, which is the *equality* relation on X, and $\mathbf{1}_X = \bigvee \mathrm{Eq}(X) = X \times X$, which is the *trivial* equivalence relation on X.

There are many sources of lattices. We discuss two lattices which come from algebras: the subalgebra lattice and the congruence lattice. We begin with a definition of an algebra. An *algebra* is a structure of the form $(A, \langle f_i : i \in I \rangle)$ such that for each $i \in I$ there is some $n < \omega$ for which $f_i : A^n \longrightarrow A$. The number n associated with the operation f_i is the *arity* of f_i. It is allowed for the arity of an operation to be 0; operations of arity 0 are sometimes called *constants*.

The functions f_i are the *operations* of A. The *type* of this algebra is the function $\tau : I \longrightarrow \omega$, where $\tau(i)$ is the arity of the operation f_i.

A lattice is an algebra. Officially, the type of a lattice is the function $\tau : \{\vee, \wedge\} \longrightarrow \omega$ where $\tau(\vee) = \tau(\wedge) = 2$. We follow the more customary practice and say that its type is $\langle 2, 2\rangle$. Some other familiar examples of algebras are any group $(G, \cdot, ^{-1}, e)$, which has type $\langle 2, 1, 0\rangle$, and any ring $(A, +, \cdot, 0)$, which has type $\langle 2, 2, 0\rangle$.

If $(A, \langle f_i : i \in I\rangle)$ and $(B, \langle g_i : i \in I\rangle)$ are algebras (of the same type τ), then B is a *subalgebra* of A if $B \subseteq A$ and, for each $i \in I$, the operations f_i and g_i agree on B. (A small point: it is allowed that a subalgebra might be empty; however, if the type has an operation of arity 0, then there are no empty algebras of that type.) We let $\mathrm{Sub}(A)$ be the set of subalgebras of A, which is to be considered as a poset ordered by the subalgebra relation. If A is an algebra and $X \subseteq A$ (by which is meant that X is a sub*set* of A but is not necessarily a sub*algebra*), then there is a smallest subalgebra $B \subseteq A$ such that $X \subseteq B$. For such X and B, we say that X *generates* B. It then follows that, not only is $\mathrm{Sub}(A)$ a lattice, but it is even a complete lattice. The lattice $\mathrm{Sub}(A)$ is the *subalgebra lattice* of A.

Not just any complete lattice can be isomorphic to some $\mathrm{Sub}(A)$. We make some definitions which are used to characterize those lattices isomorphic to subalgebra lattices.

Let $(L, \vee, \wedge)$ be a complete lattice. An element $a \in L$ is *compact* if whenever $X \subseteq L$ and $a \leq \bigvee X$, then $a \leq \bigvee Y$ for some finite $Y \subseteq X$. The compact elements of $\mathcal{P}(X)$ are the finite subsets of X (DO IT!). It is easy to identify the compact elements of a subalgebra lattice. An algebra A is *finitely generated* if there is a finite $X \subseteq A$ which generates A. The following proposition shows that the finitely generated subalgebras of A are recognizable in $\mathrm{Sub}(A)$ as an abstract lattice. Its easy proof is left as an exercise.

Proposition 4.1.1 *If A is an algebra, then the compact elements of* $\mathrm{Sub}(A)$ *are exactly the finitely generated subalgebras of A.* □

A lattice L is *algebraic* if it is complete and each element of L is the supremum of a set of compact elements. It follows from Proposition 4.1.1 that $\mathrm{Sub}(A)$ is algebraic. The reader may find it instructive to show that $\mathrm{Eq}(A)$ and $\mathrm{Eq}(A)^{\perp}$ are algebraic and to determine the compact elements of each. This definition can be refined to include a cardinal parameter. For κ an infinite cardinal, we say that L is *κ-algebraic* if L is algebraic and for each compact $x \in L$, $|\{a \in L : a \leq x \text{ and } a \text{ is compact}\}| < \kappa$. Every finite lattice is $\aleph_0$-algebraic. The following proposition, which is most relevant to models of PA in the case when $\kappa = \aleph_0$, has a routine proof.

Proposition 4.1.2 *If κ is an infinite cardinal and A is an algebra with at most κ operations, then* $\mathrm{Sub}(A)$ *is a κ^+-algebraic lattice.* □

There is a converse to Proposition 4.1.2.

Proposition 4.1.3 *Let κ be an infinite cardinal and L a κ^+-algebraic lattice. Then there is an algebra A with κ operations such that* $\mathrm{Sub}(A) \cong L$.

Proof Let A be the set of compact elements of L. For each $x \in L$, let K_x be the set of compact $a \leq x$, and then let $K_x = \{a_{x,i} : i < \kappa\}$. For each $i < \kappa$, let $f_i : A^2 \longrightarrow A$ be such that $f_i(x, y) = a_{x \vee y, i}$. Then $\mathrm{Sub}(A, \vee, \langle f_i : i < \kappa \rangle) \cong L$; in fact, the function $B \mapsto \bigvee B$ is an isomorphism (Do IT!). (See Exercise 4.8.4.) □

The subalgebra lattice is one of the important lattices associated with an algebra. We next consider another one, the congruence lattice of an algebra.

Let A be an algebra. A *congruence* of A is an equivalence relation Θ on the set A such that whenever f is an n-ary operation of A and $a_i, b_i \in A$ for $i < n$ are such that $\langle a_i, b_i \rangle \in \Theta$ for each $i < n$, then $\langle f(\bar{a}), f(\bar{b}) \rangle \in \Theta$. For a familiar example, consider a group G. If N is a normal subgroup of G, then the cosets of N are the equivalence classes of a congruence of G. Conversely, for any congruence Θ of G, the set of elements congruent to e is a normal subgroup of G whose cosets are precisely the equivalence classes of Θ. Let $\mathrm{Cg}(A)$ be the set of congruences of the algebra A. The intersection of an arbitrary collection of congruences is a congruence (Do IT!). Thus, $\mathrm{Cg}(A)$ is a complete lattice in which $\bigwedge \mathrm{Cg}(A) = \mathbf{0}_A$ and $\bigvee \mathrm{Cg}(A) = \mathbf{1}_A$, and we always think of $\mathrm{Cg}(A)$ as a lattice. Not only is $\mathrm{Cg}(A)$ a sublattice of $\mathrm{Eq}(A)$ but also, if $\mathcal{C} \subseteq \mathrm{Cg}(A)$, then $\bigvee \mathcal{C}$ and $\bigwedge \mathcal{C}$ are independent of whether they are interpreted in $\mathrm{Eq}(A)$ or $\mathrm{Cg}(A)$.

The next proposition shows that $\mathrm{Cg}(A)$ is algebraic. Before stating it, we define the product of two algebras. Given two algebras $(A, \langle f_i : i \in I \rangle)$ and $(B, \langle g_i : i \in I \rangle)$ of the same type τ, their *product*, also of type τ, is the algebra $(A \times B, \langle h_i : i \in I \rangle)$, where

$$h_i(\langle a_0, b_0 \rangle, \langle a_1, b_1 \rangle, \ldots, \langle a_{n-1}, b_{n-1} \rangle) = \langle f_i(\bar{a}), g_i(\bar{b}) \rangle.$$

Notice that the product of two lattices is a lattice.

Proposition 4.1.4 *If A is an algebra, then* $\mathrm{Cg}(A)$ *is algebraic.*

Proof It is rather straightforward to prove this proposition by a direct attack, but we take another route.

Let D be the algebra obtained from $A \times A$ by adjoining some additional operations: a 0-ary operation c_a for each $a \in A$, where $c_a = \langle a, a \rangle$; a 1-ary operation r, where $r(\langle x, y \rangle) = \langle y, x \rangle$ for each $x, y \in A$; and a 2-ary operation t, where

$$t(\langle x, y \rangle, \langle z, w \rangle) = \begin{cases} \langle x, w \rangle & \text{if } y = z, \\ \langle x, y \rangle & \text{if } y \neq z, \end{cases}$$

for $x, y, z, w \in A$. Then $\mathrm{Cg}(A) = \mathrm{Sub}(D)$ (Do IT!), so that $\mathrm{Cg}(A)$ is algebraic by Proposition 4.1.2. □

The previous proposition has a converse, the proof of which is not easy and will not be given here.

Theorem 4.1.5 *If L is an algebraic lattice, then there is an algebra A such that* $\mathrm{Cg}(A) \cong L$. □

This theorem leaves open some important and difficult questions that concern additional conditions being put on the algebra A. What is the most relevant with regard to the substructure lattice of models of PA is whether A can be finite whenever L is. This is discussed more fully in Sections 4.5 and 4.9.

4.2 Substructure lattices

A model M of Peano Arithmetic can be interpreted as an algebra with countably many operations. More generally, a model of $\mathsf{PA}^*(\mathcal{L})$ can be interpreted as an algebra with $|\mathcal{L}| + \aleph_0$ operations. Let the set of all Skolem terms be I, which is considered as an index set, and for each $i \in I$, let $f_i : M^n \longrightarrow M$ be the function that it defines. This allows us to consider $M \models \mathsf{PA}^*(\mathcal{L})$ as an algebra $(M, \langle f_i : i \in I \rangle)$, and then it makes perfect sense to consider $\mathrm{Sub}(M)$. Of course, a subalgebra of M (qua an algebra) is the same as an elementary substructure of M (qua a model of $\mathsf{PA}^*(\mathcal{L})$). If $M \models \mathsf{PA}^*$, then $\mathrm{Lt}(M) = \mathrm{Sub}(M)$, so we get the following consequence of Proposition 4.1.2.

Corollary 4.2.1 *If M is a model of* PA^*, *then* $\mathrm{Lt}(M)$ *is an $\aleph_1$-algebraic lattice.* □

More generally, consider $N \prec M$. Now let I be the set of all Skolem terms in which parameters from N are allowed, and let $f_i : M^n \longrightarrow M$ be the function that i defines. Letting M_N be the algebra $(M, \langle f_i : i \in I \rangle)$, we see that $\mathrm{Lt}(M/N) = \mathrm{Sub}(M_N)$, entailing the following generalization of the previous corollary.

Corollary 4.2.2 *If $N \prec M$ are models of* PA *and $\kappa = |N|$, then* $\mathrm{Lt}(M/N)$ *is a κ^+-algebraic lattice.* □

The fundamental question concerning substructure lattices is:

Which lattices are isomorphic to substructure lattices?

There is no known restriction on such a lattice other than what is in Corollary 4.2.1. Included in the class of $\aleph_1$-algebraic lattices are all the finite lattices. Thus, every finite lattice is a candidate for being a substructure lattice. However, there are finite lattices which are not known to be substructure lattices.

The simplest of all lattices is the rather uninteresting one-element lattice **1**. For any model M, $\mathrm{Lt}(M) \cong \mathbf{1}$ iff M is prime. The next simplest lattice is the two-element lattice **2**. (See Figure 4.1.) For any $M \models \mathsf{PA}$, $\mathrm{Lt}(M) \cong \mathbf{2}$ iff M is a minimal extension of its prime model. More generally, if $M \prec N$, then

$\mathrm{Lt}(N/M) \cong \mathbf{2}$ iff N is a minimal extension of M. Thus, there are two ways to realize $\mathbf{2}$ as a substructure lattice, corresponding to Corollaries 2.1.4 and 2.1.6, respectively.

Corollary 4.2.3 *Each model M of* PA *has an elementary end extension N such that* $\mathrm{Lt}(N/M) \cong \mathbf{2}$. □

Corollary 4.2.4 *Each nonstandard countable model M of* PA *has a cofinal extension N such that* $\mathrm{Lt}(N/M) \cong \mathbf{2}$. □

These two corollaries demonstrate how to realize the lattice $\mathbf{2}$ in two different ways: by an end extension or by a cofinal extension. These differences get lost when the lattice $\mathrm{Lt}(N/M)$ is considered as an abstract lattice. The lattice $\mathrm{Lt}(N/M)$ is not discriminating enough to reveal which of these two types of extensions actually occurs. In order to capture this difference abstractly, we are going to define a new algebra obtained from a lattice by adjoining a 1-ary operation which is a kind of rank function. Recall from Gaifman's Splitting Theorem 1.3.5 that if $M \prec N$, then there is a unique $\overline{M}$ such that $M \preccurlyeq_{\mathsf{cof}} \overline{M} \preccurlyeq_{\mathsf{end}} N$. The function $r : \mathrm{Lt}(N) \longrightarrow \mathrm{Lt}(N)$, which we call the *rank* function of N, is such that if $M \preccurlyeq N$, then $r(M)$ is that unique structure $\overline{M}$. We let $\mathrm{Ltr}(N) = (\mathrm{Lt}(N), r)$, and we call $\mathrm{Ltr}(N)$ the *ranked* substructure lattice of N. If $M \prec N$ and $\mathrm{Ltr}(N) = (\mathrm{Lt}(N), r)$, where r is the restriction of the rank function of N to $\mathrm{Lt}(N/M)$, then we let $\mathrm{Ltr}(N/M) = (\mathrm{Lt}(N/M), r)$. We call $\mathrm{Ltr}(N/M)$ the *ranked* interstructure lattice.

The important feature of the ranked substructure lattice is that it distinguishes between elementary end extensions and cofinal extensions. This is made precise by the following proposition with a routine proof (DO IT!).

Proposition 4.2.5 *Let $M \prec N$ and let r be the rank function of N. Suppose $N_0, N_1 \in \mathrm{Lt}(N/M)$ are such that $N_0 \prec N_1$. Then:*

(1) $N_0 \prec_{\mathsf{cof}} N_1$ *iff* $r(N_0) = r(N_1)$;

(2) $N_0 \prec_{\mathsf{end}} N_1$ *iff* $N_0 = r(N_0) \wedge N_1$. □

In particular, a model N is a cofinal extension of its prime submodel iff $r(0_L) = 1_L$, and it is an end extension of its prime submodel iff $r(0_L) = 0_L$. Consequently, we say that a rank function r of L is a *cofinal* rank function if $r(0_L) = 1_L$ and is an *end* rank function if $r(0_L) = 0_L$.

Definition 4.2.6 An algebra (L, r) of type $\langle 2, 2, 1\rangle$ is a *ranked lattice* if L is a lattice and $r : L \longrightarrow L$ satisfies the following for each $x, y \in L$:

(1) $x \le r(x)$;

(2) $r\big(r(x)\big) = r(x)$;

(3) $r(x) \leq r(y)$ or $r(y) \leq r(x)$;
(4) $r(x \vee y) = r(x) \vee r(y)$.

In a ranked lattice (L, r), we call the set $R = \{r(x) : x \in L\}$ the *rankset* of (L, r). The rank function r can be recovered from the rankset R by setting $r(x) = \bigwedge\{y \in R : x \leq y\}$. Therefore, when describing a ranked lattice (L, r) we may say what R is instead of saying what r is. In a finite lattice L, a subset $R \subseteq L$ is a rankset iff R is linearly ordered and $1 \in R$.

It is easily checked that if $M \prec N$, then $\mathrm{Ltr}(N/M)$ is a ranked lattice (Do it!). However, there are ranked lattices, even very small finite ones, which are not isomorphic to any ranked interstructure lattices. For an example of such a ranked lattice, consider the ranked four-element Boolean lattice $(\mathbf{B_2}, r)$ whose rankset is $R = \{0, 1\}$. Suppose that $\mathrm{Ltr}(N/M) \cong (\mathbf{B_2}, r)$ and that $\mathrm{Lt}(N/M) = \{M, M_1, M_2, N\}$. Then N is a cofinal extension of both M_1 and M_2, so that by Theorem 2.1.1, $M = M_1 \cap M_2$ is a cofinal substructure of N, which is a contradiction. The next proposition, which is just a reformulation of Blass' Theorem (Corollary 2.1.6), imposes a condition on a ranked lattice necessary for it to be an interstructure lattice.

Proposition 4.2.7 (The Blass Condition) *Suppose that $M \prec N$ and that $(L, r) \cong \mathrm{Ltr}(N/M)$. Then*

$$r(x) = r(y) \Longrightarrow r(x) = r(x \wedge y)$$

whenever $x, y \in L$ and x is compact. □

We say that a ranked lattice (L, r) satisfies the *Blass Condition* if it satisfies the condition of the previous proposition. The only purpose of the next exercise is to help the reader to gain familiarity with the notions involved.

Exercise 4.2.8 There are exactly 2^n nonisomorphic expansions of the Boolean lattice $\mathbf{B_n}$ to a ranked lattice $(\mathbf{B_n}, r)$, but only n of them satisfy the Blass condition.

Proposition 4.2.7 is useful for showing that certain lattices cannot appear as substructure lattices of models of True Arithmetic. The salient feature of True Arithmetic that gets used is that every one of its nonstandard models is an elementary end extension of its prime model. In other words, if M is a model of True Arithmetic and $\mathrm{Ltr}(M) \cong (L, r)$, then $r(0) = 0$. Thus, the following example shows that if M is a model of True Arithmetic, then $\mathrm{Lt}(M) \ncong \mathbf{M_3}$. More generally, if $M \prec N$ and $\mathrm{Lt}(N/M) \cong \mathbf{M_3}$, then $M \prec_{\mathsf{cof}} N$. Refer to Section 4.5 to see that the lattice $\mathbf{M_3}$ can actually be realized as a substructure lattice.

Proposition 4.2.9 *If $(\mathbf{M_3}, r)$ is a ranked lattice satisfying the Blass Condition, then $r(0) = 1$.*

Proof Let $\mathbf{M_3} = \{0, 1, a, b, c\}$. By (3) in Definition 4.2.6, the rankset is linearly ordered, so there are distinct $x, y \in \{a, b, c\}$ which are not in the rankset. Therefore, by (1) it must be that $r(x) = r(y) = 1$, and then by the Blass Condition, $r(0) = r(x \wedge y) = r(x) = 1$. □

Corollary 4.2.10 *If $M \prec N$ and* $\mathrm{Lt}(N/M) \cong \mathbf{M_3}$, *then $M \prec_{\mathsf{cof}} N$.* □

Another restriction that must be imposed on a ranked lattice for it to be isomorphic to a ranked interstructure lattice comes from the next lemma.

Lemma 4.2.11 *Suppose that $M \prec_{\mathsf{end}} N$ and that $N_0, N_1 \prec N$ are such that $N_0 \cap M = N_1 \cap M$ and $N = \mathrm{Scl}\big(M \cup (N_0 \cap N_1)\big)$. Then $N_0 = N_1$.*

Proof By symmetry, it suffices to show that $N_0 \preccurlyeq N_1$. Consider any $b \in N_0$. Then $N \models b = t(c, a)$, for some $c \in M$, $a \in N_0 \cap N_1$, and Skolem term $t(x, y)$. Let $d \in N$ be the least such that $N \models b = t(d, a)$. Then $d \in M$ since $d \leq c$. But also $d \in N_0$ since d is defined from a, b. Therefore, $d \in N_1$, and then $b = t(d, a) \in N_1$. □

The previous lemma yields the following necessary condition on a ranked lattice for it to be an interstructure lattice (Do it!).

Proposition 4.2.12 (The Gaifman Condition) *Suppose that $M \prec N$ and that $(L, r) \cong \mathrm{Ltr}(N/M)$. If $x, y, z \in L$ are such that $x < y < x \vee z$, $z = r(z)$, and $x \wedge z = y \wedge z$, then $x = y$.* □

We say that a ranked lattice (L, r) satisfies the *Gaifman Condition* if it satisfies the condition of the previous proposition. Following is an example of the use of the Gaifman Condition. Refer to Figure 4.1 to see what the hexagon lattice $\mathbf{H}$ is.

Proposition 4.2.13 *If r is a rank function for* $\mathbf{H}$ *satisfying both the Blass and Gaifman Conditions, then $r(0) = 1$.*

Proof Let the elements of $\mathbf{H}$ be $0, a, b, c, d, 1$, where $0 < a < b < 1$ and $0 < c < d < 1$. Since the rankset is linearly ordered, it cannot be that both $r(b) = b$ and $r(d) = d$, so without loss of generality assume that $r(b) = 1$. If $r(d) = d$, then letting $a = x$, $b = y$, and $d = z$ in Proposition 4.2.12 contradicts the Gaifman Condition. Therefore, $r(d) = 1$, but then by letting $b = x$ and $d = y$ in Proposition 4.2.7, we see that the Blass Condition implies $r(0) = 1$. □

Corollary 4.2.14 *If $M \prec N$ and* $\mathrm{Lt}(N/M) \cong \mathbf{H}$, *then $M \prec_{\mathsf{cof}} N$.* □

4.3 Finite distributive lattices, I

It is proved in this section that every finite distributive lattice is a substructure lattice. Stronger results involving ranked substructure lattices and ranked interstructure lattices are proved in the next section. This section begins with a method for constructing extensions of lattices by doubling a filter, then relates these types of extensions with substructure lattices, and culminates with Theorem 4.3.7 that every finite distributive lattice is a substructure lattice.

Any lattice L can be extended to the lattice $L \times \mathbf{2}$. Here we are letting $\mathbf{2}$ be the two-element lattice with elements $0, 1$ (see Figure 4.1), and then identifying each element $a \in L$ with the element $\langle a, 0\rangle \in L \times \mathbf{2}$. In this extension, no new element is below any old element. Another way to get an extension with no new element below any old one is by adding to the lattice L just one new element which is greater than each element of L. This one-element extension can be identified with the sublattice of $L \times \mathbf{2}$ consisting of the elements $\langle a, i\rangle$, where either $i = 0$ or $a = 1$. Extensions of L, generalizing these two types of extensions, are considered next.

A subset F of a lattice L is a *filter* of L if it is nonempty sublattice such that whenever $x \geq y \in F$ then $x \in F$. Each filter F determines an extension of L which is referred to as the lattice obtained by *doubling the filter* F. This extension is the sublattice of $L \times \mathbf{2}$ consisting of those $\langle z, i\rangle$ for which $z \in F$ or $i = 0$. To realize this lattice as an extension of L, identify each element $z \in L$ with $\langle z, 0\rangle \in L \times \mathbf{2}$. Two examples of filters are L itself and $\{1\}$ (if 1 exists in L). The lattice obtained by doubling L is $L \times \mathbf{2}$, and the one obtained by doubling $\{1\}$ is the one-element extension.

A filter is *principal* if it has the form $\{z \in L : z \geq e\}$ for some $e \in L$. All filters in a finite lattice are principal. The only filters which we use for doubling are principal filters, and we refer to the lattice obtained by doubling the principal filter $\{z \in L : z \geq e\}$ as the *e-doubling extension* of L. An extension of L is a *doubling* extension if it is an e-doubling extension for some $e \in L$. Every doubling extension of a κ-algebraic lattice is κ-algebraic (Do it!).

The reader is reminded of Corollary 2.2.12 asserting that every countable model has a conservative, superminimal extension. The next lemma gets such an extension but with an additional property.

Lemma 4.3.1 *Let M be a countable model. Then there is a type $p(x) \in S_1(M)$ such that the $p(x)$-extension $M(b) \succ M$ is a conservative, superminimal extension and, for each Skolem term $t(u, x)$ there is an $\mathcal{L}$-formula $\theta(x)$ in $p(x)$ such that each of the following hold in M:*

(1) *Every function $t(u, \cdot)$ is eventually constant or eventually one-to-one on $\theta(M)$. Specifically, the sentence*

$$\forall u[\forall x \forall x' (u \leq x \leq x' \wedge \theta(x) \wedge \theta(x') \longrightarrow t(u, x) \neq t(u, x')) \\ \vee \forall x \forall x' (u \leq x \leq x' \wedge \theta(x) \wedge \theta(x') \longrightarrow t(u, x) = t(u, x'))]$$

holds in M.

(2) *If* $u < v$ *and then the functions* $t(u,\cdot)$ *and* $t(v,\cdot)$ *eventually have disjoint images on* $\theta(M)$ *or they eventually agree on* $\theta(M)$. *Specifically, the sentence*

$$\begin{aligned}\forall u \forall v[u < v \longrightarrow & \forall x \forall x'\big(v \leq x \wedge v \leq x' \wedge \theta(x) \wedge \theta(x') \\ & \longrightarrow t(u,x) \neq t(v,x')\big) \vee \forall x\big(v \leq x \wedge \theta(x) \\ & \longrightarrow t(u,x) = t(v,x)\big)]\end{aligned}$$

holds in M.

Proof We indicate how to modify the proof of Theorem 2.1.12 to get the type $p(x)$. In the proof of Theorem 2.1.12, a decreasing sequence $X_0 \supseteq X_1 \supseteq X_2 \supseteq \cdots$ of unbounded definable subsets of M was constructed. Each X_n has the property that there is a Skolem term $s(x)$ (in the language $\mathcal{L}$) and an element $a \in M$ such that $M \models \forall x[x \in X_n \longrightarrow s(x) = a]$, and X_n is definable in M using only the parameter a. For this proof, we need to dovetail into the construction another step:

> Suppose X_n, a, and $s(x)$ are as just described. At this step we consider a Skolem term $t(u,x)$. Then there is an $\mathcal{L}$-formula $\theta(x)$ such that $\theta(M) \cap X_n$ is unbounded and (1) and (2) hold.

We show how to get such a $\theta(x)$. Let $\varphi(a,x)$ be a formula defining X_n. Let $Y \subseteq M^2$ be defined by $\varphi(u,x) \wedge s(x) = u$, and let $Y_u = \{x \in M : \langle u,x\rangle \in Y\}$. Notice that if $u \neq v$, then $Y_u \cap Y_v = \emptyset$. Then, using an inductive construction in M, there is an $\mathcal{L}$-formula $\theta(x)$ (involving no parameters since neither $t(u,x)$ and $\varphi(u,x)$ do) such that (1) and (2) hold and $\theta(M)$ has an unbounded intersection with each unbounded Y_u. Now let $X_{n+1} = X_n \cap \theta(M)$. Notice that X_{n+1} is also definable from a. □

In the statement of the next theorem, the notation $M \star N_0$ is from Definition 2.3.7. Also, since M_0 is an element of the lattice $\mathrm{Lt}(M)$, it makes sense to refer to the M_0-doubling extension of $\mathrm{Lt}(M)$.

Theorem 4.3.2 *Let* M_0 *be a countable model. Then* M_0 *has a conservative extension* N_0 *such that whenever* $M_0 \prec M$ *and* $N = M \star N_0$, *then* $\mathrm{Lt}(N)$ *is isomorphic to the* M_0*-doubling extension* L *of* $\mathrm{Lt}(M)$. *(In fact, there is an isomorphism* $\alpha : \mathrm{Lt}(N) \longrightarrow L$ *which is the identity on* $\mathrm{Lt}(M)$.*)*

Proof Suppose $M \succ M_0$. Then $M_0 \in \mathrm{Lt}(M)$, so let L be the M_0-doubling extension of $\mathrm{Lt}(M)$. The elements of L have the form $\langle K, 0\rangle$, where $K \in \mathrm{Lt}(M)$, or $\langle K, 1\rangle$, where $K \in \mathrm{Lt}(M/M_0)$. By identifying each $K \in \mathrm{Lt}(M)$ with $\langle K, 0\rangle$, we have that $\mathrm{Lt}(M)$ is a sublattice of L.

Notice that the N_0 is necessarily a superminimal extension of M_0. Apparently, not any conservative, superminimal extension will do, and that is why we need Lemma 4.3.1. Let $N_0 = M_0(b)$ be a conservative, superminimal extension as described in that lemma, and then let $N = M(b) = M{\star}N_0$. There are two crucial facts about this extension that suffice to prove the theorem:

(1) If $K \preccurlyeq N$, then either $M_0 \prec K$ or $K \preccurlyeq M$.

(2) If $M_1 \in \mathrm{Lt}(M/M_0)$, then there is a unique $K \in \mathrm{Lt}(N)\backslash \mathrm{Lt}(M)$ such that $M_1 = K \cap M$.

We first prove (1). Suppose $K \preccurlyeq N$ and $K \notin \mathrm{Lt}(M)$. Let $a \in K\backslash M$. To show that $M_0 \prec K$, it suffices to show that $b \in K$. Thus, we want a Skolem term $t_0(y)$ for which $N \models t_0(a) = b$.

There are an element $c \in M$ and a Skolem term $t(u, x)$ such that $N \models a = t(c, b)$. Let $\theta(x)$ be as in the Lemma 4.3.1, so that $N \models \theta(b)$. Let $Y = \theta(M)$. Let $t_0(y)$ be the Skolem term defining the function $f : N \longrightarrow N$ which we now describe in words how to evaluate. Let v be the least such that there is $x \in Y$ such that $t(v, x) = a$, and then let $t_0(y)$ be that uniqe element x.

We verify that $t_0(a) = b$. Clearly, $v \leq c$ since $t(c, b) = a$. If $v < c$ and $t(v, x) = a$ for some x, then, since $c \in M < a$, it must be that $t(v, \cdot)$ and $t(c, \cdot)$ eventually agree on Y, so that $x = b$. If $v = c$, then $t(c, \cdot)$ is eventually one-to-one on Y, so $x = b$.

For the proof of (2) just observe that $K = M_1(b)M_1 \star K$.

Having proved (1) and (2), we define $\alpha : \mathrm{Lt}(N) \longrightarrow L$ so that it is the identity on $\mathrm{Lt}(M)$ and $\alpha(K) = \langle K \cap M, 1\rangle$ if $K \notin \mathrm{Lt}(M)$. Then α is the isomorphism. □

The previous theorem yields a whole slew of lattices which can be realized as substructure lattices. Let $L_0, L_1, L_2, \ldots, L_n$ be a finite sequence of lattices in which L_0 is the one-element lattice and each subsequent lattice L_{i+1} is some doubling extension of L_i. In particular, L_1 is the two-element lattice. Then, not only can L_n be realized as a substructure lattice, but each completion T of PA^* has a model M such that $\mathrm{Lt}(M) \cong L_n$. So, what exactly are these lattices derivable from the one-element lattice by successively taking doubling extensions? The answer: they are precisely the finite distributive lattices. A brief introduction to distributive lattices follows.

Recall that $(\mathcal{P}(X), \cup, \cap)$ is a lattice. A lattice is *distributive* if, for some set X, it is isomorphic to a sublattice of $\mathcal{P}(X)$. Alternative definitions can be given using the distributive laws or by forbidding certain sublattices as in the next two propositions. Clearly, a lattice is distributive iff its dual is distributive (Do it!).

Proposition 4.3.3 *A lattice L is distributive iff it obeys one (or, equivalently, both) of the two distributive laws:*

$$x \wedge (y \vee z) = (x \wedge y) \vee (x \wedge z),$$
$$x \vee (y \wedge z) = (x \vee y) \wedge (x \vee z).$$

□

Proposition 4.3.4 *A lattice L is distributive iff neither of the lattices $\mathbf{N_5}$ and $\mathbf{M_3}$ is isomorphic to a sublattice of L.* □

In order to state the Representation Theorem for finite distributive lattices, we need some definitions. For a lattice L, an element $a \in L$ is *join-irreducible* if whenever $a = b \vee c$, then $a \in \{b, c\}$. Let $J(L)$ be the set of join-irreducible elements of L. If the element $0 = \bigwedge L$ exists, then 0 is join-irreducible, although in the literature 0 is not consistently regarded as being join-irreducible; we let $J_0(L)$ be the set of those join-irreducibles other than 0. Since L can be thought of as a poset and $J(L) \subseteq L$, we consider $J(L)$ and $J_0(L)$ as posets. Given a poset P, we call a subset $D \subseteq P$ a *downset* if whenever $a \leq b \in D$, then $a \in D$. The intersection and the union of an arbitrary collection of downsets of P are also downsets of P. Therefore, if we let $\mathcal{D}(P)$ be the collection of downsets of P, then $(\mathcal{D}(P), \subseteq)$ is a lattice. Furthermore, $\mathcal{D}(P)$ is a sublattice of $\mathcal{P}(P)$, so $\mathcal{D}(P)$ is a distributive lattice.

Theorem 4.3.5 (The Representation Theorem) *Let L be a finite distributive lattice. Then the function*

$$\varphi : L \longrightarrow \mathcal{D}\big(J_0(L)\big)$$

defined by $\varphi(x) = \{a \in J_0(L) : a \leq x\}$ is a lattice isomorphism. □

The next theorem characterizes the finite distributive lattices as those which can be built up from the one-element lattice by a finite sequence of doubling extensions.

Theorem 4.3.6 *Suppose L is a finite lattice. Then L is distributive iff there is a sequence $L_0, L_1, L_2, \ldots, L_n$ of lattices such that L_0 is the one-element lattice, $L_n \cong L$, and each L_{i+1} is a doubling extension of L_i.*

Proof The half of the proof that shows that L_n is distributive is an easy induction on i. Since L_i is distributive, then so is $L_i \times \mathbf{2}$. Then L_{i+1}, being a sublattice of $L_i \times \mathbf{2}$, also is.

For the other direction, let $a_0, a_1, a_2, \ldots, a_n$ be the join-irreducibles of L arranged so that if $a_i < a_j$, then $i < j$. In particular, $a_0 = 0$. Let L_i be the sublattice of L generated by $\{a_0, a_1, \ldots, a_i\}$. Then it follows from the Representation Theorem 4.3.5 that $L \cong L_n$ and that each L_{i+1} is an x-doubling extension of L_i, where $x = a_{i+1} \wedge \bigvee L_i$ (DO IT!). □

The next theorem is the principal theorem of this section. See Corollary 4.4.5 in the next section for an improvement.

Theorem 4.3.7 *Let D be a finite distributive lattice. Let M_0 be a prime model of* PA*. *Then M_0 has an elementary end extension M such that M is resolute and* $\mathrm{Lt}(M) \cong D$.

Proof Theorem 4.3.6 shows that D can be obtained by a finite sequence of doubling extensions. Then, by Theorem 4.3.2, there is a corresponding sequence of conservative minimal extensions resulting in $M \succ_{\mathsf{end}} M_0$ such that $\mathrm{Lt}(M) \cong D$. Then M is resolute by Theorem 3.4.7. □

Corollary 4.3.8 *Let D be a finite distributive lattice. Then every model M has an elementary end extension N such that* $\mathrm{Lt}(N/M) \cong D$. □

4.4 Finite distributive lattices, II

The proof of Theorem 4.3.7 in the previous section yields some information about ranked substructure lattices. Let D be a finite distributive lattice and let r be a rank function whose rankset is a maximal subchain of D. Then there are $a_0, a_1, \dots, a_n \in D$, as in the proof of Theorem 4.3.6, such that the rankset is $\{a_0 \vee a_1 \vee \cdots \vee a_i : i \le n\}$. The model M constructed in the proof of Theorem 4.3.6 is such that $\mathrm{Ltr}(M) \cong (D, r)$ (Do it!).

In this section the construction of the previous section is refined to yield a characterization of those finite ranked distributive lattices which can appear as ranked substructure lattices. There are two theorems which enter into this construction; the first is Theorem 4.3.2, which was crucial in the proof of Theorem 4.3.7, and the second is its analogue for cofinal extensions. The main results of this section are Corollaries 4.4.2 and 4.4.4.

We need the following analogue of Theorem 4.3.2 for cofinal extensions.

Theorem 4.4.1 *Let M_0 be a nonstandard countable model generated by the proper initial segment $I \subseteq_{\mathsf{end}} M_0$. Then M_0 has a cofinal I-extension N_0 such that whenever M is an I-extension of M_0 and $N = M \star N_0$, then* $\mathrm{Lt}(N)$ *is isomorphic to the M_0-doubling extension L of* $\mathrm{Lt}(M)$. *(In fact, there is an isomorphism* $\alpha : \mathrm{Lt}(N) \longrightarrow L$ *which is the identity on* $\mathrm{Lt}(M)$.*)*

Proof The model N_0 will be a superminimal, I-conservative cofinal extension of M_0 having the form $M_0(b)$. We construct a bounded type $p(x) \in S_1(M)$ that b should realize. This is done by constructing a decreasing sequence $X_0 \supseteq X_1 \supseteq X_2 \supseteq \cdots$ of bounded definable subsets M_0. These sets have to satisfy certain properties that will be stated in a moment.

Let $h(v)$ be a term which defines a very fast growing function; that is $h(v+1)$ should be much larger than $h(v)$. How big is "much larger"? The short answer is: big enough for this proof to work.

Let $Y_v = \{x \in M : h(v) \le x < h(v+1)\}$. Without loss of generality, we can assume that $a \in M$ is nonstandard and that $I \subseteq_{\mathsf{end}} \{x \in M : x < h(a)\}$. Then we let $X_0 = Y_a$. Then $h(a+1)$ should be big enough so that for each $n < \omega$, $\mathrm{card}^M(X_n) > h(a)$.

The type $p(x)$ that this sequence generates should have the following properties somewhat analogous to those in Lemma 4.3.1. For every Skolem term $t(u, x)$

there is an $\mathcal{L}$-formula $\theta(x)$ such that the following hold in M_0:

(1) If $u < h(v)$, then the function $t(u, \cdot)$ is constant or one-to-one on $\theta(M) \cap Y_v$.
(2) If $u < u' < h(v)$, then the functions $t(u, \cdot)$ and $t(u', \cdot)$ have disjoint images on $\theta(M) \cap Y_v$ or are identical on $\theta(M) \cap Y_v$.
(3) If $v' < v$, $u' < h(v')$, $u < h(v)$, and the function $t(u, \cdot)$ is one-to-one on $\theta(M) \cap Y_v$, then the image of $t(u, \cdot)$ on $\theta(M) \cap Y_v$ and the image of $t(u', \cdot)$ on $\theta(M) \cap Y_{v'}$ are disjoint.

It is also be required that:

(4) For each $n < \omega$, there are a Skolem term $s(x)$ and $a \in I$ such that X_n is definable using only the parameter a and $M_0 \models \forall x \in X_n[s(x) = a]$.
(5) For each $a \in I$, there is $n < \omega$ and a Skolem term $s(x)$ such that $M_0 \models \forall x[x \in X_n \longrightarrow s(x) = a]$.
(6) $\mathrm{card}^M(X_n)$ is large enough.

Again, the question is how large is "large enough," and the answer again is large enough for the proof to work. We see, however, that as long as h was chosen correctly, (6) takes care of itself.

Suppose we are at a stage of the construction where we have X_n and are concerned about a in (5). Let X_{n+1} be the set of those $x \in X_n$ which, for some $i \in M$, is the $\langle i, a\rangle$th element of X_n. This takes care of (5).

Next suppose that we are at a stage of the construction where we have X_n and are concerned about (1), (2), and (3) for some Skolem term $t(u, x)$. Along with X_n, we also have a and $s(x)$. Let $\varphi(a, x)$ be a formula defining X_n. Let $Z_v \subseteq Y_v \cap \varphi(v, M)$ be the biggest such that:

(1′) If $u < h(v)$, then $t(u, \cdot)$ is constant or one-to-one on Z_v.
(2′) If $u < u' < h(v)$, then the functions $t(u, \cdot)$ and $t(u', \cdot)$ have disjoint images or are identical on Z_v.
(3′) If $v' < v$, $u' < h(v')$,$u < h(v)$ and the function $t(u, \cdot)$ is one-to-one on $\theta(M) \cap Y_v$, then the image of $t(u, \cdot)$ on $\theta(M) \cap Y_v$ and the image of $t(u', \cdot)$ on $\theta(M) \cap Y_{v'}$ are disjoint.

There may be more than one biggest set satisfying (1′)–(3′), in which case let Z_v be the first in some canonical ordering of the subsets. Let $\theta(x)$ be the formula $x \in \bigcup_v Z_v$, and then let $X_{n+1} = Z_a$. The set X_{n+1} should be big enough so that $\mathrm{card}^M(X_{n+1}) > h(a)$; moreover, for each $v \geq n$, it should be that $\mathrm{card}^M(Z_v) > h(v)$.

It is now clear that, with this construction, (1)–(6) will hold. The proof can now be completed just as the proof of Theorem 4.3.2 was. □

From Theorems 4.3.2 and 4.4.1, we can obtain the following theorem characterizing those ranked finite distribute lattices which appear as ranked interstructure lattices for extensions which are not end extensions. Notice that the condition on the rank is just the Blass Condition, the Gaifman Condition being redundant. The following theorem does remain true if the hypothesis that $r(0) > 0$ is deleted; however, when $r(0) = 0$, then the extension must be an elementary end extension, and for such extensions a stronger result appears as Corollary 4.4.5.

Corollary 4.4.2 *Let (D, r) be a ranked finite distributive lattice such that $r(0) > 0$ and*

$$r(x) = r(y) \Longrightarrow r(x) = r(x \wedge y)$$

whenever $x, y \in D$. Then every nonstandard countable model M has an elementary extension N such that $\mathrm{Ltr}(N/M) \cong (D, r)$.

Proof It suffices to assume that M is a prime model, for if it is not, then just expand the language to include constant symbols for all elements of M.

Let $a_0, a_1, \ldots, a_n$ be a sequence of the join-irreducibles arranged as in the proof of Theorem 4.3.6 (that is, $a_i < a_j \Longrightarrow i < j$) but with the additional property that for each $x \in D$, there is $i \leq n$ such that $r(x) = a_0 \vee a_1 \vee \cdots \vee a_i$. It follows from the Representation Theorem 4.3.5 that there is such a sequence. For each $i \leq n$, let D_i be the sublattice of D generated by the elements $a_0, a_1, \ldots, a_i$ and let $1_i = a_0 \vee a_1 \vee \cdots \vee a_i = \bigvee D_i$. We let r_i be the rank function on D_i defined by $r_i(x) = r(x) \wedge 1_i$. We inductively obtain a sequence $M_0 \prec M_1 \prec \cdots \prec M_n$ of models such that $\mathrm{Ltr}(M_i) \cong (D_i, r_i)$. Then, since $(D_n, r_n) = (D, r)$, we can let $N = M_n$.

Let $M_0 = M$. Now suppose that $i < n$ and that we have M_i. We get M_{i+1} to be isomorphic to a $(a_i \wedge 1_i)$-doubling extension of D_i using Theorem 4.3.2 if $r_{i+1}(1_i) = 1_i$ and Theorem 4.4.1 otherwise. The argument that $\mathrm{Lt}(N) \cong D$ is just like the argument in Theorem 4.3.2. □

This is an appropriate point to make some more definitions. Let $p(x)$ be a type and L a lattice. We say that $p(x)$ *produces* L if, whenever M is a $p(x)$-extension of its prime submodel, then $\mathrm{Lt}(M) \cong L$. Similarly, $p(x)$ *produces* (L, r) if $\mathrm{Ltr}(M) \cong (L, r)$. This definition is useful in dealing with substructure lattices but is less useful for interstructure lattices. Resolute types, which are defined in Definition 3.4.3 and discussed in Section 3.4 of the previous chapter, are tailor-made for realizing lattices as interstructure lattices. Recall that if $p(x) \in S_1(T)$ is a resolute type, then $p(x)$ is definable, so we can refer to the $p(x)$-extension of a model M of T and to the ranked lattice it produces. The following proposition identifies the key feature of resolute types.

Proposition 4.4.3 *Let T be a completion of* PA^* *and let $p(x) \in S_1(T)$ be a resolute type which produces the ranked lattice (L, r). If $M \models T$ and N is a $p(x)$-extension of M, then* $\mathrm{Ltr}(N/M) \cong (L, r)$.

Proof Let $M(b)$ be a $p(x)$-extension of M, and let M_0 be the prime model of T. We can say exactly what the isomorphism is: it is $N \mapsto N \cap M_0(b)$. It follows right from Proposition 3.4.2 that this is an isomorphism of the lattices (Do IT!) which preserves ranksets (Do IT!). □

The next corollary follows in the manner of Corollaries 4.3.7 and 4.4.2. The resoluteness of the type depends on Theorem 3.4.8 (Do IT!).

Corollary 4.4.4 *Let T be a completion of* PA^*, *and let (D, r) be a ranked finite distributive lattice such that $r(0) = 0$ and*

$$r(x) = r(y) \Longrightarrow r(x) = r(x \wedge y)$$

whenever $x, y \in D$. Then there is a resolute type $p(x) \in S_1(T)$ producing (D, r). □

By invoking Proposition 4.4.3, we get the following corollary.

Corollary 4.4.5 *Let (D, r) be a ranked finite distributive lattice such that $r(0) = 0$ and*

$$r(x) = r(y) \Longrightarrow r(x) = r(x \wedge y)$$

whenever $x, y \in D$. Then every model M has an elementary end extension N such that $\mathrm{Ltr}(N/M) \cong (D, r)$. □

4.5 Finite lattices

One of the results of the previous section is that for any finite distributive lattice D, every countable nonstandard model M of PA^* has a cofinal extension N such that $\mathrm{Lt}(N/M) \cong D$. In this section we see that there are many finite nondistributive lattices L for which the same is true. For example, if $3 \le n \le 15$, then every countable nonstandard model M of PA^* has a cofinal extension N such that $\mathrm{Lt}(N/M) \cong \mathbf{M_n}$. Other such lattices are $\mathbf{N_5}$ and $\mathbf{H}$ depicted in Figure 4.1. The current state of our knowledge does not rule out the possibility that every finite lattice has this property. However, there are finite lattices which are not even known to be interstructure lattices. The lattice $\mathbf{M_{16}}$ is perhaps the simplest instance of our ignorance here.

This section begins with a discussion about representations of finite lattices. Recall from Section 4.1 that $\mathrm{Eq}(A)$ is the lattice of equivalence relations on the set A and that its extreme elements are $\mathbf{1}_A = \bigvee \mathrm{Eq}(A)$ and $\mathbf{0}_A = \bigwedge \mathrm{Eq}(A)$.

Definition 4.5.1 If L is a finite lattice and A any set, then $\alpha : L \longrightarrow \mathrm{Eq}(A)$ is a *representation* of L if α is an injection for which each of the following holds for $x, y \in L$:

$$\begin{aligned}\alpha(x \vee y) &= \alpha(x) \wedge \alpha(y);\\ \alpha(0) &= \mathbf{1}_A;\\ \alpha(1) &= \mathbf{0}_A.\end{aligned}$$

The representation $\alpha : L \longrightarrow \mathrm{Eq}(A)$ is *finite* if A is finite.

Notice that a representation of the lattice L is an embedding of the $\vee$-semilattice L into the $\wedge$-semilattice $\mathrm{Eq}(A)$. The reader is cautioned that it is not required of a representation α that it satisfies

$$\alpha(x \wedge y) = \alpha(x) \vee \alpha(y)$$

for all $x, y \in L$, but if α does satisfy this, then we say that that α is a *lattice* representation.

Let L be a finite lattice and $\alpha : L \longrightarrow \mathrm{Eq}(A)$ a representation. If $B \subseteq A$, then $\alpha|B : L \longrightarrow \mathrm{Eq}(B)$ is the function such that for any $x \in L$, $(\alpha|B)(x) = \alpha(x) \cap B^2$. There is no guarantee that $\alpha|B$ is also a representation of L, for it may be that B is just too small. But if $C \subseteq B \subseteq A$ and $\alpha|C$ is a representation, then so is $\alpha|B$.

Let $\alpha : L \longrightarrow \mathrm{Eq}(A)$ and $\beta : L \longrightarrow \mathrm{Eq}(B)$ be two representations of L. Then α and β are *isomorphic* if there is a bijection $h : A \longrightarrow B$ such that for any $x \in L$ and $a, b \in A$, $\langle a, b\rangle \in \alpha(x)$ iff $\langle h(a), h(b)\rangle \in \beta(x)$. The bijection h is said to *confirm* that α and β are isomorphic.

Every finite lattice L has a finite representation $\alpha : L \longrightarrow \mathrm{Eq}(L)$. For each $r \in L$, let $\alpha(r) \in \mathrm{Eq}(L)$ be such that $\langle x, y\rangle \in \alpha(r)$ iff either $x, y \leq r$ or $x = y$ (DO IT!). It follows from Theorem 4.1.5 that every finite lattice has a lattice representation, but there are no guarantees that this will give a finite representation. Much more difficult is the following important theorem of Pudlák–Tůma [155]

Theorem 4.5.2 *Every finite lattice has a finite lattice representation.* □

Finite lattice representations of finite distributive lattices are easy to come by. Here is one way to get them. Let D be a finite distributive lattice and let $A = J(D) \subseteq D$. (See Theorem 4.3.5 and the paragraph preceding it.) Define $\alpha : D \longrightarrow \mathrm{Eq}(A)$ so that for each $r \in D$, the equivalence relation $\alpha(r)$ has only singletons for its equivalence classes with the (possible) exception of the equivalence class containing 0_D, which is $\{x \in J(D) : x \leq r\}$.

Exercise 4.5.3 This function $\alpha : D \longrightarrow \mathrm{Eq}(A)$ is a lattice representation of D.

The lattice $\mathbf{M_3}$ has a very simple finite lattice representation since, in fact, $\mathbf{M_3} \cong \mathrm{Eq}(3)^\perp$ (DO IT!). More generally, each lattice $\mathrm{Eq}(n)^\perp$ has a finite lattice representation, namely the identity function α on $\mathrm{Eq}(n)^\perp$. (Reminder: $L^\perp$ is the dual lattice of L obtained by interchanging the roles of $\wedge$ and $\vee$.)

If you take a peek at Theorem 4.5.32 later in this section, you will see that substructure lattices are universal in the sense that every finite lattice is a isomorphic to a sublattice of some substructure lattice. The proof of Theorem 4.5.32 makes use of the difficult Theorem 4.5.2. There is a weaker version of universality having a proof avoiding Theorem 4.5.2 and which is conceptually simpler than the proof of Theorem 4.5.32. The proof of this theorem can serve as an easy introduction to the ideas presented later in this section. Readers not interested in such an introduction may skip right to Definition 4.5.7.

The following definition introduces a partition property, reminiscent of Ramsey's Theorem, which representations may have.

Definition 4.5.4 Let $\alpha : L \longrightarrow \mathrm{Eq}(A)$ and $\beta : L \longrightarrow \mathrm{Eq}(B)$ be representations of the lattice L. Then $\alpha \longrightarrow (\beta)_2$ if whenever $A = C \cup D$, there is $X \subseteq B$ such that $\beta \cong \alpha|X$ and either $X \subseteq C$ or $X \subseteq D$. More generally, if $2 \leq k < \omega$, then $\alpha \longrightarrow (\beta)_k$ if whenever $A = C_0 \cup C_1 \cup \cdots \cup C_{k-1}$, then there are $i < k$ and $X \subseteq C_i$ such that $\beta \cong \alpha|X$.

Theorem 4.5.5 *Let $2 \leq k < \omega$ and let $\alpha : L \longrightarrow \mathrm{Eq}(A)$ be a finite lattice representation of the finite lattice L. Then there is $n < \omega$ such that $\alpha^n \longrightarrow (\alpha)_k$.* □

The above theorem is, in a somewhat disguised form, what is arguably the most fundamental theorem in Ramsey Theory, that branch of finite combinatorics which evolved from FRT. For $2 \leq k < \omega$ and $1 \leq m < \omega$, let $|A| = m$ and then let $\gamma : \mathrm{Eq}(A)^\perp \longrightarrow \mathrm{Eq}(A)$ be the identity represenatation of the lattice $\mathrm{Eq}(A)^\perp$. This representation is certainly a lattice representation, so an instance of Theorem 4.5.5 is that there is $n < \omega$ such that $\gamma^n \longrightarrow (\gamma)_k$. This special case of Theorem 4.5.5 is the Hales–Jewett Theorem, stated in a way which may not appear to some readers to be what they know as that theorem. We see that it is.

If A is a finite set and $1 \leq n < \omega$, then a subset $B \subseteq A^n$ is a *combinatorial line* of A^n if there is a nonempty subset $I \subseteq n$ and a function $c : n\backslash I \longrightarrow A$ such that

$$B = \{a \in A^n : a_i = c_i \text{ if } i \in n\backslash I \text{ and } a_i = a_j \text{ if } i, j \in I\}.$$

If B is a combinatorial line as just given, then there is a natural bijection $f : A \longrightarrow B$ where $f(a)_i = a_i$ if $i \in I$. The following is the more traditional form of the Hales–Jewett Theorem.

> THE HALES–JEWETT THEOREM: *Suppose that $1 \leq m, r < \omega$, and $|A| = m$. Then there is $n < \omega$ such that if $A^n = C_0 \cup C_1 \cup \cdots \cup C_{k-1}$, then some C_i contains a combinatorial line.*

We use the Hales–Jewett Theorem to prove the special case of Theorem 4.5.5. Given k and $\gamma : \mathrm{Eq}(A)^{\perp} \longrightarrow \mathrm{Eq}(A)$, let n be as in the Hales–Jewett Theorem, and let $A^n = C_0 \cup C_1 \cup \cdots \cup C_{k-1}$. Then, letting $B \subseteq C_i$ be a combinatorial line, we see that $\alpha \cong \alpha^n | B$ as confirmed by the natural bijection $f : A \longrightarrow B$.

Conversely, the Hales–Jewett Theorem follows from this special case of Theorem 4.5.5 using the same value of n. For, letting $A^n = C_0 \cup C_1 \cup \cdots \cup C_{k-1}$, we get $B \subseteq C_i$ such that $\alpha^n | B \cong \alpha$ as confirmed by the function f. But then B is a combinatorial line with $f : A \longrightarrow B$ being the natural bijection (Do it!).

We easily prove Theorem 4.5.5. If $1 \leq k < \omega$ and $\alpha : L \longrightarrow \mathrm{Eq}(A)$ is a finite lattice representation, then let n be as in the Hales–Jewett Theorem. This n is easily seen to work (Do it!).

By considering the lattice $L = \mathbf{M_3} \cong \mathrm{Eq}(3)^{\perp}$ in the next proposition, we see that there are substructure lattices that are not distributive.

Proposition 4.5.6 *If L is a finite lattice having a finite lattice representation and M is a countable nonstandard model, then M has a cofinal extension N such that L is isomorphic to a sublattice of* $\mathrm{Lt}(N/M)$.

Proof Let $\alpha : L \longrightarrow \mathrm{Eq}(A)$ be a finite lattice representation of the lattice L. That α is such a representation can be formalized in M. By overspill and Theorem 4.5.5, let $n \in M$ be nonstandard so that, in M, $\alpha : L \longrightarrow \mathrm{Eq}(A)$ is a representation with the property that for each $k < \omega$, $\alpha^n \longrightarrow (\alpha)_k$. Let $X_0 = A^n$. For each $i \in L$, let $f_i : X_0 \longrightarrow X_0$ be the definable function for which $f_i(x) = y$ iff $y = \min\{z \in X_0 : \langle x, z\rangle \in \alpha^n(i)\}$.

Let $D_0, D_1, D_2, \ldots$ be list all the definable subsets of M. We obtain inductively a sequence $X_0 \supseteq X_1 \supseteq X_2 \supseteq \cdots$ of definable subsets of M such that if $i, k < \omega$, then $\alpha^n | X_i \longrightarrow (\alpha)_k$. Now suppose we have X_i. Then, by overspill, there is a nonstandard even $2r \in M$ such that $\alpha^n | X_i \longrightarrow (\alpha)_{2r}$, and therefore, letting X_{i+1} be either $X_i \cap D_i$ or $X_i \backslash D_i$, we get that $\alpha^n | X_{i+1} \longrightarrow (\alpha)_r$. Then, since r is nonstandard, $\alpha^n | X_{i+1} \longrightarrow (\alpha)_k$ for each $k < \omega$. Clearly, the sequence $X_0 \supseteq X_1 \supseteq X_2 \supseteq \cdots$ determines a unique type $p(x)$, which is bounded since X_0 is. Let N be the $p(x)$-extension of M generated by the element $a \in N$. For each $i \in L$, let $b_i = f_i(a)$ and let M_i be the elementary extension of M generated by the element b_i. To prove that $i \mapsto M_i$ is an embedding of L in $\mathrm{Lt}(N/M)$, we prove the following four things:

(1) $M_0 = M$ *and* $M_1 = N$. For the first equality, notice that the function f_0 is constant on X_0, mapping each $x \in X_0$ to $\min(X_0) \in M$. Thus, $b_0 = f_0(a) \in M$, so that $M_0 = M$. For the second equality, just observe that f_1 is the identity function on X_0. Thus, $b_1 = f_1(a) = a$, so that $M_1 = N$.

(2) $M_{i \vee j}$ *is generated by* $M_i \cup M_j$. First we show that $b_i \in M_{i\vee j}$ (which, by symmetry, also shows that $b_j \in M_{i \vee j}$). Since $\alpha^n(i \vee j)$ refines $\alpha^n(i)$, it follows that $f_i(x) = f_i(f_{i\vee j}(x))$ for all $x \in X_0$. Therefore, $b_i = f_i(b_{i\vee j})$, so $b_i \in M_{i \vee j}$.

For the converse, observe that $f_{i\vee j}(x) = \min\{z \in X_0 : f_i(z) = f_i(x)$ and $f_j(z) = f_j(x)\}$. Therefore, $b_{i\vee j} = \min\{z \in X_0 : f_i(z) = b_i$ and $f_j(z) = b_j(x)\}$, so that $b_{i\vee j}$ is in the model generated by $\{b_i, b_j\}$.

(3) $M_i \cap M_j = M_{i\wedge j}$. The inclusion $M_{i\wedge j} \wedge M_i \cap M_j$ follows from (2) since, for example, $M_{i\wedge j} \subseteq M_{i\vee(i\wedge j)} \subseteq M_i$. For the reverse inclusion, consider some $b \in M_i \cap M_j$. Then there are M-definable functions $g_i, g_j : N \longrightarrow N$ such that $b = g_i(b_i) = g_j(b_j)$. There is some $k < \omega$ such that $D_k = \{x \in X_0 : g_i(f_i(x)) = g_j(f_j(x))\}$. Since $N \models b \in D_k$, it must be that $X_{k+1} \subseteq D_k$. Thus, we might as well assume that $g_i(f_i(x)) = g_j(f_j(x))$ for all $x \in X_0$.

We claim: if $\langle x, y\rangle \in \alpha^n(i \wedge j)$, then $g_i(f_i(x)) = g_j(f_j(x))$. Working in M, consider such $x, y \in X_0$. Since α^n is a lattice representation, there are $x = x_0, x_1, x_2, \dots, x_t = y$ such that $\langle x_r, x_{r+1}\rangle \in \alpha^n(i)$ for even $r \leq t$ and $\langle x_r, x_{r+1}\rangle \in \alpha^n(j)$ for odd $r \leq t$. Thus $f_i(x_r) = f_i(x_{r+1})$ for even r and $f_j(x_r) = f_j(x_{r+1})$ for odd r. Then $g_i(f_i(x)) = g_i(f_i(x_0)) = g_i(f_i(x_1)) = g_j(f_j(x_1)) = g_j(f_j(x_2)) = g_i(f_i(x_2)) = \cdots = g_i(f_i(x_t)) = g_i(f_i(y))$, proving the claim. Thus, $b = g_i(f_{i\wedge j}(a)) = g_i(b_{i\wedge j})$, so $b \in M_{i\wedge j}$.

(4) *If $i \neq j$, then $M_i \neq M_j$.* It suffices to show that if $i < j$, then $b_j \notin M_i$. For a contradiction, suppose that $b_j = g(b_i)$, where $g : N \longrightarrow N$ is M-definable. Since $\alpha^n(j)$ is a refinement of $\alpha^n(i)$, it follows that $f_i(x) = f_i(f_j(x))$ for all $x \in X_0$, so that $b_i = f_i(b_j)$. Thus, $b_j = g(f_i(b_j))$. Therefore, there is $k < \omega$ such that $f_j(x) = f_j(g(f_i(x)))$ for all $x \in X_k$. Then, $(\alpha^n|X_k)(i)$ is a refinement of $(\alpha^n|X_k)(j)$. This implies that $(\alpha^n|X_k)(i) = (\alpha^n|X_k)(j)$, contradicting that $\alpha^n|X_k$ is a representation of L. □

Recall that if A is an algebra, then $\mathrm{Cg}(A)$ is the lattice of congruences of A.

Definition 4.5.7 Let $\alpha : L \longrightarrow \mathrm{Eq}(A)$ be a representation of the finite lattice L. We say that the representation is a *congruence* representation if there is an algebra A such that α is an isomorphism of L and $\mathrm{Cg}(A)^{\perp}$.

Simple examples of lattices having finite congruence representations are the finite lattices $\mathrm{Eq}(A)^{\perp}$. The identity function is a congruence representation since, if A is considered as an algebra with no operations, then $\mathrm{Eq}(A) = \mathrm{Cg}(A)$.

Every congruence representation is a lattice representation. Theorem 4.1.5 implies that every finite lattice has a congruence representation. Unfortunately, arbitrary congruence representations do not appear to be so useful for showing that a given finite lattice is isomorphic to a substructure lattice. However, finite congruence representations turn out to be very useful. All known examples of finite lattices which are isomorphic to intermediate structure lattices come from the following theorem. Of course, with our current state of knowledge, it is still possible that every finite lattice has a finite congruence representation.

Theorem 4.5.8 *Let L be a finite lattice which has a finite congruence representation. Then every countable nonstandard model M has an elementary cofinal extension N such that* $\mathrm{Lt}(N/M) \cong L$. □

Before starting the proof of this theorem, we take a look at some more examples of finite lattices and their finite congruence representations.

Example 4.5.9 *Every finite distributive lattice has a finite congruence representation.*

In fact, the representations given in Exercise 4.5.3 are congruence representations. Let $\{f_i : i \in I\}$ be the set of all functions $f : A \longrightarrow A$ such that $f(x) \leq x$ for all $x \in A$. The congruences of $(A, \langle f_i : i \in I \rangle)$ are exactly those which are in the range of α (DO IT!). Much more difficult to prove is that for any finite distributive lattice there is a finite lattice L such that $D \cong \mathrm{Cg}(L)$.

Example 4.5.10 *The lattice* $\mathbf{N_5}$ *has a finite congruence representation.*

Perhaps the simplest such representation comes from an algebra having four elements and two 1-ary operations. Let $A = \{1, 2, 3, 4,\}$ and let $f, g : A \longrightarrow A$ be the functions such that $f(1) = 2,\ f(2) = 1,\ f(3) = 4,\ f(4) = 3,\ f(5) = 1$, and $g(1) = g(3) = 2, g(2) = g(4) = 2$. Consider the algebra (A, f, g) of type $\langle 1, 1\rangle$. Besides the two extreme congruences $\Theta_0 = \mathbf{1}_A$ and $\Theta_1 = \mathbf{0}_A$, there are three others, $\Theta_a, \Theta_b, \Theta_c$, whose corresponding partitions are:

$$\{\{1,2\},\{3,4\}\}, \quad \{\{1,2\},\{3\},\{4\}\}, \quad \{\{1,3\},\{2,4\}\}.$$

It can be checked that $\mathbf{N_5} \cong \mathrm{Cg}(A)^{\perp}$ (DO IT!). In fact, the function $r \mapsto \Theta_r$ is a congruence representation.

Example 4.5.11 *The lattice* $\mathbf{H}$ *has a finite congruence representation.*

There is an algebra A having 16 elements such that $\mathbf{H} \cong \mathrm{Cg}(A)^{\perp}$. It is convenient to let $A = \mathbb{Z}_4 \times \mathbb{Z}_4$, where $\mathbb{Z}_4$ is the additive cyclic group of order 4. Let $\gamma : A \longrightarrow A$ be the permutation defined by $\gamma(\langle i, j\rangle) = \langle j+1, i\rangle$. Besides the two extreme congruences $\Theta_0 = \mathbf{1}_A$ and $\Theta_1 = \mathbf{0}_A$, there are four others: $\Theta_a, \Theta_b, \Theta_c$, and Θ_d. We define Θ_a and Θ_b so that if $i, j, r, s \in \mathbb{Z}_4$, then:

$$\langle i, j\rangle \Theta_a \langle r, s\rangle \iff i = r;$$

$$\langle i, j\rangle \Theta_b \langle r, s\rangle \iff i = r \text{ and either } j = i = s \text{ or } j \neq i \neq s.$$

Then let $\Theta_c = \gamma(\Theta_a)$ and $\Theta_d = \gamma(\Theta_b)$. These six equivalence relations form a sublattice of $\mathrm{Eq}(A)^{\perp}$ which is isomorphic to $\mathbf{H}$. This sublattice is also a congruence lattice. (See Exercise 4.8.10.)

Example 4.5.12 *If $q = p^k$, where p is a prime and $1 \leq k < \omega$, then* $\mathbf{M_{q+1}}$ *has a finite congruence representation.*

Let $\mathbb{F}_q$ be the Galois field of order q, and let $A = \mathbb{F}_q \times \mathbb{F}_q$. For each $a \in \mathbb{F}_q$, let Θ_a be the equivalence relation on A such that $\langle x, y\rangle \Theta_a \langle u, v\rangle$ iff $a(x-u) = y - v$, and let $\Theta_\infty \in \mathrm{Eq}(A)$ be such that $\langle x, y\rangle \Theta_\infty \langle u, v\rangle$ iff $x = u$. Then $\{\mathbf{0}_A, \mathbf{1}_A, \Theta_\infty\} \cup \{\Theta_a : a \in \mathbb{F}_q\}$ is a sublattice of $\mathrm{Eq}(A)$ which is isomorphic to $\mathbf{M_{q+1}}$ (Do it!). We can think of A as the affine plane over $\mathbb{F}_q$ and the defined equivalence relations as partitions of this plane into parallel lines. To see that this lattice is actually a congruence lattice, consider A as an algebra whose operations are the linear functions $f : A \longrightarrow A$, meaning that there are $a, b, c, d \in \mathbb{F}_q$ such that $f(\langle x, y\rangle) = \langle ax + by, cx + dy\rangle$. Then, $\mathbf{M_{q+1}} \cong \mathrm{Cg}(A)^\perp$ (Do it!).

The reader may have observed that in Examples 4.5.9, 4.5.10, and 4.5.12 all of the operations of the given algebras have arity 1. The following exercise, which is used in the proof of Theorem 4.5.27, shows that this is not an accident.

Exercise 4.5.13 Let $(A, \langle f_i : i \in I\rangle)$ be an algebra. A function $p : A^n \longrightarrow A$ is a *polynomial* of A if it can be obtained as the composition of operations of A and constant functions $c : A \longrightarrow A$. If $\{p_j : j \in J\}$ is the set of all 1-ary polynomials of $(A, \langle f_i : i \in I\rangle)$, then $\mathrm{Cg}(A, \langle f_i : i \in I\rangle) = \mathrm{Cg}(A, \langle p_j : j \in J\rangle)$.

In the following definition, we define the Canonical Partition Property which is a property that a representation of a finite lattice might have. The abbreviation CPP is used.

Definition 4.5.14 Let L be any lattice and $\alpha : L \longrightarrow \mathrm{Eq}(A)$ a representation of L.

(1) If $\Theta \in \mathrm{Eq}(A)$ and $B \subseteq A$, then we say that Θ is *canonical* for α on B if there is $r \in L$ such that $\Theta \cap B^2 = (\alpha|B)(r)$.
(2) Let L be a finite lattice. Using induction on $n < \omega$, we define when α is an n-CPP representation of L. We say that α is a 0-CPP representation if $\alpha(r)$ has more than two equivalence classes whenever $0 < r \in L$, and we say that α is an $(n+1)$-representation if whenever Θ is an equivalence relation on A, then there is $B \subseteq A$ such that $\alpha|B$ is an n-CPP representation and Θ is canonical on B.

The converse to the following exercise is also true. It is easily deduced from Lemma 4.5.19.

Exercise 4.5.15 A representation $\alpha : \mathbf{2} \longrightarrow \mathrm{Eq}(A)$ is n-CPP if $|A| > 2^{2^n}$.

Lemma 4.5.16 *Suppose that L is a finite lattice and $m < n < \omega$. Then every n-CPP representation of L is also an m-CPP representation.*

Proof It suffices to consider only the case in which $m = 0$ as then the general case easily follows (Do it!). The proof is by induction on n, and the basis step $n = 0$ is trivial. Assume that every n-CPP representation is a 0-CPP representation, and then suppose that α is an $(n+1)$-CPP representation which is not

0-CPP. Let $\alpha(y)$ be an offending equivalence relation. Let $B \subseteq A$ be such that $\alpha|B$ is an n-CPP representation and for some $x \in L$, $\alpha(x) \cap B^2 = \alpha(y) \cap B^2$. Necessarily, $x = y$. By the inductive hypothesis, $\alpha|B$ is 0-CPP, so $(\alpha|B)(y)$ does not have exactly two equivalence classes, so it must have just one, and therefore $y = 0$. But then $\alpha(y) = \mathbf{1}_A$, contradicting that it has two equivalence classes. □

The following easy exercise is used in the proof of the lemma following it.

Exercise 4.5.17 If $\alpha : L \longrightarrow \mathrm{Eq}(A)$ is a representation of the finite lattice L and $B \subseteq A$ is such that $\alpha|B$ is an n-CPP representation of L, then α is an n-CPP representation.

Lemma 4.5.18 *Suppose that L is a finite lattice and $\alpha : L \longrightarrow \mathrm{Eq}(A)$ is an $(n+1)$-CPP representation of L. If $A = A_0 \cup A_1$, then $\alpha|A_0$ or $\alpha|A_1$ is an n-CPP representation of L.*

Proof We can assume that $A_0, A_1 \neq \emptyset$ and, by the previous exercise, that $A_0 \cap A_1 = \emptyset$. Let Θ be the equivalence relation having just A_0 and A_1 for equivalence classes. Let $B \subseteq A$ be such that $\alpha|B$ is an n-CPP representation and Θ is canonical on B. By Lemma 4.5.16 $\alpha|B$ is 0-CPP, so that either $B \subseteq A_0$ or $B \subseteq A_1$. Then, by Exercise 4.5.17, either $\alpha|A_0$ or $\alpha|A_1$ is n-CPP. □

Lemma 4.5.19 *Let $\alpha : L \longrightarrow \mathrm{Eq}(A)$ be an n-CPP representation. If $0 < r \in L$, then $\alpha(r)$ has more than 2^{2^n} equivalence classes.*

Proof The proof is by induction on n. If $n = 0$, then this follows from the definition of a 0-CPP representation. For the inductive step, let $\alpha : L \longrightarrow \mathrm{Eq}(A)$ be an $(n+1)$-CPP representation. Consider $0 < r \in L$, assuming that $\alpha(r)$ has no more than $2^{2^{n+1}}$ equivalence classes. Let $\Theta \in \mathrm{Eq}(A)$ be such that it is refined by $\alpha(r)$ and that it has no more than 2^{2^n} equivalence classes each containing no more than 2^{2^n} equivalence classes of $\alpha(r)$. Let $B \subseteq A$ be such that $\alpha|B$ is n-CPP and Θ is canonical on B. By the inductive hypothesis, $\Theta \cap B^2$ has only one equivalence class, contradicting that $r > 0$. □

Lemma 4.5.20 *If $\alpha : L \longrightarrow \mathrm{Eq}(A)$ is an n-CPP representation, then there is a finite $B \subseteq A$ such that $\alpha|B$ is an n-CPP representation of L.*

Proof The proof is by induction on n. The case $n = 0$ is easy. Let $\alpha : L \longrightarrow \mathrm{Eq}(A)$ be an $(n+1)$-CPP representation, and suppose that for no finite $B \subseteq A$ is $\alpha|B$ an $(n+1)$-CPP representation. For each finite $B \subseteq A$, let Θ_B be an equivalence relation on B demonstrating that $\alpha|B$ is not n-CPP. By a compactness argument, there is an equivalence relation Θ on A such that for any finite $C \subseteq A$ there is a finite B such that $C \subseteq B \subseteq A$ and $\Theta \cap B^2 = \Theta_B$. Now let $C \subseteq A$ and $x \in L$ be such that $\alpha|C$ is an n-CPP representation of L and $(\alpha|C)(x) = \Theta \cap C^2$. By the inductive hypothesis, we can assume that C is finite. Let $B \supseteq C$ be finite such that $\Theta \cap B^2 = \Theta_B$. By Exercise 4.5.17 $\alpha|B$ is an n-CPP representation of L, and also $(\alpha|B)(x) = \Theta \cap B^2$. But this contradicts the choice of Θ_B. □

A useful consequence of the previous lemma is that for any finite lattice L there is a Σ_1 $\mathcal{L}_{\mathsf{PA}}$ formula $cpp(L, x)$ such that for any $n < \omega$, the sentence $cpp(L, n)$ expresses the fact that L has an n-CPP representation. Thus, recalling that $\mathbb{N}$ is the standard model, $\mathbb{N} \models cpp(L, n)$ iff L has an n-CPP representation. Since each sentence $cpp(L, n)$ is Σ_1, if $\mathbb{N} \models cpp(L, n)$, then $\mathsf{PA} \vdash cpp(L, n)$.

The next theorem is the principal way to get cofinal extensions having a specified finite interstructure lattice.

Theorem 4.5.21 *Let M be a countable nonstandard model of PA^* and let L be a finite lattice. If $M \models cpp(L, n)$ for every $n < \omega$, then M has a cofinal extension N such that* $\mathrm{Lt}(N/M) \cong L$.

Proof The extension N is obtained by constructing a complete type $p(x) \in S_1(M)$ and then letting N be a $p(x)$-extension of M. Since the theorem applies to models of PA^*, we can expand M by adjoining constant symbols denoting each of its elements. In this way, we can assume, without loss of generality, that *M is a nonstandard prime model of* PA^*.

Since M is nonstandard, by overspill there is a nonstandard $c \in M$ such that $M \models cpp(L, c)$. Working in M, we can find a finite c-CPP representation $\alpha : L \longrightarrow \mathrm{Eq}(A)$. Thus A is a bounded definable set and α is definable. We now define *large* sets to be those definable subsets $X \subseteq A$ for which there is some nonstandard $d \in M$ such that $\alpha|X$ is a d-CPP representation.

Let $\Theta_0, \Theta_1, \Theta_2, \ldots$ be an enumeration of all definable equivalence relations on A. Then we can get a decreasing sequence $A = X_0 \supseteq X_1 \supseteq X_2 \supseteq \cdots$ of large subsets such that, for each n, Θ is canonical on X_{n+1}.

Let $p(x)$ be the type consisting of all formulas $\varphi(x)$ such that $M \models \forall x[x \in X_n \longrightarrow \varphi(x)]$ for some $n < \omega$.

The type $p(x)$ is complete. For, consider any formula $\varphi(x)$, and consider the equivalence relation Φ defined by the formula $\varphi(x) \longleftrightarrow \varphi(y)$. Then Φ is canonical on some X_n; that is, $\Phi \cap X_n^2 = \alpha(r) \cap X_n^2$ for some $r \in L$. Since Φ has only two equivalence classes and $\alpha|X_n$ is 0-CPP, it is clear that $\Phi \cap X_n^2 = \Theta_0 \cap X_n^2$, and therefore either $M \models \forall x[x \in X_n \longrightarrow \varphi(x)]$ or $M \models \forall x[x \in X_n \longrightarrow \neg\varphi(x)]$. Thus, either $\varphi(x) \in p(x)$ or $\neg\varphi(x) \in p(x)$, so $p(x)$ is complete.

Let N be a $p(x)$-extension of M generated by the element d realizing $p(x)$. For each $r \in L$, let $t_r(x)$ be a Skolem term such that whenever $x \in A$, then $t_r(x)$ is the least element of the Θ_r-equivalence class to which x belongs. Let $M_r = \mathrm{Scl}(t_r(d))$.

Each model M_r just defined is in $\mathrm{Lt}(N)$. In fact, these are the only models in $\mathrm{Lt}(N)$. For, suppose that $M \preccurlyeq N_0 \preccurlyeq N$ and that N_0 not one of the M_r. We can assume, without loss of generality, that $N_0 = \mathrm{Scl}(b)$ for some $b \in N$. Thus, there is a Skolem term $t(x)$ such that $b = t(d)$. Let Ψ be the equivalence relation that $t(x)$ induces; that is, $x \Psi y \Longleftrightarrow t(x) = t(y)$. Then there is some X_n on which Ψ

is canonical, so we can let $r \in L$ be such that $\Psi \cap X_n^2 = \alpha(r) \cap X_n^2$. There are Skolem terms $t'(x)$ and $t''(x)$ such that

$$\forall x \in X_n[t'(t_r(x)) = t(x) \ \wedge \ t''(t(x)) = t_r(x)]$$

holds in M. But then $N \models t'(t_r(d)) = t(d) \ \wedge \ t''(b) = t_r(d)$, so that $N_0 = M_r$.

Finally, to complete this proof, we show that if $q, r \in L$, then $q \leq r$ iff $M_q \preccurlyeq M_r$. First, suppose $q \leq r$. Then $M \models \forall x \in A[t_q(t_r(x)) = t_q(x)]$, so that $t_q(t_r(d)) = t_q(d)$, thereby implying that $M_q \preccurlyeq M_r$. Conversely, suppose that $M_q \preccurlyeq M_r$. Then there is a Skolem term $t(x)$ such that $N \models t(t_r(d)) = t_q(d)$. Let Ψ be the equivalence relation that $t(x)$ induces, and let X_n be such that Ψ is canonical on X_n. Then $\alpha(r) \cap D_n^2$ refines $\Theta_q \cap X_n^2$, so that $q \leq r$. □

There is a strong converse to Theorem 4.5.21 which is the next theorem. Notice that in this theorem there are no restrictions on the cardinality of M nor on the nature of the extension $N \succ M$.

Theorem 4.5.22 *Let L be a finite lattice and suppose that* $\mathrm{Lt}(N/M) \cong L$. *Then,* $M \models \mathit{cpp}(L, n)$ *for every* $n < \omega$.

Proof Let the function $s \mapsto M_s$ be an isomorphism of L onto $\mathrm{Lt}(N/M)$. For each $s \in L$, let $a_s \in N$ be such that $M_s = \mathrm{Scl}(M \cup \{a_s\})$. Then, there are Skolem terms $t_s(x)$ (allowing parameters from M) such that $N \models t_s(a_1) = a_s$ for each $s \in L$. Let Θ_s be the equivalence relation on M induced by $t_s(x)$; that is, $x\Theta_s y$ iff $t_s(x) = t_s(y)$.

The function $\alpha : L \longrightarrow \mathrm{Eq}(M)$, where $\alpha(s) = \Theta_s$ is a 0-CPP representation of L (DO IT!). The proof, for each $n < \omega$, that $M \models \mathit{cpp}(L, n)$ is like the proof of Lemma 4.5.20 but carried out inside of M. It can be shown that there is a large enough $b \in M$ such that $\alpha|[0, b]$ is an n-CPP representation of L (DO IT!). □

Theorems 4.5.21 and 4.5.22 jointly yield the following corollary, which, at least for nonstandard countable models, asserts that it is easier to realize a finite lattice by a cofinal extension than it is by an end extension.

Corollary 4.5.23 *Suppose that L is a finite lattice and* $\mathrm{Lt}(N/M) \cong L$. *Then every nonstandard countable $M_0 \equiv M$ has a cofinal extension N_0 such that* $\mathrm{Lt}(N_0/M_0) \cong L$. □

The remaining ingredient in the proof of Theorem 4.5.8 is the proof that every lattice having a finite congruence representation has an n-CPP representation for each n. To this end, we make some more definitions.

Definition 4.5.24 Let L be a finite lattice, and let $\alpha : L \longrightarrow \mathrm{Eq}(A)$ and $\beta : L \longrightarrow \mathrm{Eq}(B)$ be two representations of L. We say that α *arrows* β [notation: $\alpha \longrightarrow \beta$] if, whenever $\Theta \in \mathrm{Eq}(A)$, then there are $C \subseteq A$ and $x \in L$ such that $(\alpha|C)(x) = \Theta \cap C^2$ and $\alpha|C$ is isomorphic to β.

One way to show that the finite lattice L has an n-CPP representation is to show that there are representations $\alpha_0, \alpha_1, \alpha_2, \ldots, \alpha_n, \ldots$ such that α_0 is a 0-CPP representation and $\alpha_{n+1} \longrightarrow \alpha_n$ for each $n < \omega$. For such representations, α_n is n-CPP. This is exactly the approach that we will use for lattices having finite congruence representations.

Let $\alpha : L \longrightarrow \mathrm{Eq}(A)$ be a representation of the finite lattice L. If $1 \leq n < \omega$, then we let $\alpha^n : L \longrightarrow \mathrm{Eq}(A^n)$ be the function such that whenever $a_0, a_1, \ldots, a_{n-1}, b_0, b_1, \ldots, b_{n-1} \in A$ and $x \in L$, then $\langle \bar{a}, \bar{b} \rangle \in \alpha^n(x)$ iff $\langle a_i, b_i \rangle \in \alpha(x)$ for each $i < n$. If $\alpha : L \longrightarrow \mathrm{Eq}(A)$ is a representation of the finite lattice L and $1 \leq m, n < \omega$, then α^n is a representation of L and $(\alpha^m)^n$ and α^{mn} are isomorphic representations (Do it!).

Exercise 4.5.25 Suppose that $1 \leq n < \omega$. If α is a lattice representation of the finite lattice L, then so is α^n.

Lemma 4.5.26 *Let α be a finite congruence representation of the finite lattice L. Then:*

(1) *α^2 is a* 0-CPP *representation;*

(2) *if $1 \leq n < \omega$, then α^n is a finite congruence representation.*

Proof (1) Clearly, if $x \in L$, then $|\alpha^2(x)| = |\alpha(x)|^2 \neq 2$.

(2) By Example 4.5.9, we can let $\alpha : L \longrightarrow \mathrm{Cg}(A)^{\perp}$ be an isomorphism, where A is a finite algebra all of whose operations are 1-ary. Without loss of generality, we can assume that among the operations of A are all the constant functions and the identity function. For, any of these which happen not to be among the operations can be adjoined without altering $\mathrm{Cg}(A)$.

We wish to turn A^n into an algebra. The operations of A^n will be 1-ary; that is, each will be a function $g : A^n \longrightarrow A^n$. The function $g : A^n \longrightarrow A^n$ will be among the operations of A^n iff one of the following holds:

- there are operations $f_0, f_1, \ldots, f_{n-1}$ of A such that

$$g(\langle a_0, a_1, \ldots, a_n \rangle) = \bar{f}(\bar{a})$$

  for every $a_0, a_1, \ldots, a_{n-1} \in A$;
- for some $i < n$, $g = h_i$ where

$$h_i(\bar{a}) = \langle a_i, a_i, \ldots, a_i \rangle$$

  for every $a_0, a_1, \ldots, a_{n-1} \in A$.

Notice that in the items above, we have used abbreviations such as $\bar{f}(\bar{a})$ for $\langle f_0(a_0), f_1(a_1), \ldots, f_{n-1}(a_{n-1}) \rangle$. We continue with this practice.

Verifying that $\alpha^n : L \longrightarrow \mathrm{Cg}(A^n)^\perp$ is an isomorphism will complete the proof. Since we know from Exercise 4.5.25 that α^n is a lattice representation, it suffices to prove that $\alpha^n(L) = \mathrm{Cg}(A^n)$.

We first prove that $\alpha^n(L) \subseteq \mathrm{Cg}(A^n)$. Consider $x \in L$. To show that $\alpha^n(x) \in \mathrm{Cg}(A^n)$, we need to show that whenever $\bar{c}, \bar{d} \in A^n$ and g is an operation of A^n, then

$$\langle \bar{c}, \bar{d} \rangle \in \alpha^n(x) \Longrightarrow \langle g(\bar{c}), g(\bar{d}) \rangle \in \alpha^n(x)\,.$$

Since g can be one of two kinds of operations, two arguments are required.

If g is the first kind of operation defined from the operations $f_0, f_1, \ldots, f_{n-1}$, then

$$\begin{aligned}
\langle \bar{c}, \bar{d} \rangle \in \alpha^n(x) &\Longrightarrow \langle c_i, d_i \rangle \in \alpha(x) \quad \text{for all } i < n \\
&\Longrightarrow \langle f_i(c_i), f_i(d_i) \rangle \in \alpha(x) \quad \text{for all } i < n \\
&\Longrightarrow \langle \bar{f}(\bar{c}), \bar{f}(\bar{d}) \rangle \in \alpha^n(x) \\
&\Longrightarrow \langle g(\bar{c}), g(\bar{d}) \rangle \in \alpha^n(x).
\end{aligned}$$

If g is the second kind of operation, where $g = h_i$, then

$$\begin{aligned}
\langle \bar{c}, \bar{d} \rangle \in \alpha^n(x) &\Longrightarrow \langle c_i, d_i \rangle \in \alpha(x) \\
&\Longrightarrow \langle \langle c_i, c_i, \ldots, c_i \rangle, \langle d_i, d_i, \ldots, d_i \rangle \rangle \in \alpha^n(x) \\
&\Longrightarrow \langle g(\bar{c}), g(\bar{d}) \rangle \in \alpha^n(x).
\end{aligned}$$

We next show that $\mathrm{Cg}(A^n) \subseteq \alpha^n(L)$. Consider some $\Theta \in \mathrm{Cg}(A^n)$, with the aim of finding some $x \in L$ such that $\alpha^n(x) = \Theta$. Let $\Theta_0 \in \mathrm{Eq}(A)$ be such that

$$\langle a, b \rangle \in \Theta_0 \Longleftrightarrow \langle \langle a, a, \ldots a \rangle, \langle b, b, \ldots, b \rangle \rangle \in \Theta\,.$$

Then $\Theta_0 \in \mathrm{Cg}(A)$. For, if $a, b \in A$ and f is an operation of A, then

$$\begin{aligned}
\langle a, b \rangle \in \Theta_0 &\Longrightarrow \langle \langle a, a, \ldots, a \rangle, \langle b, b, \ldots, b \rangle \rangle \in \Theta \\
&\Longrightarrow \langle \langle f(a), f(a), \ldots, f(a) \rangle, \langle f(b), f(b), \ldots, f(b) \rangle \rangle \in \Theta \\
&\Longrightarrow \langle f(a), f(b) \rangle \in \Theta_0\,.
\end{aligned}$$

Since $\Theta_0 \in \mathrm{Cg}(A)$, there is some $x \in L$ such that $\alpha(x) = \Theta_0$. We finish the proof by showing that $\alpha^n(x) = \Theta$. We first show that $\Theta \subseteq \alpha^n(x)$. If $\bar{c}, \bar{d} \in A^n$, then

$$\begin{aligned}
\langle \bar{c}, \bar{d} \rangle \in \Theta &\Longrightarrow \langle h_i(\bar{c}), h_i(\bar{d}) \rangle \in \Theta && \text{for each } i < n \\
&\Longrightarrow \langle \langle c_i, c_i, \ldots, c_i \rangle, \langle d_i, d_i, \ldots, d_i \rangle \rangle \in \Theta && \text{for each } i < n \\
&\Longrightarrow \langle c_i, d_i \rangle \in \Theta_0 && \text{for each } i < n \\
&\Longrightarrow \langle c_i, d_i \rangle \in \alpha(x) && \text{for each } i < n \\
&\Longrightarrow \langle \bar{c}, \bar{d} \rangle \in \alpha^n(x).
\end{aligned}$$

To show that $\alpha^n(x) \subseteq \Theta$, suppose $\langle \bar{c}, \bar{d} \rangle \in \alpha^n(x)$. Then $\langle c_i, d_i \rangle \in \alpha(x)$ for each $i < n$. But $\Theta_0 = \alpha^n(x)$, so $\langle c_i, d_i \rangle \in \Theta_0$ for each $i < n$ and, therefore, $\langle \langle c_i, c_i, \ldots, c_i \rangle, \langle d_i, d_i, \ldots, d_i \rangle \rangle \in \Theta$ for each $i < n$. Now let $g_0, g_1, \ldots, g_{n-1} : A^n \longrightarrow A^n$ be such that whenever $i < n$ and $\bar{a} \in A^n$, then $g_i(\bar{a}) = \bar{b}$, where

$$b_j = \begin{cases} d_j & \text{if } j < i, \\ a_j & \text{if } j = i, \\ c_j & \text{if } j > i. \end{cases}$$

Clearly, each g_i is an operation of A^n, so $\langle g_i(\langle c_i, c_i, \ldots, c_i \rangle), g_i(\langle d_i, d_i, \ldots, d_i \rangle) \rangle \in \Theta$ for each $i < n$. But since

$$\begin{aligned}
\bar{c} &= g_0(\langle c_0, c_0, \ldots, c_0 \rangle), \\
g_i(\langle d_i, d_i, \ldots, d_i \rangle) &= g_{i+1}(\langle c_{i+1}, c_{i+1}, \ldots, c_{i+1} \rangle), \\
g_{n-1}(\langle d_{n-1}, d_{n-1}, \ldots, d_{n-1} \rangle) &= \bar{d},
\end{aligned}$$

we can conclude that $\langle \bar{c}, \bar{d} \rangle \in \Theta$. □

Theorem 4.5.27 *If α is a finite congruence representation of the finite lattice L, then there is $n < \omega$ such that $\alpha^n \longrightarrow \alpha$.*

Proof The proof of this theorem uses a generalization of the Hales–Jewett Theorem, which has been referred to earlier in this chapter. The Hales–Jewett Theorem implies an extension of itself. Combinatorial lines of A^r, which were defined earlier, could also be called one-dimensional combinatorial subsets of A^r. We extend this definition to higher dimensions. If $s \leq r$, then $B \subseteq A$ is an *s-dimensional combinatorial* subset of A^r if there are a subset $I \subseteq r$ partitioned into s nonempty sets $I_0, I_1, \ldots, I_{s-1}$ and a function $c : r \backslash I \longrightarrow A$ such that

$$B = \{a \in A^r : a_i = c_i \text{ if } i \in n \backslash I \text{ and } a_i = a_j \text{ if } i, j \in I_k \text{ for some } k < r\}.$$

If B is an s-dimensional combinatorial subset of A^r as just described, there is a natural bijection $f : A^s \longrightarrow B$ where $f(a)_i = a_j$ if $i \in I_j$.

THE EXTENDED HALES–JEWETT THEOREM: *Suppose that* $1 \leq m, s, k < \omega$ *and* $|A| = m$. *Then there is* $r < \omega$ *such that if* $A^r = C_0 \cup C_1 \cup \cdots \cup C_{k-1}$, *then some* C_i *contains an* s*-dimensional combinatorial subset of* A^r.

The Extended Hales–Jewett Theorem involves partitioning A^r into some fixed finite number k of sets. There is a further generalization of the Hales–Jewett Theorem, due to Prömel and Voigt, in which A^r is partitioned into an arbitrary number of sets.

THE PRÖMEL–VOIGT THEOREM: *Suppose that* $1 \leq m, s < \omega$ *and* $|A| = m$. *Then there is* $r < \omega$ *such that if* $\Theta \in \mathrm{Eq}(A^r)$, *then there is an* s*-dimensional combinatorial subset* $B \subseteq A^r$ *with a natural bijection* $h : A^s \longrightarrow B$ *and* $\Psi \in \mathrm{Eq}(A)$ *such that* $h(\Psi)^s) = h(A^s) \cap \Theta$.

The following is an immediate corollary of the Prömel–Voigt Theorem.

Corollary 4.5.28 *Suppose that* $1 \leq m, s < \omega$, $|A| = m$, *and* $\gamma : \mathrm{Eq}(A)^{\perp} \longrightarrow \mathrm{Eq}(A)$ *is the identity representation. There is* $r < \omega$ *such that* $\gamma^r \longrightarrow \gamma^s$. □

The representation γ in the previous corollary is a finite congruence representation since A can be considered as an algebra having no operations. Then, by Lemma 4.5.26(2), γ^s is also a finite congruence representation. Thus, by choosing n so that $ns \geq r$, we see that $(\gamma^s)^n \longrightarrow \gamma^r$, thereby obtaining the special case of the theorem for $\alpha = \gamma^r$. We use this special case of Theorem 4.5.27 to prove the theorem for arbitrary α.

Continuing with the proof of the theorem, let $\alpha : L \longrightarrow \mathrm{Cg}(A)^{\perp}$ be an isomorphism. With thanks to Exercise 4.5.13, we assume that all the operations of A are 1-ary; moreover, we assume that the set of operations is closed under composition and includes the identity function. Let $f_0, f_1, \ldots, f_m$ be the operations of A, and let $F : A^{m+1} \longrightarrow A$ be such that $F(a) = \langle f_0(a), f_1(a), \ldots, f_m(a)\rangle$. Letting $\gamma : \mathrm{Eq}(A)^{\perp} \longrightarrow \mathrm{Eq}(A)$ be the identity representation, we can get from the Prömel–Voigt Theorem some $n < \omega$ such that $\gamma^n \longrightarrow \gamma^{m+1}$. We claim that $\alpha^n \longrightarrow \alpha$.

To prove this claim, consider some $\Theta \in \mathrm{Eq}(A^n)$. Let $B \subseteq A^n$ be such that $\gamma^n|B$ is isomorphic to γ^{m+1} (as confirmed by $g : A^{m+1} \longrightarrow B$) and let $z \in \mathrm{Eq}(A)$ be such that $(\gamma^n|B)(z) = \Theta \cap B^2$. Now define $h : A \longrightarrow A^n$ so that that $h(a) = g(F(a))$, and let $C \subseteq A^n$ be its image. To finish the proof, it suffices to show that (1) $\alpha^n|C$ is isomorphic to α, and (2) $(\alpha^n|C)(x) = \Theta \cap C^2$ for some $x \in L$.

We prove (1) by showing that h confirms the isomorphism. To this end, consider $x \in L$ and $a, b \in A$. Then:

$$\begin{aligned}\langle a,b\rangle \in \alpha(x) &\Longleftrightarrow \langle f_i(a), f_i(b)\rangle \in \alpha(x) \quad \text{for each } i \leq m\\ &\Longleftrightarrow \langle F(a), F(b)\rangle \in \alpha^{m+1}(x)\\ &\Longleftrightarrow \langle g(F(a)), g(F(b))\rangle \in \alpha^{n}(x)\\ &\Longleftrightarrow \langle h(a), h(b)\rangle \in \alpha^{n}(x).\end{aligned}$$

To complete the proof of the theorem, it remains to prove (2). Let $\Psi \in \mathrm{Eq}(A)$ be such that

$$\langle a,b\rangle \in \Psi \Longleftrightarrow \langle h(a), h(b)\rangle \in \Theta\,.$$

Since h confirms that $\alpha^n|C$ and α are isomorphic, it suffices to show that $\Psi \in \mathrm{Cg}(A)$. Thus, we show that for any $j \leq m$, if $\langle a,b\rangle \in \Psi$, then $\langle f_j(a), f_j(b)\rangle \in \Psi$. First notice that for any $a, b \in A$,

$$\begin{aligned}\langle a,b\rangle \in \Psi &\Longleftrightarrow \langle h(a), h(b)\rangle \in \Theta\\ &\Longleftrightarrow \langle g(F(a)), g(F(b))\rangle \in \Theta\\ &\Longleftrightarrow \langle F(a), F(b)\rangle \in \gamma^{m+1}(z)\\ &\Longleftrightarrow \langle f_i(a), f_i(b)\rangle \in \gamma(z) \quad \text{for each } i \leq m.\end{aligned}$$

Now suppose that $\langle a,b\rangle \in \Psi$. Then $\langle f_i(a), f_i(b)\rangle \in \gamma(z)$ for each $i \leq m$. Since the set of operations of A is closed under composition, we have that $\langle f_i(f_j(a)), f_i(f_j(b))\rangle \in \gamma(z)$ for each $i \leq m$, so that $\langle f_j(a), f_j(b)\rangle \in \Psi$, thereby proving (2). □

Corollary 4.5.29 *If L is a finite lattice which has a finite congruence representation, then for each $n < \omega$, L has an n-CPP representation.*

Proof Let α be a finite congruence representation of the lattice L. Let $\alpha_0 = \alpha^2$, which, by Lemma 4.5.26(1), is a 0-CPP representation of L. Applying Lemma 4.5.26(2) repeatedly, we can get finite congruence representations $\alpha_0, \alpha_1, \alpha_2, \alpha_3 \ldots$ such that

$$\cdots \longrightarrow \alpha_3 \longrightarrow \alpha_2 \longrightarrow \alpha_1 \longrightarrow \alpha_0.$$

Then, α_n is an n-CPP representation of L. □

The pieces of the proof of Theorem 4.5.8 can now be assembled. Let M be a countable nonstandard model and L a finite lattice having a finite congruence representation. By the previous corollary, L has an n-CPP representation for each $n < \omega$. Thus, each of the sentences $cpp(L, n)$ is true in the standard model $\mathbb{N}$, and so each is true in M. Then, by Theorem 4.5.21, M has a cofinal extension N such that $\mathrm{Lt}(N/M) \cong L$, completing the proof of Theorem 4.5.8.

The following is a sample corollary.

Corollary 4.5.30 *If $q = p^k$, where p is a prime and $1 \leq k < \omega$, then every countable nonstandard model M has a cofinal extension N such that* $\mathrm{Lt}(N/M) \cong \mathbf{M_{q+1}}$.

Proof Refer to Theorem 4.5.21 and Example 4.5.12. □

The method of proof of Theorem 4.5.21 can also be used for some infinite lattices. For an infinite cardinal κ, let $\mathbf{M}_\kappa$ be the lattice of cardinality κ in which all elements, other than 0 and 1, are pairwise incomparable. Example 4.5.12 can be extended to the case where $q \in M$ is nonstandard, resulting in the following.

Corollary 4.5.31 *Every countable nonstandard model M has a cofinal extension N such that* $\mathrm{Lt}(N/M) \cong \mathbf{M}_{\aleph_0}$. □

While the question of whether every finite lattice is isomorphic to a substructure lattice is still unresolved, there is nothing to prevent a finite lattice from being embeddable into a finite substructure lattice.

Corollary 4.5.32 *If L is a finite lattice and M is a countable nonstandard model, then M has a cofinal extension N such that* $\mathrm{Lt}(N/M)$ *is finite and L is isomorphic to a sublattice of* $\mathrm{Lt}(N/M)$.

Proof By Theorem 4.5.2, $L^\perp$ has a finite lattice representation; that is, there is a finite set A and a lattice embedding $\alpha : L^\perp \longrightarrow \mathrm{Eq}(A)^\perp$. Then α is an embedding of L in $\mathrm{Eq}(A)$. The lattice $\mathrm{Eq}(A)$ has a finite congruence representation since $\mathrm{Eq}(A) \cong \mathrm{Cg}(A)$, where A is the algebra having no operations. Thus, Theorem 4.5.8 implies that there is a cofinal extension N such that $\mathrm{Lt}(N/M) \cong \mathrm{Eq}(A)$. Then L is isomorphic to a sublattice of $\mathrm{Lt}(N/M)$. □

In all cases that we know of, that is for any finite lattice L and $n < \omega$, if $M \models cpp(L, n)$ for some model M, then in fact $\mathsf{PA} \vdash cpp(L, n)$. Moreover, if the sentence $\forall x[cpp(L, x)]$ is true in the standard model of PA, then it is a consequence of PA. This section ends with some results concerning models in which $\forall x[cpp(L, x)]$ holds.

An important feature of Lemma 4.5.19, when restricted to finite representations, is that it and its proof can be formalized in PA. We make use of this feature in the next theorem which is an analogue of Theorem 4.5.21 for the case that $\forall x[cpp(L, x)]$ holds.

Theorem 4.5.33 *Let M be a countable nonstandard model of* PA^* *and L a finite lattice such that $M \models \forall x[cpp(L, x)]$. Then for each $a \in M$, there is a cofinal $[0, a]$-extension N such that* $\mathrm{Lt}(N/M) \cong L$.

Proof We can assume that a is nonstandard. To construct the extension, proceed as in the proof of Theorem 4.5.21 but requiring that the nonstandard c at the beginning of the proof be large enough, say $c > 2a$. Then the sets X_n that we get are such that $\alpha|X_n$ is a $(c-n)$-CPP representation. Consequently, by Lemma 4.5.19 (or, more precisely, its formalization in PA) each $\alpha(r)$, where $0 < r \in L$, has at least a equivalence classes. We obtain a cofinal extension $N = M(d)$ just as in the proof of Theorem 4.5.21. It remains to show that N is an $[0,a]$-extension.

Consider a Skolem term $t(x)$ such that $N \models t(d) < a$. Then there is some n such that $M \models \forall x[x \in X_n \longrightarrow t(x) < a]$. Then the formula $t(x) = t(y)$ defines an equivalence relation Θ on X_n, so there is some $m > n$ such that Θ is canonical on X_m. But, as $\Theta \cap X_m^2$ has no more than a equivalence classes, it must be that $(\alpha|X_m)(0) = \Theta \cap X_m^2$. But then $t(x)$ is constant on X_m, so that $t(d) \in M$. □

Some results which concern the producing of lattices by cofinal extensions have consequences about the producing of lattices by end extensions. In particular, we will see that certain finite lattices are produced by resolute end-extensional types.

For finite lattices L_1 and L_2 (assumed to be disjoint), let $L_1 \oplus L_2$ be their *linear sum*, which is the lattice simultaneously extending L_1 and L_2 and having elements $L_1 \cup L_2$, except that 1_{L_1} and 0_{L_2} are identified. Thus, $L \oplus \mathbf{2}$ is a 1-doubling extension of L, and $\mathbf{2} \oplus L$ is L with a new 0 added to it. If N is a $p(x)$-extension of M, where $p(x)$ is an end-extensional type, and $\mathrm{Lt}(N/M)$ is finite, then, by Theorem 2.1.1 (Do it!), $\mathrm{Lt}(N/M)$ is isomorphic to some $\mathbf{2} \oplus L$. The following is a sort of converse to this.

Corollary 4.5.34 *Let T be a completion of* PA^*. *Let L be a finite lattice such that $T \vdash \forall x[cpp(L,x)]$. Then there is an end-extensional resolute type $p(x) \in S_1(T)$ which produces $\mathbf{2} \oplus L$.*

Proof It is a little easier to work in the prime model M_0 of T and then construct the end-extensional resolute type $p(x) \in S_1(M_0)$. Since $M_0 \models \forall x[cpp(L,x)]$, there are definable sequences $\langle A_x : x \in M_0 \rangle$ and $\langle \alpha_x : x \in M_0 \rangle$ such that $\langle A_x : x \in M_0 \rangle$ is a sequence of pairwise disjoint, bounded subsets of M_0, and each $\alpha_x : L \longrightarrow \mathrm{Eq}(A_x)$ is an x-CPP representation of L. Let $A = \bigcup_x A_x$. We say that the unbounded definable subset $Y \subseteq A$ is *large* if there is $n < \omega$ such that $\alpha_x|(A_x \cap Y)$ is an $(x-n)$-CPP representation whenever $A_x \cap Y \neq \emptyset$. Thus, A is large. There are two lemmas which are used in constructing the type. The second of the lemmas should be compared with Theorem 4.5.21. Their proofs are omitted, but (Do it!).

> Lemma: *Suppose that Y is large and that $\langle f_a : a \in M_0 \rangle$ is a definable sequence of functions on M_0. Then there is a large $Z \subseteq Y$ such that for each $a \in M_0$, either f_a is eventually constant on $\{x \in M_0 : A_x \cap Z \neq \emptyset\}$ or is eventually one-to-one on that set.*

LEMMA: *Suppose that Y is large and that $\langle \Theta_a : a \in M_0 \rangle$ is a definable sequence of equivalence relations on M_0. Then there is a large $Z \subseteq Y$ such that for each $a \in M_0$ there is $r \in L$ such that for all sufficiently large $x \in M_0$, if $A_x \cap Z \neq \emptyset$, then $\Theta_a \cap (A_x \cap Z)^2 = \alpha_x(r) \cap (A_x \cap Z)^2$.*

These lemmas are used to construct a type by forming a decreasing sequence $A = Y_0 \supseteq Y_1 \supseteq Y_2 \supseteq \cdots$ of large sets such that:

(1) whenever $\langle f_a : a \in M_0 \rangle$ is a definable sequence of functions $f : M_0 \longrightarrow M_0$, then there is $m < \omega$ such that for each $a \in M_0$, either f_a is eventually constant on $\{x \in M_0 : A_x \cap Y_m \neq \emptyset\}$ or is eventually one-to-one on that set;

(2) whenever $\langle \Theta_a : a \in M_0 \rangle$ is a definable sequence of equivalence relations on M_0, then there is $m < \omega$ such that for each $a \in M_0$ there is $r \in L$ such that for all sufficiently large $x \in M_0$, if $A_x \cap Y_m \neq \emptyset$, then $\Theta_a \cap (A_x \cap Y_m)^2 = \alpha_x(r) \cap (A_x \cap Y_m)^2$.

The sequence $\langle Y_m : m < \omega \rangle$ determines a type $p(x)$, where $\varphi(x) \in p(x)$ iff for some $m < \omega$,

$$M_0 \models \forall x[x \in Y_m \longrightarrow \varphi(x)]\,.$$

This type is definable (DO IT!) and produces $\mathbf{2} \oplus L$ (DO IT!).

Finally, $p(x)$ is resolute. In fact, $p(x)$ is $(1+|L|)$-resolute (DO IT!) from which it follows by Proposition 3.4.4 that it is resolute. □

We saw in Corollary 4.5.31 that there is a type producing $\mathbf{M}_{\aleph_0}$. This type cannot be used get a resolute type producing $\mathbf{2} \oplus \mathbf{M}_{\aleph_0}$. In fact, what we do get is a definable type $p(x)$ such that whenever N is a $p(x)$-extension of M and $|M| = \kappa$, then $\mathrm{Lt}(N/M) \cong \mathbf{M}_\kappa$. In particular, this $p(x)$ is not resolute. See Exercise 4.8.17.

4.6 The pentagon lattice

The pentagon lattice, also known as $\mathbf{N_5}$, is an important nondistributive lattice, being one of the two nondistributive lattices which are in the forbidden sublattice characterization of distributive lattices in Proposition 4.3.4. The other lattice is $\mathbf{M_3}$. By means of cofinal extensions, both of these lattices were shown in the previous section to be substructure lattices. Corollary 4.2.10 is the converse to this for $\mathbf{M_3}$: if $M \prec N$ and $\mathrm{Lt}(N/M) \cong \mathbf{M_3}$, then $M \prec_{\mathsf{cof}} N$. The bulk of this section is devoted to proving Wilkie's Theorem 4.6.2 that the lattice $\mathbf{N_5}$ is, in this respect, different from $\mathbf{M_3}$ in that it can be realized by an elementary end extension. This theorem is historically the first proving the existence of a finite nondistributive substructure lattice.

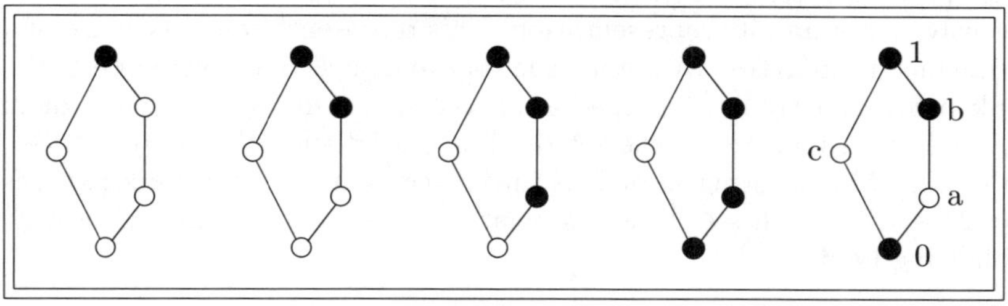

FIG. 4.2. Five ranked pentagon lattices

The lattice $\mathbf{N_5}$ has ten different rank functions, five of which satisfy both the Blass and Gaifman Conditions (DO IT!). These five are displayed in Figure 4.2. Note that both $\circ$ and $\bullet$ denote points of the lattice, but the latter is used for points in the rankset. The first of these ranked lattices has a cofinal rank function, and the last two have end rank functions. We saw in the previous section that the first is a ranked substructure lattice, and the proof of Theorem 4.6.2 shows that the last one also is. The first result of this section, Theorem 4.6.1, implies that the second depicted lattice cannot be a ranked substructure lattice. It will not be shown here, but the remaining two ranked lattice are also ranked substructure lattices. Throughout this section, when referring to $\mathbf{N}_5$, we assume that it consists of the points $0, a, b, c, 1$ as indicated in Figure 4.2.

Theorem 4.6.1 *Suppose that $M \prec N$ and that* $\mathrm{Ltr}(N/M) \cong (\mathbf{N}_5, r)$. *Then* $r(0) \neq b$.

Proof Let $F : (\mathbf{N}_5, r) \longrightarrow \mathrm{Ltr}(N/M)$ be an isomorphism, and suppose that $r(0) = b$. Thus, $(\mathbf{N}_5, r)$ is the second of the depicted ranked lattices in Figure 4.2. We assume there are elements $a, b, c \in N$ such that $F(a) = M(a), F(b) = M(b)$, and $F(c) = M(c)$. Let $e \in M$ be such that $a, b < e$. Then there is an M-definable function $G : N \times N \longrightarrow N$ such that $G(c, a) = b$, and let $g \in M(c)$ be (a code for) the function $\{\langle x, y\rangle : x, y < e \text{ and } G(c, a) = b\}$. Since $M \prec_{\mathsf{end}} M(c)$ and $g < 2^{2^e}$, it follows that $g \in M$. Thus $b = g(a) = G(c, a) \in M(a)$, which is a contradiction. □

Theorem 4.6.2 *Every countable model M of* PA *has an elementary end extension N such that* $\mathrm{Lt}(N/M) \cong \mathbf{N_5}$. □

Proving Theorem 4.6.2 in the case that M is the standard model is notationally much easier than in the general case, although conceptually there's not a whole lot of difference. We take the expedient approach of presenting the proof just for the standard model leaving it up to the reader to make the appropriate modifications for the general proof.

The proof involves the study of a particular representation of $\mathbf{N}_5$ which, necessarily, is not a finite representation. This representation makes use of a tree $(T, \lhd)$, which is an ω-tree with a unique root and such that each node of the tree of rank n has exactly 2^{n+1} immediate successors. One way to get a definitive such tree is by letting T be the set of all $s \in \omega^{<\omega}$ such that $\ell(s) = 2^n - 1$ for some $n < \omega$. The ordering $\lhd$ of T is initial segment, and the rank $rk(s)$ is n if $\ell(s) = 2^n - 1$. To each $s \in T$, we associate a subset $X_s \subseteq \omega$ in such a way that the following hold:

(1) If $s \lhd t$, then $X_s \supseteq X_t$.

(2) If s, t are incomparable, then $X_s \cap X_t = \emptyset$.

(3) There are infinitely many elements of X_s which, for any $t \rhd s$ are not in X_t.

(4) For every $k < \omega$ there is a rank n such that $k \notin X_s$ whenever s has rank k.

These conditions determine $\langle X_s : s \in T \rangle$ up to a bijection. That is, if (1)–(4) hold for of $\langle X_s : s \in T \rangle$ and $\langle Y_s : s \in T \rangle$, then there is a bijection $f : X_\emptyset \longrightarrow Y_\emptyset$ such that for any $s \in T$, $Y_s = \{f(k) : k \in X_s\}$.

Let $A = \{\langle s, k \rangle : s \in T, k \in X_s \rangle\}$. We now define the representation $\alpha : \mathbf{N}_5 \longrightarrow \mathrm{Eq}(A)$. There is no question about what $\alpha(0)$ and $\alpha(1)$ are; we define $\alpha(a), \alpha(b), \alpha(c)$ by identifying the equivalence classes of each of these equivalence relations.

The $\alpha(c)$-classes are the *columns* of A, the kth column being the set of those $\langle s, k \rangle$ in A. Each X_s is an $\alpha(b)$-class. For each $s \in T$, the union of all the X_t, where t is an immediate successor of s, is an $\alpha(a)$-class. This determines $\alpha(a)$ as long as we also let $X_\emptyset$ be an $\alpha(a)$-class.

It is clear that α is a 0-CPP representation of $\mathbf{N}_5$ (Do IT!). The next lemma is the key combinatorial fact about this representation.

Lemma 4.6.3 $\alpha \longrightarrow \alpha$.

What is this lemma saying? Whenever Θ is an equivalence relation on A, there is $B \subseteq A$ such that $\alpha | B \cong \alpha$ and Θ is canonical on B. Thus, to prove the lemma, we start with Θ and then find the requisite B. This is done by a rather slow process of gradually thinning down A until we get to B.

In the definition of α we made use of a tree which had very specific branching. This was done just to be definitive; it would have been just as good if we had required that every point of rank n had at least 2^{n+1} immediate successors. For, if $\beta : \mathbf{N}_5 \longrightarrow \mathrm{Eq}(B)$ is such a representation, then there are subsets $A' \subseteq B$ and $B' \subseteq A'$ such that $\beta | A' \cong \alpha$ and $\alpha | B' \cong \beta$. We say that $B \subseteq \omega$ is *correct* if $\alpha | B$ is a representation satisfying this weaker requirement. If B is correct, then the set $(\alpha | A)(b)$-classes form a tree, and then for $n < \omega$, we let $\Lambda_n(B)$ be the subset of B which is the union of all the $(\alpha | B)(b)$-classes of rank n in this tree.

In the following series of claims we assume that $B \subseteq A$ is correct and $\Theta \in \mathrm{Eq}(A)$.

Claim 1: *Let $X \subseteq \Lambda_n(B)$ be an $(\alpha|B)(b)$-class. Then there is a correct $D \subseteq B$ such that:*

(1) $\Lambda_i(D) = \Lambda_i(B)$ *for each* $i < n$;
(2) $X \cap D \neq \emptyset$;
(3) *If Y is an $(\alpha|B)(b)$-class of rank n and $Y \neq X$, then $Y \subseteq D$;*
(4) $\Theta \cap (B \cap D)^2$ *is either trivial or discrete.*

By repeated applications of Claim 1, we can get the following claim.

Claim 2: *There is a correct $D \subseteq B$ such that whenever X is an $(\alpha|D)(b)$-class, then $\Theta \cap X^2$ is either trivial or discrete.*

In the light of Claim 2, we can assume that each point in the tree is "trivial" or "discrete." By taking an appropriate subtree, we can get that either all points are "trivial" or all are "discrete," resulting in the following improvement to Claim 2.

Claim 3: *There is a correct $D \subseteq B$ such that either:* (1) *whenever X is an $(\alpha|D)(b)$-class, then $\Theta \cap X^2$ is trivial; or* (2) *whenever X is an $(\alpha|D)(b)$-class, then $\Theta \cap X^2$ is discrete.*

We have reduced Lemma 4.6.3 to the two cases of Claim 3. Let us assume that the original $\Theta \in \mathrm{Eq}(A)$ is as in (1) or (2) of Claim 3.

Suppose (1) occurs; that is $\Theta \cap X^2$ is trivial for each $\alpha(b)$-class. We find a correct $B \subseteq A$ such that $\Theta \cap B^2$ is either trivial, $\alpha(a)|B$ or $\alpha(b)|B$. Since Θ induces an equivalence relation the tree T, this problem of finding such a B reduces to a problem about trees. This problem is solved by the right lemma about the trees. Since we are just concerned about the tree T, we will state the lemma for that tree only.

Let Φ be the equivalence relation on T for which any two elements are equivalent iff they have the same immediate predecessor. A warning about notation: if $s \in T$, then $rk(s)$ always is the rank of s in the sense of T. Now let us say that a subset $S \subseteq T$is a *strong subtree* if $S \neq \emptyset$ and whenever $s \in S$ and $rk(s) = m$, then there is $n > m$ such that whenever $s \lhd s' \in T$ and $rk(s') = m + 1$, then there is a unique $t \in T'$ such that $rk(t) = n$, $s' \unlhd t$ and there is no $t' \in S$ for which $s' \lhd t' \lhd t$.

Lemma 4.6.4 *Suppose that $E \in \mathrm{Eq}(T)$. Then there is a strong subtree $S \subseteq T$ such that $E \cap S^2$ is either trivial, discrete, or $\Phi \cap S^2$.* □

No formal proof will be given, but here is how one would go about constructing S. First try to construct S such that $S \cap E^2$ is discrete, working your way up the tree. If this succeeds, fine. If not, then it is for one of two reasons: either you always have to choose a point in an equivalence class in which there already is a point, or you have to choose two points in the same class. If it is for the first reason, then there is a strong subtree $S' \subseteq T$ such that $E \cap S'^2$ is finite, and then it is easy to get a strong subtree $S \subseteq S'$ such that $E \cap S^2$ is trivial. If it is for the second reason, then there is a strong subtree $S' \subseteq T$ such that $\Phi \cap S'^2 \subseteq E \cap S'^2$. Next, try to construct a strong subtree $S \subseteq S'$ such that $S \cap S'^2$, again working your way up the tree. If this succeeeds, fine. Otherwise, there is $S \subseteq S'$ such that $E \cap S^2$ is trivial.

So we can suppose that (2) of Claim 3 occurs; that is $\Theta \cap A^2$ is discrete for each $\alpha(b)$-class. We will obtain a correct $D_0 \subseteq A$ such that $\Theta \cap D_0^2 \subseteq \alpha(c)$. This can be done in the following way: let $B_0, B_1, B_2, \ldots$ be a list of all $\alpha(b)$-class with each block appearing infinitely often. Successively pick $x_n \in B_n$ which is not in the same column as any previously chosen x_m and no y which is in the same column as x_n is in the same Θ-class of some z in the same column as a previously selected x_m. Let C_n be the set of points y in the same column as x_n that are below x_n, by which we mean that there are $i \leq j < \omega$ such that $y \in \Lambda_i(A)$ and $x_n \in \Lambda_j(A)$. Let $D_0 = \bigcup_n C_n$. Clearly, D_0 is correct and has the required property.

The trouble is that in C_n there may be some points in the same Θ-class and others that are not. It is now possible (exactly how is left to be worked out) to find a correct $D_2 \subseteq D_1$ such that if C_n meets D_2, then $C_n \subseteq D_2$ and such that if $y \in \Lambda_i(D_1) \cap C_n$, $z \in \Lambda_j(D_1) \cap C_n$, $y' \in \Lambda_i(D_1) \cap C_m$, and $z' \in \Lambda_i(D_1) \cap C_m$, then $\langle y, z\rangle \in \Theta$ iff $\langle y', z'\rangle \in \Theta$. This allows us to define an equivalence relation R on ω as follows: $\langle i, j\rangle \in R$ iff for every (or, equivalently, some) $y \in \Lambda_i(D_1)$ and $z \in \Lambda_j(D_1)$, $\langle y, z\rangle \in \Theta$. Let $I \subseteq \omega$ be infinite such that $R \cap I^2$ is discrete or trivial, and let $D = \bigcup_{i \in I} \Lambda_i(D_1)$. Clearly, D is correct. If $R \cap I^2$ is discrete, then $\Theta \cap D^2$ is discrete, and if $R \cap I^2$ is trivial, then $\Theta \cap D^2 = \alpha(c) \cap D^2$.

Theorem 4.6.2 was stated for countable M only and for good reason. The next theorem implies that there are uncountable M having no elementary end extension N such that $\mathrm{Lt}(N/M) \cong \mathbf{N}_5$. In fact, no rather classless M does.

Theorem 4.6.5 *If* $\mathrm{Lt}(N/M) \cong \mathbf{N}_5$, *then* N *is not a conservative extension of* M.

Proof The submodels $M(a), M(b)$, and $M(c)$ are as in the proof of Theorem 4.6.1. Suppose N is a conservative extension of M. Then $M \prec_{\mathsf{end}} N$, so that $M \prec_{\mathsf{end}} M(c)$ and $M(b) \prec_{\mathsf{end}} N$ (Do it!).

Let $g : N \longrightarrow N$ be an $M(c)$-definable function such that $g(a) = b$. Since N is a conservative extension of M, there is an M-definable $G \subseteq N^2$ which agrees with g on M; that is, whenever $x, y \in M$, then $\langle x, y\rangle \in G$ iff $g(x) = y$. Since $G \cap M^2$ is a function, so is G. Let D be its domain. Then D is M-definable and G and g agree on $M \cap D$ (Do it!). We investigate whether or not $a \in D$, deriving a contradiction in either case.

Suppose $a \in D$. Then $G(a) \in M(a)$, so that $G(a) \neq b = g(a)$, and G and g do not agree D. Let $m \in D$ be the least such that $G(m) \neq g(m)$. Then $m \leq a$ and $m \in M(c)$, so that $m \in M$, which is a contradiction.

Suppose $a \notin D$. Let

$$I = \{z \in N : \forall x \leq z[x \notin D \longrightarrow g(x) > z]\}.$$

Then I is an $M(c)$-definable initial segment of N which includes M, so by overspill, there is $m \in (I \cap M(c))\backslash M$. Since $a < m$, it follows that $g(a) > m$, contradicting that $g(a) = b < m$. □

Corollary 4.6.6 *There is no resolute type producing* $\mathbf{N}_5$. □

4.7 Infinite distributive lattices

It was seen in Corollary 4.2.1 that for any model M, $\mathrm{Lt}(M)$ is an $\aleph_1$-algebraic lattice. Whether each $\aleph_1$-algebraic lattice is isomorphic to some $\mathrm{Lt}(M)$ is still unknown. Indeed, there are even finite lattices for which this is not known. However, for distributive lattices, nothing more than its $\aleph_1$-algebraicity is required. This is the content of the following theorem.

Theorem 4.7.1 (Mills' Theorem) *For any completion T of* PA* *and for any $\aleph_1$-algebraic distributive lattice D, there is a resolute model $M \models T$ such that* $\mathrm{Lt}(M) \cong D$. □

The proof of Mills' Theorem, which involves some highly technical details, will not be presented here. The key to the proof is a vast generalization of Theorem 4.4.1 for which a generalization of the notion of doubling extension is needed.

Let L and D be two lattices, each having a 0 element. For this construction, the lattice D need not be distributive, but in the applications it will be. In particular, if $D = \mathbf{2}$, then the extension of L that is about to be defined is a doubling extension. Let $F : D \longrightarrow L$ be such that $F(0) = 0$ and $F(x \vee y) = F(x) \vee F(y)$ for all $x, y \in D$. The (F, D)-bling extension of L is the sublattice of $L \times D$ consisting of those pairs $\langle z, x \rangle$ such that $z \geq F(x)$. This is indeed a lattice (DO IT!), and by identifying $z \in L$ with $\langle z, 0 \rangle$ (which we will do), it is an extension of L (DO IT!). Notice that if $D = \mathbf{2}$, then the (F, D)-bling extension is just the $F(1)$-doubling extension of L. An extension of L is a D-bling extension if it is an (F, D)-bling extension for some F.

Theorem 4.7.2 *Let M be a countable resolute model, let D be a distributive algebraic lattice having at most countably many compact elements, and let L be a D-bling extension of* $\mathrm{Lt}(M)$. *Then M has a conservative extension N which is* *resolute such that* $\mathrm{Lt}(N) \cong L$. *(In fact, there is an isomorphism* $\alpha : \mathrm{Lt}(N) \longrightarrow L$ *which is the identity on* $\mathrm{Lt}(M)$.*)* □

This theorem implies its own generalization obtained by weakening the hypothesis that D has at most countably many compact elements to requiring that D be $\aleph_1$-algebraic. Then Theorem 4.7.1 results by letting M be the prime model of T.

There is a variant of Mills' Theorem which applies to cofinal extensions: *if* M *is a countable, nonstandard model and* D *is an* $\aleph_1$*-algebraic distributive lattice* D*, then* M *has a cofinal extension* N *such that* $\mathrm{Lt}(N/M) \cong D$. The proof of this involves some of the same highly technical details that the proof of Mills' Theorem does. The proof of the following weaker theorem avoids some of these and will be presented here.

Theorem 4.7.3 *Let* M *be a countable nonstandard model, and let* D *be an algebraic distributive lattice having at most countably many compact elements. Then* M *has a cofinal extension* N *such that* $\mathrm{Lt}(N/M) \cong D$. □

The lattice D in this theorem is complete, so $\bigvee D$ exists. This element may not be compact, but without loss of generality we can assume it is by adjoining to D a new element $1 > \bigvee D$. We make use of the representations of finite distributive lattices D found in Exercise 4.5.3. So as to be able to refer to such representations, let us say that the representation $\alpha : D \longrightarrow \mathrm{Eq}(A)$ is *normal* if it is isomorphic to a power of the ones described in that exercise. The next lemma is rather technical in its appearance, but its significance is clear.

Suppose that you are constructing a cofinal extension using the method of Theorem 4.5.21 to produce an interstructure lattice isomorphic to D_1. The representations of D_1 that you are using are the α_1^n for nonstandard n. You are proceeding merrily along in the construction, and then suddenly you change your mind and decide to produce D_2 instead. No problem, as Lemma 4.7.4 says that you can switch in midstream to the represenations α_2^m of D_2.

Lemma 4.7.4 *Let* $(D_1, \vee, \wedge)$ *and* $(D_2, \vee, \wedge)$ *be finite distributive lattices, with* $(D_1, \vee)$ *being a subsemilattice of* $(D_2, \vee)$*, and* $\bigvee D_1 = \bigvee D_2$*,* $\bigwedge D_1 = \bigwedge D_2$*. Then for each normal* $\alpha_1 : D_1 \longrightarrow \mathrm{Eq}(A_1)$ *there is a normal* $\alpha_2 : D_2 \longrightarrow \mathrm{Eq}(A_2)$*, with* $A_2 \subseteq A_1$ *such that* $\alpha_1 | A_2 = \alpha_2 \restriction D_1$.

Proof We first assume that $\alpha_1 : D_1 \longrightarrow \mathrm{Eq}(J(D_1))$. Instead of actually getting $A_2 \supseteq J(D_1)$, we find a large enough n and an embedding $f : J(D_1)) \longrightarrow (J(D_2))^n$. Define $\mu : D_2 \longrightarrow D_1$ so that $\mu(r) = \bigwedge\{x \in D_1 : r \leq x\}$. Then $\bigvee D_1 = \bigvee D_2$ implies that $r \leq \mu(r)$ for each $r \in D_2$. We choose a function

$f : J(D_2) \longrightarrow (J(D_1))^n$ such that:

(1) if $r \in J(D_2)$ and $i < n$, then $f(r)_i \leq \mu(r)$;

(2) if $r, s \in J(D_2)$ are distinct and $x \in J(D_1)$ is such that $x \leq \mu(r)$, then there is $i < n$ such that $f(r)_i = x$ and $f(s)_i = 0$.

By letting $n = |J(D_2)|^2 \times |J(D_1)|$, we have n large enough to get such a function.

To verify the required condition, let us consider $x \in D_1$ and $r, s \in J(D_2)$. If $r = s$, then there is no problem, so we assume that $r \neq s$.

$$\begin{aligned}\langle r, s\rangle \in \alpha_2(x) \Longrightarrow r, s \leq x \quad &\Longrightarrow \quad \mu(r), \mu(s) \leq x \\ &\Longrightarrow f(r)_i, f(s)_i \leq x \quad \text{for all } i < n \\ &\Longrightarrow \langle f(r), f(s)\rangle \in \alpha_1^n(x).\end{aligned}$$

Conversely, we have

$$\begin{aligned}\langle r, s\rangle \notin \alpha_2(x) \Longrightarrow r \not\leq x \text{ (say)} \quad &\Longrightarrow \mu(r) \not\leq x \\ &\Longrightarrow f(r)_i \not\leq x, f(s)_i = 0 \quad \text{for some } i < n \\ &\Longrightarrow \langle f(r)_i, f(s)_i\rangle \notin \alpha_1(x) \quad \text{for some } i < n \\ &\Longrightarrow \langle f(r), f(s)\rangle \notin \alpha_1^n(x),\end{aligned}$$

completing the proof for $m = 1$. (Notice that the third implication in the converse direction follows from the Representation Theorem 4.3.5 and the fourth implication from the hypothesis $\bigwedge D_1 = \bigwedge D_2$ (DO IT!).)

For arbitrary normal $\alpha_2 : D_2 \longrightarrow (J(D_2))^m$, let $f_1 : J(D_2) \longrightarrow (J(D_1))^{n_1}$ be a function which works for $m = 1$, and then let $n = mn_1$ and $f : (J(D_2))^m \longrightarrow (J(D_1))^n$ be such that $f(r)_{jn_1+i} = f_1(r_j)_i$ where $j < m$ and $i < n_1$. That this f also works can easily be checked (DO IT!). □

The following fact about algebraic distributive lattices enables us to implement the previous lemma in the proof of Theorem 4.7.3. For a lattice L, we let $K(L)$ be the set of its compact elements. Note that if $x, y \in K(L)$, then $x \vee y \in K(L)$ (DO IT!). In other words, $(K(L), \vee)$ is a subsemilattice of $(L, \vee)$.

Lemma 4.7.5 *Let D be an algebraic distributive lattice. For each finite $X \subseteq K(D)$ there is a finite $D_0 \subseteq K(D)$ such that $X \subseteq D_0$, $x \vee y \in D_0$ whenever $x, y \in D_0$, and D_0 is a distributive lattice (whose meet might not be the same as the meet of D).*

Proof Let $F \subseteq D$ be the sublattice of D generated by X. Then F is also a distributive lattice and, being finitely generated, is finite. If it happens that

$F \subseteq K(D)$, then we can let $D_0 = F$, and we are done. But in general F contains elements that are not compact. Let $J \subseteq F$ be the join-irreducibles of F. For each $x \in F$, there is a $J_x \subseteq J$ such that $x = \bigvee J_x = \bigvee\{y \in K(L) : y \leq j \text{ for some } j \in J_x\}$. Therefore, by compactness, there is, for each $j \in J$, a compact $k_j \leq j$ such that for each compact $x \in F$, $x = \bigvee\{k_j : x \geq j \in J\}$. By choosing the k_j's large enough, we can also have that if $x \not\geq k_j$, then $x \not\geq j$. Now let D_0 be the upper semilattice that the k_j's generate; that is, $D_0 = \{\bigvee\{k_j : j \in J'\} : J' \subseteq J\}$.

Clearly, $x \vee y \in D_0$ whenever $x, y \in D_0$. To see that D_0 has the other required properties, define $\varphi : F \longrightarrow D_0$ so that $\varphi(x) = \bigvee\{k_j : x \geq j \in J\}$. Then $\varphi(x) = x$ for $x \in X$, so $X \subseteq D_0$. To finish the proof it suffices to observe that if $x, y \in F$, then $x \leq y \Longleftrightarrow \varphi(x) \leq \varphi(y)$. If $x \leq y$, then $\varphi(x) = \bigvee\{k_j : x \geq j \in J\} \leq \bigvee\{k_j : y \geq j \in J\} = \varphi(y)$. Conversely, suppose $x \not\leq y$, then there is some $k_j \leq x$ such that $k_j \not\leq y$, and then $k_j \leq \varphi(x)$ and $k_j \not\leq \varphi(y)$ so that $\varphi(x) \not\leq \varphi(y)$. □

Proof of Theorem 4.7.3 Let M be countable and nonstandard. Assume that D is as in the theorem and also that 1_D is compact. Let $D_0 = \{0_D, 1_D\}$. Using Lemma 4.7.5, we can get a sequence $D_0 \subseteq D_1 \subseteq D_2 \subseteq \cdots$ of finite distributive subsemilattices of $K(D)$ such that each is a distributive lattice and that $K(D) = \bigcup_i D_i$. The idea is to start the construction as if you are producing D_0, and then after a while switch to D_1 (as in Lemma 4.7.4), and still later to D_2, and so on. Then, when all is said and done, you will have produced D.

Let $g_0, g_1, g_2, \ldots$ be a list of all definable functions $g : M \longrightarrow M$. We will obtain a sequence of normal representations $\alpha_i : D_i \longrightarrow \mathrm{Eq}(A_i)$, each one being n-CPP for some nonstandard $n \in M$. Let $\alpha_0 : D_0 \longrightarrow \mathrm{Eq}(A_0)$ be such that A_0 is a bounded subset of M of nonstandard cardinality. At stage i, we have a normal α_i which is n-CPP for some nonstandard $n \in M$. First let $B \subseteq A_i$ be such that $\alpha_i|B$ is $(n-1)$-CPP and normal and such that g_i is canonical on B. Then use Lemma 4.7.4 to get $A_{i+1} \subseteq A_i$ and $\alpha_i : D_{i+1} \longrightarrow \mathrm{Eq}(A_{i+1})$ which is m-CPP for nonstandard m.

The desceneding sequence $A_0 \supseteq A_1 \supseteq A_2 \supseteq \cdots$ determines a type over M. Let $N = M(c)$ be an extension generated by c realizing that type. Then $\mathrm{Lt}(N/M) \cong D$. The argument verifying this is just like the argument at the end of the proof of Theorem 4.5.21 and is left for the reader to supply. □

4.8 Exercises

A *semilattice* is an algebra $(S, \vee)$ which satisfies the commutative, associative, and idempotency laws for $\vee$. If S is a semilattice, then define $\leq$ on S so that $x \leq y$ iff $x \vee y = y$. If L is a lattice, then let $K(L)$ be the set of its compact elements.

♣4.8.1 If S is a semilattice, then $(S, \leq)$ is a poset.

♣**4.8.2** If $(L, \vee, \wedge)$ is a lattice, then $(K(L), \vee)$ is a subsemilattice of $(L, \vee)$.

♣**4.8.3** Let L_1, L_2 be algebraic lattices and let $f : K(L_1) \longrightarrow K(L_2)$ be a semilattice isomorphism (see 4.8.2). Then there is a unique isomorphism $g : L_1 \longrightarrow L_2$ which extends f.

If $(S, \vee)$ is a semilattice, then $I \subseteq S$ is an *ideal* of S if whenever $x, y \in I$ and $z \in S$ are such that $z \leq x \vee y$, then $z \in I$. Let $\mathcal{I}(S)$ be the set of ideals of S.

♦**4.8.4** If S is a semilattice, then $(\mathcal{I}(S), \subseteq)$ is an algebraic lattice and $K(\mathcal{I}(S)) \cong S$. Define the isomorphism.

♦**4.8.5** If κ is an infinite cardinal and A is an algebra which has no more than κ operations, then $\mathrm{Cg}(A)$ is κ^+-algebraic.

♣**4.8.6** Let A be a finitely generated algebra having a proper subalgebra. Then it has a proper subalgebra $B \subsetneq A$ such that A is a *minimal* extension of B. (That is, whenever $B \subseteq C \subseteq A$, then either $C = B$ or $C = A$.)

If L_0 is a sublattice of L_1 and whenever $a, c \in L_0$ and $b \in L_1$ are such that $a < b < c$, then $b \in L_0$, then L_0 is a *convex* sublattice of L_1.

♦**4.8.7** Let L_0 be a finite convex sublattice of L_1. If M is a countable nonstandard model which has an elementary extension N_1 such that $\mathrm{Lt}(N_1/M) \cong L_1$, then M has a cofinal extension N_0 such that $\mathrm{Lt}(N_0/M) \cong L_0$.

♣**4.8.8** If p be a prime and $\mathbb{Z}_p$ is the cyclic group of order p, then $\mathrm{Cg}(\mathbb{Z}_p \times \mathbb{Z}_p) \cong \mathbf{M}_{p+1}$.

♥**4.8.9** If L is a lattice, then $\mathrm{Cg}(L)$ is a distributive lattice.

♣**4.8.10** Show that the lattice in Example 4.5.11 is a congruence lattice.

♣**4.8.11** Let L be a lattice having $\mathbf{H}$ as a sublattice with $0_{\mathbf{H}} = 0_L$ and $1_{\mathbf{H}} = 1_L$. If r is a rank function on L satisfying the Blass and Gaifman Conditions, then $r(0) = 1$. (This generalizes Proposition 4.2.13.)

The next two exercises concern the lattice of convex subsets of a three-element chain ordered by reverse inclusion, which we refer to as $\mathbf{K}_3$.

♦**4.8.12** Every countable nonstandard M has a cofinal extension N such that $\mathrm{Lt}(N/M) \cong \mathbf{K}_3$.

♠**4.8.13** Is there a model M of True Arithmetic such that $\mathrm{Lt}(M) \cong \mathbf{K}_3$?

The next three exercises are all modifications of Theorem 4.3.2.

♦**4.8.14** Show that in Theorem 4.3.2, if both occurrences of the word "resolute" are omitted, then the theorem remains true.

♦**4.8.15** Show that in Theorem 4.3.2, if both occurrences of the word "resolute" are replaced with the word "definable," then the theorem remains true.

♣**4.8.16** Suppose M is a countable model, L is a lattice for which $1 < |L| < \aleph_0$, and L' is a doubling extension of L. If M has a cofinal extension N such that $\mathrm{Lt}(N/M) \cong L$, then M has a cofinal extension N' such that $\mathrm{Lt}(N'/M) \cong L'$.

♥**4.8.17** There is a definable type producing $\mathbf{2} \oplus \mathbf{M}_{\aleph_0}$ which is not resolute.

♥**4.8.18** There an end-extensional type producing $\mathbf{3}$ which is not resolute.

♥**4.8.19** There is a resolute type producing an infinite lattice.

♠**4.8.20** Is there a resolute type producing $\mathbf{2} \oplus \mathbf{M}_{\aleph_0}$?

If $a \in M \preccurlyeq N = Scl(M \cup [0,a]_N)$, then we say that N is an *a-cofinal* extension of M.

♦**4.8.21** Let M be a countable nonstandard model and L a finite lattice. Then the following are equivalent: (1) For every nonstandard $a \in M$ there is an a-cofinal extension N of M such that $\mathrm{Lt}(N/M) \cong L$; (2) L has an n-CPP representation of each $n < \omega$.

♥**4.8.22** Suppose that L is a finite lattice which has an n-CPP repesentation for each $n < \omega$. Let M_0 be a nonstandard prime model. Then M_0 has a cofinal extension M such that:

(1) $\mathrm{Lt}(M) \cong \mathbf{2} \oplus L$;
(2) $\mathrm{SSy}(M) = \mathrm{SSy}(M_0)$;
(3) M is generated by a such that $a < c$ whenever $\omega < c \in M_0$.

♣**4.8.23** A converse to Theorem 4.5.33: Let M be any model of PA^* and L a finite lattice. If for each $a \in M$ there is an elementary a-end extension such that $\mathrm{Lt}(N/M) \cong L$, then $M \models \forall x[cpp(L,x)]$.

♥**4.8.24** If α is a finite representation and $\alpha^m \longrightarrow \alpha$, then α is a congruence representation.

4.9 Remarks & References

Lattices have their roots in the nineteenth century, but it was not until the publication of the first edition of Birkhoff [11] that there was a subject of lattice theory. There are now very many books available on this subject. We mention, besides [11], [3] which is devoted just to distributive lattices and [129] which is about both lattices and algebras.

Theorem 4.2.1 was proved by G. Grätzer & E.T. Schmidt [48]. A much simpler proof can be found in Pudlák [153].

The first result on interstructure lattices was Gaifman's improvement of the MacDowell-Specker Theorem to minimal end extensions. It was discussed in detail by Gaifman in [45] and in some earlier papers as well. Blass' Theorem on minimal cofinal extensions appeared in [13]. The paper [45] also contains a proof of Theorem 4.3.7 (but without requiring M to be resolute) as well as some additional cases in which D is an infinite distributive lattice. Corollary 4.4.5 was first proved by Schmerl in [162]. Theorem 4.7.3 was proved by Paris [144], and the complete charaterization for distributive lattices was later proved by Mills [131].

It is unknown whether or not every uncountable model has a minimal cofinal extension. See Question 2 in Chapter 12.

Two papers concerning substructure lattices appeared in the same issue of *Fundamenta Mathematicae* in 1977. In one paper, Paris [145] proved Proposition 4.5.6 using a completely different method, thereby showing that a substructure lattice need not be distributive. In the other, Alex Wilkie [210] proved that a substructure lattice did not have to be modular by showing that the pentagon lattice could be realized. The proof of Theorem 4.6.2 is only slightly different from the one in [210]. Theorem 4.6.5 is from Schmerl [178]. The notion of n-CPP representations was first isolated in Schmerl [167], and Theorems 4.5.21 and 4.5.22 were proved there. It was also proved in [167] that $\mathbf{M_3}$ is a substructure lattice. Theorem 4.5.8 and Corollaries 4.5.29 and 4.5.30 are from Schmerl [169].

Theorem 4.3.5 is attributable to G. Birkhoff. It is the finite case of the fundamental representation theorem for arbitrary distributive lattices due to Birkhoff [10] and M.H. Stone [201]. This can be found in any book on lattice theory.

We make some remarks about the status of the problem of determining which finite lattices can appear as substructure lattices. It is possible that for every finite lattice L, every nonstandard countable model M has a cofinal extension N such that $\mathrm{Lt}(N/M) \cong L$. On the other hand, it is possible that some finite lattices, one such being $\mathbf{M_{16}}$, are not isomorphic to any $\mathrm{Lt}(N/M)$. All the finite lattices which are known to be interstructure lattices have finite congruence representations, and so Theorem 4.5.8 applies. There is nothing now known which would preclude any finite lattice from having a finite congruence representation. But there is also nothing known which precludes a finite lattice from being a substructure lattice and having no finite congruence representation.

It is still an open problem to find a finite lattice not having a finite congruence representation. Although it is possible that there are none, the most popular conjecture is that they do exist. The following theorem has been proved by Pálfy & Pudlák [143].

Theorem 4.9.1 *The following are equivalent:*

(1) *Every finite lattice has a finite congruence representation.*

(2) *Every finite lattice is isomorphic to a convex sublattice of the lattice of subgroups of a finite group.*

The implication (2) $\Longrightarrow$ (1) is true locally in that any convex sublattice of the lattice of subgroups of a finite group has a finite congruence representation (DO IT!). The proof of the converse (1) $\Longrightarrow$ (2) does not apply locally: it could be that there is a finite lattice which has a finite congruence representation but is not isomorphic to any convex sublattice of the lattice a subgroup lattice.

A. Lucchini [122] has given some remarkable improvements to Example 4.5.12, vastly extending earlier results of Walter Feit [39] for $\mathbf{M_7}$ and $\mathbf{M_{11}}$. If either $n = q + 2$, where q a power of a prime, or $n = ((q^t + 1)/(q + 1)) + 1$, where q is a power of a prime and t is an odd prime, then $\mathbf{M_n}$ is isomorphic to a convex sublattice of the lattice of subgroups of some finite group. Thus, each such $\mathbf{M_n}$ has a finite congruence representation. For no other values of n is this known to hold.

Three lattices that we know to be substructure lattices are $\mathbf{2}$, $\mathbf{N}_5$, and $\mathbf{M}_3$. The variety $\mathcal{V}(\mathbf{2})$ generated by $\mathbf{2}$ is the class of distributive lattices, and we know that every finite distributive lattice has a finite congruence representation so is a substructure lattice. By results in [154], every finite *finitely fermentable* lattice has a finite congruence representation so is a substructure lattice. Among these lattices is $\mathbf{N}_5$, so that all finite lattices in $\mathcal{V}(\mathbf{N}_5)$ are finitely fermentable and thus are substructure lattices. While $\mathbf{M}_3$ is not finitely fermentable, John W. Snow [200] proved every finite lattice in $\mathcal{V}(\mathbf{M}_3)$ has a finite congruence representation and, consequently, is a substructure lattice.

5

HOW TO CONTROL TYPES

The method of model construction using indiscernibles is sometimes referred to as the EM-technology to honor Ehrenfeucht and Mostowski, who introduced it in their seminal 1956 paper. One of its features is its use of Ramsey's Theorem. This chapter is devoted to an important extension of the EM-technology that can be used to construct models having no (or only very small) sets of indiscernibles. We were tempted to call this chapter "AH-Technology" in honor of Abramson and Harrington who intoduced the technique and also proved the needed combinatorial theorem generalizing Ramsey's Theorem. We will use that as this chapter's unofficial subtitle.

5.1 Solid bases and AH-sets

We make some definitions concerning a subset A of some fixed model $M \models \mathsf{PA}^*$. Recall that A is a set of generators if M is the only $N \preccurlyeq M$ for which $A \subseteq N$. We say that the set A is a *basis* for M if for any $X \subseteq A$ there is exactly one $N \preccurlyeq M$ such that $X = N \cap A$. In particular, a basis is a set of generators (DO IT!). When dealing with a basis A for M, if $X \subseteq A$, then we let M_X be that unique model for which X is a set of generators. Finally, a basis A is *solid* if whenever $X, Y \subseteq A$ are finite subsets and $f : M_X \longrightarrow M_Y$ is an isomorphism, then $f{\restriction}X : X \longrightarrow Y$ is a bijection. Alternatively, A is solid iff if $a \in A$ and $b \in M$ are such that $\mathrm{tp}(a) = \mathrm{tp}(b)$, then $b \in A$ (DO IT!).

There are some observations to be made about bases. Let B be a basis for M, and let $X = \{a_0, a_1, \ldots, a_{n-1}\} \subseteq A$, where $a_0 < a_1 < \cdots < a_{n-1}$. Suppose also that $b \in M_X$ generates M_X. Then there are Skolem terms $t(x_0, x_1, \ldots, x_{n-1})$ and $s_i(y)$ for $i < n$ such that

$$N \models b = t(a_0, a_1, \ldots, a_{n-1}) \wedge \bigwedge_{i<n} a_i = s_i(b)\,.$$

This notion of a solid basis appeared implicitly in Proposition 3.3.15, from which it follows that if N is an I-canonical extension of a prime model M, then I is a solid basis. It is possible to construct solid bases exhibiting much more diversity. The method for obtaining such solid bases is discussed in detail in this section, the key theorem being Theorem 5.1.3.

5.1.1 *Controlling indiscernibles and automorphisms*

The first proposition of this subsection shows that solid bases are useful in controlling the amount of indiscernibility that exists in a model. A set of indiscernibles in a model is an ordered set and, therefore, has an order type, by which we mean the isomorphism type of this ordered set. A finite order type is just the same as a finite ordinal. If λ is an order type, then λ^* is the reverse order type. Thus, $n^* = n$ for finite n. Every infinite order type has either ω or ω^* as a suborder type.

Proposition 5.1.1 *Suppose B is a nonempty solid basis for $M \models \mathsf{PA}^*$ and λ is an order type. If there is $I \subseteq M$ which is a set of indiscernibles having order type λ, then there is $J \subseteq B$ which is a set of indiscernibles having order type λ or λ^*.*

Proof If $\lambda = 0$, then there is nothing to prove, and if $\lambda = 1$, then any singleton $J \subseteq B$ will do. So we assume that λ is at least 2.

Consider $i < j$ in I. Since I is a set of indiscernibles, i, j realize the same type q. Since B is a basis, we can let $X, Y \subseteq B$ be such that M_X, M_Y are the submodels generated by i, j respectively, and since M_X, M_Y are finitely generated, the sets X and Y are finite. Notice that by Ehrenfeucht's Lemma, i and j are the only elements of M_X and M_Y, respectively, realizing q. Therefore $M_X \neq M_Y$, so that $X \neq Y$. Since i, j realize the same type, the models M_X and M_Y are isomorphic, so we can let $f : M_X \longrightarrow M_Y$ be an isomorphism, which is unique since $f(i) = j$. As B is a solid basis, $f{\restriction}X$ maps X onto Y, and then there is some $a \in X$ such that $a \neq f(a)$. Let a be the kth element of X. Thus, we have seen that the kth element of X is not the same as the kth element of Y.

Now let $s(y)$ be a Skolem term such that $M \models s(i) = a$. Then $s(j)$ is the kth element of Y via the isomorphism f. By the indiscernibility of I (actually, only 2-indiscernibilty is needed here), $s(i_0) \neq s(i_1)$ for any distinct $i_0, i_1 \in I$. Thus, $s(y)$ defines a one-to-one function on I into B. Let $J = \{s(i) : i \in I\} \subseteq B$. If $s(i) < s(j)$, then (again by the 2-indiscernibility of I) the set J has order type λ, and if $s(i) > s(j)$, then J has order type λ^*. Finally, since I is a set of indiscernibles, then also J is. □

The next proposition shows that solid bases are useful in controlling the automorphisms of a model. If $M \models \mathsf{PA}^*$ and $B \subseteq M$, then there is a natural way of expanding $(B, <)$ to a structure $\mathfrak{B} = (B, <, \ldots)$. For each $n < \omega$ and $p \in S_n(\mathrm{Th}(M))$, let R_p be the n-ary relation on B consisting of those n-tuples realizing p. Let $\mathfrak{B}$ be the expansion of $(B, <)$ by adjoining all possible R_p, and call this structure $\mathfrak{B}$ the *type expansion* of B.

Proposition 5.1.2 *Suppose B is a nonempty solid basis for $M \models \mathsf{PA}^*$, and let $\mathfrak{B}$ be the type expansion of B. Then, $\mathrm{Aut}(M) \cong \mathrm{Aut}(\mathfrak{B})$.*

Proof There is a very natural isomorphism $\alpha : \mathrm{Aut}(M) \longrightarrow \mathrm{Aut}(\mathfrak{B})$ that is defined by letting $\alpha(f) = f\restriction B$ for $f \in \mathrm{Aut}(M)$.

Observe that $\alpha(f) : B \longrightarrow B$ because B is a solid basis, and then also $\alpha(f)^{-1} : B \longrightarrow B$. Then, $\alpha(f)$ is an automorphism of $\mathfrak{B}$ since $f \in \mathrm{Aut}(M)$ (DO IT!). Hence, α does indeed map $\mathrm{Aut}(M)$ into $\mathrm{Aut}(\mathfrak{B})$, and clearly it is a homomorphism. It then follows, just from the fact that B generates M, that α is one-to-one and onto. Therefore, α is an isomorphism. □

5.1.2 *AH-sets*

Just a reminder: For a set B and $n < \omega$, we let $[B]^n$ be the set of n-element subsets of B, and $[B]^{<\omega}$ be the set of all finite subsets of B. An n-type (for a completion T of PA^*) is a set of formulas whose free variables are among $x_0, x_1, \ldots, x_{n-1}$. If $I \in [\omega]^{<\omega}$, then an I-ary formula has its free variables among $\{x_i : i \in I\}$, and an I-type is a consistent set of I-ary formulas. Given $J \subseteq I \in [\omega]^{<\omega}$ and an I-type p, we let $p|J$ be the J-type consisting of all J-ary formulas in p. If $I = \{i_0, i_1, \ldots, i_{n-1}\} \in [\omega]^n$, where $i_0 < i_1 < \cdots < i_{n-1}$, then we say that an I-ary type p is *solid* if it is complete, the formula $x_{i_0} < x_{i_1} < \cdots < x_{i_{n-1}}$ is in p, and whenever $b_0 < b_1 < \cdots < b_{i_{n-1}}$ are elements of M that realize p, and M is generated by $B = \{b_0, b_1, \ldots, b_{i_{n-1}}\}$, then B is a solid basis for M.

Theorem 5.1.3 *Let $T \supseteq \mathsf{PA}^*$ be a completion having a nonstandard prime model. Then there is a set $\mathcal{P}$ such that:*

(1) *For each $p \in \mathcal{P}$, there is $I \in [\omega]^{<\omega}$ for which p is a bounded solid I-type.*
(2) *T is (the unique $\emptyset$-type) in $\mathcal{P}$.*
(3) *If $p \in \mathcal{P}$ is an I-type and $J \subseteq I$, then $p|J \in \mathcal{P}$.*
(4) *If $i_0 < i_1 < \cdots < i_{n-1} < \omega$ and if $p(x_{i_0}, x_{i_2}, \ldots, x_{i_{n-1}})$ is an I-type, then $p(x_{i_0}, x_{i_2}, \ldots, x_{i_{n-1}})$ is in $\mathcal{P}$ iff the n-type $p(x_0, x_1, \ldots, x_{n-1})$ is in $\mathcal{P}$.*
(5) *Suppose that $1 \le n < \omega$ and $p_I \in \mathcal{P}$ is an I-type whenever $I \subsetneq n$, and that $p_J \subseteq p_I$ whenever $J \subseteq I \subsetneq n$. Then there are $2^{\aleph_0}$ n-types $p \in \mathcal{P}$ such that $p \supseteq \bigcup\{p_I : I \subsetneq n\}$.* □

Conditions (1)–(4) are rather routine. Condition (5), the most interesting, implies that an n-type in $\mathcal{P}$ is not determined by its proper subtypes. In fact, for any given n-type in $\mathcal{P}$, there are $2^{\aleph_0}$ n–types in $\mathcal{P}$ having exactly the same proper subtypes. The cardinal $2^{\aleph_0}$ is not especially crucial: $\aleph_0$ would work just about as well.

All the 1-types in $\mathcal{P}$ are selective, and they all fail to be 2-indiscernible. They even fail to be n-weakly Ramsey for any $n < \omega$ and, furthermore, would fail to be $\aleph_0$-weakly Ramsey had that notion been defined.

A set $\mathcal{P}$ satisfying the conditions (1)–(5) is referred to as an AH-*set* for T.

Before getting into the proof of this theorem, we take a look at how it is used to build solid bases. Some further terminology is needed.

Let $(B, <)$ be any linearly ordered set, and let $X \in [B]^n$. Then X inherits the ordering from B, so that $X = \{x_0, x_1, \ldots, x_{n-1}\}$, where x_i is the ith element of X in this ordering. If $I \subseteq n = \{0, 1, 2, \ldots, n-1\}$, then define $X \circ I = \{x_i \in X : i \in I\}$.

Now let $(B, <)$ be a linearly ordered set. For example, we might let B be a subset of some model $M \models \mathsf{PA}^*$. A function $f : [B]^{<\omega} \longrightarrow W$ is *compatible* if, whenever $I \subseteq n < \omega$ and $X, Y \in [B]^n$ are such that $f(X) = f(Y)$, then $f(X \circ I) = f(Y \circ I)$.

Example 5.1.4 *Let $M \models \mathsf{PA}^*$ and let $B \subseteq M$. We define f on $[B]^{<\omega}$ as follows. Given $X \in [B]^{<\omega}$, we let $X = \{b_0, b_1, \ldots, b_{n-1}\}$, where the b_i's are in increasing order. Then let $f(X) = \mathrm{tp}(b_0, b_1, \ldots, b_{n-1})$. It is clear that f is compatible. We could also say that* tp *is compatible on $[B]^{<\omega}$.*

Example 5.1.5 *Let $\mathcal{L}$ be any language comprising only relation symbols, including the binary relation symbol $<$. Let $\mathfrak{A} = (A, <, \ldots)$ be a linearly ordered $\mathcal{L}$-structure. Define f on $[A]^{<\omega}$ so that $f(X)$ is the isomorphism type of the substructure $\mathfrak{A}|X$. Clearly, this function f is compatible.*

Example 5.1.6 *Let $\mathfrak{A}$ be as in the previous example, and let $G = \mathrm{Aut}(\mathfrak{A})$ be its automorphism group. Then, not only does G act on A, but also G acts on A^n and on $[A]^n$ for each $n < \omega$. For $X \in [A]^n$, let $orb(X)$ be the orbit of X under the action of G on $[A]^n$. Let $\mathcal{O}$ be the set of all orbits for all n. Then $orb : [A]^{<\omega} \longrightarrow \mathcal{O}$ is a compatible function. This example works if $\mathfrak{A}$ is replaced with a model $M \models \mathsf{PA}^*$.*

When considering compatible functions on $[B]^{<\omega}$, it is the equivalence relations that the functions induce on each $[B]^n$ that are important. We say that two functions $f : [B]^{<\omega} \longrightarrow W$ and $g : [B]^{<\omega} \longrightarrow V$ are *equivalent* if they induce identical equivalence relations on each of the $[B]^n$. For example, if M is a countable recursively saturated model of PA^*, then tp and *orb* (see Examples 5.1.4 and 5.1.6) are equivalent. Any function equivalent to a compatible function is compatible.

Proposition 5.1.7 *Let $T \supseteq \mathsf{PA}^*$ be a completion and M its nonstandard prime model. Suppose that $(B, <)$ is a linearly ordered set, $|W| \leq 2^{\aleph_0}$ and $f : [B]^{<\omega} \longrightarrow W$ is compatible. Then there exists $N \succ_{\mathsf{cof}} M$ such that B is a solid basis for N and f and* tp *are equivalent functions on $[B]^{<\omega}$.*

REMARK The set W plays no role in this Proposition except for its cardinality, which is at most $2^{\aleph_0}$ so as to conform with (5) of Theorem 5.1.3.

Proof Let $\mathcal{P}$ be an AH-set for T as in Theorem 5.1.3. We assume that the elements of B are new constant symbols with the aim of finding an appropriate complete $(\mathcal{L}^*_{\mathsf{PA}} \cup B)$-theory Σ extending T. We want a compatible function $g : [B]^{<\omega} \longrightarrow \mathcal{P}$ that is equivalent to f such that the following hold whenever $X \in [B]^k$:

- $g(X)$ is an $|X|$-type.
- Suppose $I \in [k]^s$, where $I = \{i_0, i_1, \ldots, i_{s-1}\}$ and $i_0 < i_1 < \cdots < i_{s-1} < k$. Then $g(X \circ I)$ is the type $p(x_0, x_1, \ldots, x_{s-1})$, where $p(x_{i_0}, x_{i_1}, \ldots, x_{i_{s-1}})$ is the I-subtype of $g(X)$.

Let's suppose for now that we have such a function g. The function g can be used to obtain a complete theory Σ in the language $\mathcal{L}^*_{\mathsf{PA}} \cup B$. If $n < \omega$, $X \in [B]^n$, and $\varphi(\bar{x})$ is an n-ary formula, let $\varphi(X)$ be the sentence obtained from $\varphi(\bar{x})$ by replacing each free variable x_i by the ith element of X. Then let $\bar{g}(X) = \{\varphi(X) : \varphi(\bar{x}) \in g(X)\}$. Let $\Sigma = \bigcup\{\bar{g}(X) : X \in [B]^{<\omega}\}$. Clearly, Σ is a consistent extension of T. Let N' be a model of Σ, and then let N be the $\mathcal{L}^*_{\mathsf{PA}}$-reduct of the submodel generated by B. It can now be checked (DO IT!) that $N \succ_{\mathsf{cof}} M$, B is a solid basis, and that tp is equivalent to g which in turn is equivalent to f.

To get the function g, define $g_0 \subseteq g_1 \subseteq g_2 \subseteq \cdots$ inductively, where $g_i : [B]^{\leq i} \longrightarrow \mathcal{P}$, and then let g be their union. To start, let g_0 be such that $g_0(\emptyset) = T \in \mathcal{P}$ (by (2)). Now suppose that $n > 0$ and that we have $g_0, g_1, \ldots, g_{n-1}$ satisfying the necessary conditions. Using (3)–(5), it is easy (DO IT!) to see how to get an appropriate g_n. □

5.1.3 *The proof*

We now turn to the proof of Theorem 5.1.3. The proof of this theorem makes use of a somewhat weakened variation of the Nešetřil–Rödl Theorem that previously appeared at the end of Subsection 3.1.2. We need to consider finite $\mathcal{L}$-structures, where $\mathcal{L}$ is a finite language consisting only of relation symbols among which is the binary relation symbol $<$, which always denotes a linear ordering of the universe of the $\mathcal{L}$-structures. Suppose that $\mathfrak{A} = (A, <, \ldots)$ is a finite ordered $\mathcal{L}$-structure. Notice that $\mathcal{P}(A) = [A]^{<\omega}$. Suppose f is a function whose domain includes $[A]^{<\omega}$. We say that f is *homogeneous* on $\mathfrak{A}$ if whenever $X, Y \subseteq A$ and $\mathfrak{A}|X \cong \mathfrak{A}|Y$, then $f(X) = f(Y)$. We say that f is *canonical* on $\mathfrak{A}$ if for any $X \subseteq A$, there is $K \subseteq X$ such that whenever $K_1 \subseteq X_1 \subseteq A$ and $K_2 \subseteq X_2 \subseteq A$ are such that $(\mathfrak{A}|X_1, K_1) \cong (\mathfrak{A}|X_2, K_2) \cong (\mathfrak{A}|X, K)$, then $f(X_1) = f(X_2)$ iff $K_1 = K_2$. For such a canonical f, this subset $K \subseteq X$ is unique and is called the *f-core* of X.

It might help to look at the two extremes for a function f that is canonical on $\mathfrak{A} = (A, <, \ldots)$. Let $X \subseteq A$, and let $\mathfrak{X} = \{Y \subseteq A : \mathfrak{A}|Y \cong \mathfrak{A}|X\}$. Then f is constant on $\mathfrak{X}$ iff the f-core of X is $\emptyset$, and f is one-to-one on $\mathfrak{X}$ iff X is the f-core of X (DO IT!). Thus, if f is canonical and every $X \subseteq A$ has $\emptyset$ as its f-core, then f is homogeneous. For a still stronger property, f is *solidly canonical* on $\mathfrak{A}$ if f is canonical and whenever $K \subseteq X \subseteq A$, where K is the f-core of X, then either $f(X) \in K$ or $f(X) \notin A$.

THE AH/NR THEOREM: *Suppose* $\mathfrak{A} = (A, <, \ldots)$ *is a finite ordered* $\mathcal{L}$*-structure. Then there is a finite ordered* $\mathcal{L}$*-structure* $\mathfrak{B} = (B, <, \ldots)$ *such that whenever* $f : [B]^{<\omega} \longrightarrow \{0, 1\}$*, then there is* $\mathfrak{A}' \subseteq \mathfrak{B}$ *such that* $\mathfrak{A}' \cong \mathfrak{A}$ *and* f *is homogeneous on* $\mathfrak{A}'$.

This theorem is both formalizable and provable in PA. However, for the proof of Theorem 5.1.3 it suffices that it is provable in TA, that is, true in the real world.

Proof of Theorem 5.1.3 Let M be the prime model of T, and let $c \in M$ be nonstandard. Let $f_0, f_1, f_2, \ldots$ be a list of all (codes of) definable functions $f : [M]^{\leq c} \longrightarrow \{0, 1\}$. [We are adopting a convention (that is not completely spelled out) by identifying functions $f : [M]^n \longrightarrow M$ with certain other functions $f : M^n \longrightarrow M$.] Working in M, let $\mathcal{L} = \{<\} \cup \{R_{ij} : i, j < c\}$ be a relational language, where each R_{ij} is an i-ary relation symbol. For each $k \leq c$ let $\mathcal{L}_k = \{<\} \cup \{R_{ij} : i, j < k\}$. Clearly, for each standard k there is an ordered $\mathcal{L}_k$-structure $\mathfrak{A} = (A, <, \ldots)$ such that every ordered $\mathcal{L}_k$-structure of cardinality k is isomorphic to a substructure of $\mathfrak{A}$. By overspill, there is such a nonstandard $k \in M$ and such an $\mathfrak{A}$. Let $e_0 = k$ and $\mathfrak{A}_0 = \mathfrak{A}$.

Working in the real world, we construct a decreasing sequence $e_0 \geq e_1 \geq e_2 \geq \ldots$ of nonstandard elements of M, and a sequence $\mathfrak{A}_0, \mathfrak{A}_1, \mathfrak{A}_2, \ldots$ such that for each $n < \omega$:

(a) $\mathfrak{A}_n = (A_n, <, \ldots)$ is an ordered finite (still in the sense of M) $\mathcal{L}_n$-structure;

(b) $\mathfrak{A}_n$ is large enough so that every ordered $\mathcal{L}_n$-structure of cardinality e_n is isomorphic to a substructure of $\mathfrak{A}_n$;

(c) f_n is homogeneous on $\mathfrak{A}_{n+1}$;

(d) $\mathfrak{A}_{n+1} \restriction \mathcal{L}_n \subseteq \mathfrak{A}_n$.

To obtain e_{n+1} and $\mathfrak{A}_{n+1}$, assume we have e_n and $\mathfrak{A}_n = (A_n, <, \ldots)$. Expand $\mathfrak{A}_n$ to an $\mathcal{L}_{n+1}$-structure $\mathfrak{A}'_n$ having the property that for some nonstandard $e' \leq e_n$, every ordered $\mathcal{L}_{n+1}$-structure of cardinality e' is isomorphic to a substructure of $\mathfrak{A}'_n$.

[We interject a remark here that there are other ways that this construction might proceed. For example, we might have started with $\mathfrak{A}$ that embeds every $\mathcal{L}_c$-structure of cardinality c. We have adopted the approach we did because with it we do not need to know that the full AH/NR Theorem is true in M.]

By the AH/NR Theorem, for each standard $e \leq e'$ there is a substructure $\mathfrak{B} \subseteq \mathfrak{A}'_n$ on which f_n is homogeneous such that every ordered $\mathcal{L}_e$-structure of cardinality e is isomorphic to a substructure of $\mathfrak{B}$. By overspill, there is a nonstandard such e; let e_{n+1} some such e, and let $\mathfrak{A}_{n+1}$ be a corresponding $\mathfrak{B}$.

One of the beauties of this construction is that you get much more than you originally bargained for. The following are all bonuses:

(e) If $f : [M]^{\leq c} \longrightarrow M$ is definable, then there is $n < \omega$ such that f is canonical on $\mathfrak{A}_n$.

(f) If $f : [M]^{\leq c} \longrightarrow M$ is definable, then there is $n < \omega$ such that f is solidly canonical on $\mathfrak{A}_n$.

Still working in the real world, let $\mathcal{L}_\omega = \{<\} \cup \{R_{ij} : i, j < \omega\}$. For each $I \in [\omega]^{<\omega}$, let $\mathcal{Q}_I$ be the set of $\mathcal{L}_\omega$-structures $(I, <, \ldots)$, where the ordering $<$ on I coincides with usual ordering of the elements of I. Then let $\mathcal{Q} = \bigcup\{\mathcal{Q}_I : I \in [\omega]^{<\omega}\}$. This set $\mathcal{Q}$ behaves in a way that is very close to the way we want $\mathcal{P}$ to behave. Specifically, the following (which should be compared to the corresponding requirements in Theorem 5.1.3) hold:

(2′) The empty structure is the unique member of $\mathcal{Q}_0$.

(3′) If $\mathfrak{A} \in \mathcal{Q}_I$ and $J \subseteq I$, then $\mathfrak{A}|J \in \mathcal{Q}_J$.

(4′) If $\mathfrak{A} \in \mathcal{Q}_I$ and $\mathfrak{A}$ is isomorphic to $\mathfrak{B} = (J, <, \ldots)$ (by an order-preserving isomorphism), then $\mathfrak{B} \in \mathcal{Q}_J$.

(5′) If $1 \leq n < \omega$ and $\{\mathfrak{A}_I : I \subsetneq n\} \subseteq \mathcal{Q}$, where $\mathfrak{A}_I \in \mathcal{Q}_I$ and $\mathfrak{A}_J \subseteq \mathfrak{A}_I$ whenever $J \subseteq I \subsetneq n$, then there are $2^{\aleph_0}$ different $\mathfrak{A} \in \mathcal{Q}_n$ such that $\mathfrak{A} \supseteq \bigcup\{\mathfrak{A}_I : I \subsetneq n\}$.

Given $I \in [\omega]^{<\omega}$ and $\mathfrak{A} \in \mathcal{Q}_I$, we associate with them an I-type $p_{\mathfrak{A}}$. For simplicity, let $I = n$, and let $p = p_{\mathfrak{A}}$. We have to decide whether a given n-ary formula $\varphi(x_0, x_1, \ldots, x_{n-1})$ does or does not belong in p. Consider the definable function $f : [M]^n \longrightarrow 2$ such that whenever $a_0 < a_1 < \cdots < a_{n-1}$ are n elements of M, then $f(\{a_0, a_1, \ldots, a_{n-1}\}) = 0$ iff $M \models \varphi(a_0, a_1, \ldots, a_{n-1})$. Then there is some $k < \omega$ such that f is homogeneous on $\mathfrak{A}_k$. Thus, f is constant on the set of substructures of $\mathfrak{A}_k$ that are isomorphic to $\mathfrak{A}$. Then $\varphi(x_0, x_1, \ldots, x_{n-1}) \in p$ iff this constant value is 0. Thus, we get that p is an n-type.

With an eye towards (1), we will show that p is a bounded solid n-type. It is easily seen that p is bounded since $A \subseteq M$ is bounded. To show that it is solid, consider a model N generated by a set $B = \{b_0, b_1, \ldots, b_{n-1}\}$, where $\langle b_0, b_1, \ldots, b_{n-1}\rangle$ realizes p. It suffices to show that B is a solid basis for N.

For each $X \subseteq B$, let $N_X \preccurlyeq N$ be the submodel generated by X. We show for each $X \subseteq B$ that (i) $N_X \cap B = X$ and (ii) if $N' \preccurlyeq N$ and $X = N' \cap B$, then $N' = N_X$. These two facts imply that B is a basis for N. For notational simplicity, let's assume that $k \leq n$ and $X = \{b_i : i < k\}$.

(i) Suppose, without loss of generality, that $b_k \in N_X$. Then there is a k-ary Skolem term $t(x_0, x_1, \ldots, x_{k-1})$ such that $N \models b_k = t(b_0, b_1, \ldots, b_{k-1})$. It follows from (2′)–(5′) that there are $\mathfrak{B} \in \mathcal{Q}_{2n}$ such that $\mathfrak{A} = \mathfrak{B}|n \cong \mathfrak{B}|(n+1)\{k\}$. But then $t(x_0, x_1, \ldots, x_{k-1}) = x_k$ and $t(x_0, x_1, \ldots, x_{k-1}) = x_{k+1}$ are both in $p_{\mathfrak{B}}$, a contradiction.

(ii) Suppose $N' \preccurlyeq N$ and $X = N' \cap B$. Suppose $b \in N'$. Then there is an n-ary Skolem term $t(x_0, x_1, \ldots, x_{n-1})$ such that $N \models b = t(b_0, b_1, \ldots, b_{n-1})$. By (e), and with some abuse of terminology, there is a t-core $K \subseteq B$. If $b_i \in K$, then there is a Skolem term $s(y)$ such that $N \models s(b) = b_i$, so that $b_i \in X$. Therefore, $b \in N_X$.

We have shown that B is a basis. We now show that it is in fact a solid basis. Consider an n-ary Skolem term $t(x_0, x_1, \ldots, x_{n-1})$ and let $N \models b =$

$t(b_0, b_1, \ldots, b_{n-1})$, where $b \notin X$. By (f), there some A_m such that $M \models \forall x_0, x_1, \ldots, x_{n-1} \in A_m[x_0 < x_1 < \cdots < x_{n-1} \to t(x_0, x_1, \ldots, x_{n-1}) \notin A_m]$. Thus, b does not realize the same type as any element of B, thereby proving that B is a solid basis.

Finally, let $\mathcal{P}$ be the set of all those $p_{\mathfrak{A}}$. It needs to be checked that $\mathcal{P}$ satisfies (1)–(5). As we have just seen, each n-ary type in $\mathcal{P}$ is solid and, similarly, each I-type in $\mathcal{P}$ also is. Thus (1) holds. It is easily checked that (2)–(4) follow from (2′)–(4′). Since different $\mathfrak{A}$'s yield different types, (5) follows from (5′). This completes the proof of Theorem 5.1.3. □

We make two comments about Theorem 5.1.3 and its proof:

1. It would be easy to add in the proof the requirement that $\omega = \inf\{\text{card}^M(A_n) : n < \omega\}$. If this were done, then for each 1-type $p(x)$ in $\mathcal{P}$, if N is a $p(x)$-extension of the prime model M, then $\omega = \text{GCIS}(M, N)$.
2. It is also possible to modify the proof in the opposite direction of the previous comment. Consider some $d \in M$. Then it can be arranged that the AH-set $\mathcal{P}$ is such that for any n-type $p(\bar{x})$ in $\mathcal{P}$, if N is a $p(\bar{x})$-extension of the prime model M, then N is a cofinal d-end extension of M. For this modification, we use that the AH/NR Theorem is provable in PA.

5.1.4 *True Arithmetic*

A defect in Theorem 5.1.3 is that it requires the prime model of T to be nonstandard. Thus, it does not apply if $T = \mathsf{TA}$. There is way to modify the theorem so as to cover TA.

If $M \prec N$, then N is a *wasp* extension of M if whenever $N_0 \prec N$, then either $N_0 \preccurlyeq M$ or $M \preccurlyeq N_0$. For example, a minimal extension is a wasp extension iff it is superminimal.

Theorem 5.1.8 *Let M be a countable nonstandard model of* PA^* *generated by an element b, and let $T = \text{Th}(M, b)$. Then there is an AH-set $\mathcal{P}$ for T having the following additional property: whenever $n < \omega$, $p \in \mathcal{P}$ is an n-type, and (N, b) is a p-extension of (M, b), then N is a wasp extension of M.*

Proof If the additional condition about the wasp extensions were not included, then this theorem would say nothing beyond Theorem 5.1.3, and the proof of Theorem 5.1.3 would work. But with the additional requirement, the construction of $\mathcal{P}$ needs to be modified by incorporating both Comment 2 following the proof of Theorem 5.1.3 and the method of constructing a superminimal extension in Theorem 2.1.12. We will give a heuristic description of how the proof should go.

Let $T_0 = \text{Th}(M)$ and let M_0 be its prime model. Comment 2 referred to constructing what it will be convenient to refer to here as a d-end AH-set for T_0. This construction involves constructing a sequence $A_0 \supseteq A_1 \supseteq A_2 \supseteq \cdots$. Such a construction can be done uniformly for all $d \in M_0$. Thus, at stage n

of the construction, instead of having A_n, we will have a definable sequence $\langle A_{u,n} : u \in M_0 \rangle$, in which $A_{d,n} = A_n$, with each $A_{u,n}$ being a bounded, definable subset of M_0. A new step will be interlaced into the construction.

We would expect that, in general, $\mathrm{card}(A_{u+1,n})$ is much bigger than $\mathrm{card}(A_{u,n})$. But now we require that $\mathrm{card}(A_{u+1,n})$ be much, much bigger than $\mathrm{card}(A_{u,n})$. This is to accommodate the new step. At some stage n we consider a definable sequence $\langle f_u : u \in M_0 \rangle$ of definable functions $f : M_0^{<M_0} \longrightarrow M_0$, where each function f_u is canonical on $\mathfrak{A}_{u,n}$. At the next stage we have $\langle A_{u,n+1} : u \in M_0 \rangle$ such that whenever $u < v \in M_0$ and the f_v-core of $A_{v,n+1}$ is not $\emptyset$, then $f_u([A_{u,n+1}]^{<M_0}) \cap f_v([A_{v,n+1}]^{<M_0}) = \emptyset$.

This uniform construction not only produces a u-end AH-set for each $u \in M_0$, but can be applied in any model. For example, it can be applied in M, and since b generates M, the construction will produce a b-end AH-set $\mathcal{P}$ for $\mathrm{Th}(M, b)$. This AH-set $\mathcal{P}$ will have the additional property called for by the theorem.

To see that $\mathcal{P}$ has this additional property, let $p(\bar{x})$ be an m-type in $\mathcal{P}$, and let (N, b) be a $p(\bar{x})$-extension of (M, b) generated by the m-tuple $\bar{c}$. We show that N is a wasp extension of M. Suppose that $t(u, \bar{x})$ is a Skolem term and $d \in N \backslash M$ is such that $N \models d = t(b, \bar{c})$. It suffices to show that there is a Skolem term $s(y)$ such that $N \models b = s(d)$.

There is some stage $n < \omega$ of the construction at which we considered the definable sequence $\langle f_u : u \in M_0 \rangle$, where, for all $u \in M_0$ and $x_0 < x_1 < \cdots < x_{m-1} \in M_0$, $t(u, \bar{x}) = f_u(\{x_0, x_1, \ldots, x_{m-1}\})$ and each f_u is canonical on $\mathfrak{A}_{u,n}$. Then, for all $u < b$, $d \notin f_u([A_{u,n+1}]^{<M_0})$. Therefore, in M_0 we can let $s(y)$ be the least v such that $y \in f_v([A_{v,n+1}]^{<M_0})$. Clearly, $N \models b = s(d)$. □

The following lengthy proposition should be compared with Proposition 5.1.1.

Proposition 5.1.9 *Let $(M, b) \models \mathsf{PA}^*$, let $T = \mathrm{Th}(M, b)$, and let (M_0, b) be the prime model of T. Suppose B is a solid basis for $(M, b) \models \mathsf{PA}^*$ having the following additional property: whenever $n < \omega$, $p \in \mathcal{P}$ is an n-type, and (N, b) is a p-extension of (M_0, b), then N is a wasp extension of M_0. Let λ be an order type. If there is $I \subseteq M \backslash M_0$ which is a set of indiscernibles for M having order type λ, then there is $J \subseteq B$ which is a set of indiscernibles for M having order type λ or λ^*.*

Proof By the proof of Proposition 5.1.1, we get a Skolem term $s(u, y)$ such that $s(b, y)$ defines a one-to-one function on I into B. Consider some $j \in I$. There is a Skolem term $t(y)$ such that $M \models t(j) = b$. Therefore, for any $i \in I$, $t(j)$ realizes the same type as b. But it follows (DO IT!) from Ehrenfeucht's Lemma that b is the only element of M realizing its type. Hence $t(i) = b$ for all $i \in I$, and then the Skolem term $s(t(y), y)$ defines a one-to-one function on I into B. Then, as at the end of the proof of Proposition 5.1.1, $J = \{s(t(i), i) : i \in I\} \subseteq B$ is a set of indiscernibles having order type λ or λ^*. □

An important feature of Proposition 5.1.9 is that it can be applied to TA by letting M_0 be a minimal extension of the standard model generated by b.

5.2 Omitting indiscernibles

Recall that the infinite cardinal numbers $\beth_\alpha$, for α an ordinal, are defined by transfinite recursion. More generally, if κ is any cardinal then: $\beth_0(\kappa) = \kappa$; $\beth_{\alpha+1}(\kappa) = 2^{\beth_\alpha(\kappa)}$; $\beth_\beta(\kappa) = \sup\{\beth_\alpha(\kappa) : \alpha < \beta\}$ for limit ordinal β. Then let $\beth_\alpha = \beth_\alpha(\aleph_0)$.

Theorem 5.2.1 *Let $\mathfrak{A} = (A, \ldots)$ be any structure for a countable language. If $|A| > \beth_n$, then $\mathfrak{A}$ has an indiscernible set $I \in [A]^{n+1}$.*

Theorem 5.2.2 *Let $M \models \mathsf{PA}^*$, and let $1 \leq n < \omega$. Then there is $N \equiv M$ such that $|N| = \beth_n$ and there is no indiscernible set $I \in [N]^{n+1}$.*

The first of the above theorems is an immediate consequence of the Erdős–Rado Theorem. For the record, here it is.

> The Erdős–Rado Theorem: *Suppose $n < \omega$ and κ is an infinite cardinal. If $|A| > \beth_n(\kappa)$ and $f : [A]^{n+1} \longrightarrow \kappa$, then there is $H \subseteq A$ such that $|H| = \kappa^+$ and f is constant on $[H]^{n+1}$.*

The proof of Theorem 5.2.2 uses the the following converse of the ER Theorem.

> The Erdős–Hajnal–Rado Theorem: *Suppose $n < \omega$ and κ is an infinite cardinal. If $|A| \leq \beth_n(\kappa)$, then there is $f : [A]^{n+1} \longrightarrow \kappa$ for which there is no $H \in [A]^{n+2}$ such that f is constant on $[H]^{n+1}$.*

Proof of Theorem 5.2.2 The construction of the model relies on the technology of the previous section. Subsection 5.1.4 is needed only in case the prime model of $\mathrm{Th}(M)$ is a standard model.

Let $M \models \mathsf{PA}^*$, and let $1 \leq n < \omega$. If $n = 1$, then this theorem reduces to Exercise 3.3.18, so we can assume that $n \geq 2$. If M is not a prime model, then we replace it with its prime elementary submodel. The case that M is a standard model will be handled at the end of this proof; for now, assume that M is nonstandard. The model N that we will construct is a cofinal extension of M.

Let $(A, <)$ be a linearly ordered set of cardinality $\beth_n$; thus $|A| = \beth_{n-1}(2^{\aleph_0})$. Let $f : [A]^n \longrightarrow \mathbb{R}$ be as in the EHR Theorem. This function f easily extends to a compatible function $g : [A]^{<\omega} \longrightarrow \mathbb{R}$ (Do it!). Apply Proposition 5.1.7 to get $N \succ_{\mathsf{cof}} M$ such that A is a solid basis for N and g and tp are equivalent on $[A]^{<\omega}$. It follows from Proposition 5.1.1 that N contains no indiscernible set of cardinality $n + 1$. For, if it did, then there would be such an indiscernible set from A. But as g and tp are equivalent, this would contradict the characteristic property of f, finishing the proof for nonstandard M.

Now suppose that the prime model is a standard model. So, instead of taking M to be the prime model, let it be an elementary extension of the prime model generated by an element b realizing a minimal type. Proceed in a very similar manner to what was just done, using Theorem 5.1.8 and Proposition 5.1.9 instead Theorem 5.1.3 and Proposition 5.1.1. □

The technique for proving Theorem 5.2.2 extends easily to other related situations. Here is a typical one. If κ, λ are infinite cardinals, then $\kappa \longrightarrow (\omega)^{<\omega}_{\lambda}$ means: for every function $f : [\kappa]^{<\omega} \longrightarrow \lambda$ there is an infinite $H \subseteq \kappa$ such that f is constant on $[H]^n$ for each $n < \omega$. If $\kappa \longrightarrow (\omega)^{<\omega}_{\lambda}$ fails, then $\kappa \not\longrightarrow (\omega)^{<\omega}_{\lambda}$. If $\kappa \not\longrightarrow (\omega)^{<\omega}_{\lambda}$ and $\lambda < \kappa$, then $\kappa \not\longrightarrow (\omega)^{<\omega}_{2}$. The reader should have no difficulty (DO IT!) supplying the details of the proof of the following theorem.

Theorem 5.2.3 *Let M be a model of* PA*, *and let κ be an infinite cardinal. Then there is $N \equiv M$ of cardinality κ having no infinite set of indiscernibles iff $\kappa \not\longrightarrow (\omega)^{<\omega}_{2}$.* □

5.3 Hanf numbers

Theorem 5.2.2 has a consequence concerning Hanf numbers. For the moment, let us fix an arbitrary complete theory T for some countable language $\mathcal{L}$. Let $\Phi(x)$ be set of $\mathcal{L}$-formulas with x being the only free variable. Then $h(\Phi(x))$, the *Hanf number* of $\Phi(x)$, is the least cardinal κ such that every model $\mathfrak{A}$ of T having cardinality at least κ realizes $\Phi(x)$. There is no guarantee that there is such a κ, but if none exist, then we let $h(\Phi(x)) = \infty$. We let $H(T)$, the *Hanf number for omitting types*, be the supremum of all $(h(\Phi(x))^+$, as $\Phi(x)$ ranges over all sets of unary formulas in the language of T for which $h(\Phi(x)) < \infty$.

Theorem 5.3.1 (Morley) (1) *For any complete theory T for a countable language, $H(T) \leq \beth_{\omega_1}$.*

(2) *There is T (for a language with just a binary relation symbol) such that for every $\alpha < \omega_1$, there is some $\Phi_\alpha(x)$ such that $h(\Phi_\alpha(x)) = (\beth_\alpha)^+$.* □

For any completion T of PA*, it is very easy to get $\Phi(x)$ for which $h(\Phi(x)) = (\beth_0)^+ = \aleph_1$: for example, let $\Phi(x)$ assert "x is not definable." Thus, an immediate consequence of Theorems 5.2.1 and 5.2.2 is:

Corollary 5.3.2 *Let T be any completion of* PA*, *and let $n < \omega$. Then there is a type $\Phi_n(x)$ such that $h(\Phi_n(x)) = (\beth_n)^+$.* □

Thus, for any completion T of PA*, we get from Theorem 5.3.1(1) and Corollary 5.3.2 that $\beth_\omega \leq H(T) \leq \beth_{\omega_1}$. Just where within this interval does $H(T)$ lie? For completions T of PA, the answer depends on whether or not T is True Arithmetic.

Theorem 5.3.3 *Let $T \supseteq$* PA* *be a completion having a standard model. Then $H(T) = \beth_\omega$.*

Proof The inequality $H(T) \geq \beth_\omega$ comes from Corollary 5.3.2. We need to prove $H(T) \leq \beth_\omega$, which means: if $\Phi(x)$ is a set of formulas and for each $n < \omega$, there is a model $N \models T$ such that $|N| > \beth_n$ and N omits $\Phi(x)$, then there are arbitrarily large models of T omitting $\Phi(x)$. So let us assume that $\Phi(x)$ is such a set of formulas.

Let $\mathbb{N}^*$ be the prime model of T. Throughout this proof, *formula* refers to a formula in the language of PA^* whose free variables are among $x_0, x_1, x_2, \ldots$, and *model* refers to a model of T. If M is a model, $I \subseteq M$ and $\varphi(x_0, x_1, \ldots, x_{n-1})$ (or simply $\varphi(\bar{x})$) is an n-ary formula, then we say that $\varphi(x_0, x_1, \ldots, x_{n-1})$ is *valid on* I if, whenever $c_0 < c_1 < \cdots < c_{n-1}$ are elements of I, then $M \models \varphi(c_0, c_1, \ldots, c_{n-1})$, and we say that $\varphi(\bar{x})$ is ω*-valid* if there are arbitrarily large finite subsets of $\mathbb{N}^*$ on which $\varphi(\bar{x})$ is valid.

By Ramsey's Theorem, for every formula $\varphi(\bar{x})$, either $\varphi(\bar{x})$ or $\neg\varphi(\bar{x})$ is ω-valid. (There is a stronger statement that we will not need: either $\varphi(\bar{x})$ or $\neg\varphi(\bar{x})$ is valid on an unbounded definable subset of $\mathbb{N}^*$.) But there is a stronger statement that we will need:

(1) If $\theta(\bar{x})$ is ω-valid and $\varphi(\bar{x})$ is any formula, then either $\theta(\bar{x}) \wedge \varphi(\bar{x})$ or $\theta(\bar{x}) \wedge \neg\varphi(\bar{x})$ is ω-valid.

The above statement (1) also is a consequence of Ramsey's Theorem (DO IT!).

The key to this proof of the theorem is the following:

(2) If $\theta(\bar{x})$ is ω-valid and $t(\bar{x})$ is a Skolem term, then there is a formula $\varphi(x)$ in $\Phi(x)$ such that $\theta(\bar{x}) \wedge \neg\varphi(t(\bar{x}))$ is ω-valid.

We prove (2). Suppose that $t(\bar{x})$ is an $(n+1)$-ary Skolem term. Let M be a model omitting $\Phi(x)$ such that $|M| > (\beth_n)^+$. Then there is $a \in M$ such that $|[0, a]| \geq (\beth_n)^+$. Since $\theta(\bar{x})$ is ω-valid and $M \models T$, there is a bounded definable $A \subseteq M$ such that $\mathrm{card}^M(A) = a$ and $\theta(\bar{x})$ is valid on A. We next define a function $f : [A]^{n+1} \longrightarrow \omega$. Let $\Phi(x) = \{\varphi_i(x) : i < \omega\}$. If $c_0 < c_1 < \cdots < c_n$ are elements of A, then let $f(\{c_0, c_1, \ldots, c_n\}) = i$, where $i < \omega$ is chosen so that $M \models \neg\varphi_i(t(\bar{c}))$. By the Erdős–Rado Theorem, there are $i < \omega$ and an infinite $H \subseteq A$ such that f is constantly i on $[H]^{n+1}$. Letting $\varphi(x) = \varphi_i(x)$, we see that $\theta(\bar{x}) \wedge \neg\varphi(t(\bar{x}))$ is ω-valid. This proves (2).

By repeated applications of (1) and (2), we can get a set Γ of ω-valid formulas such that Γ is closed under conjunction, and (from (1)) for every $\theta(\bar{x})$ either $\theta(\bar{x})$ or $\neg\theta(\bar{x})$ is in Γ, and (from (2)) for any $\theta(\bar{x}) \in \Gamma$ and Skolem term $t(\bar{x})$, there is $\varphi(x) \in \Phi(x)$ such that $\neg\varphi(t(\bar{x})) \in \Gamma$.

Now consider an arbitrary infinite (large) cardinal κ. By compactness, there is a model N that is generated by a set I of indiscernibles such that I (and thus also N) has cardinality κ and such that each $\theta(\bar{x})$ in Γ is valid on I. It is easily checked (DO IT!) that N omits $\Phi(x)$. □

Theorem 5.3.4 *If* $T \supseteq \mathsf{PA}^*$ *is a completion having a nonstandard prime model, then* $H(T) = \beth_{\omega_1}$. □

The inequality $H(T) \leq \beth_{\omega_1}$ comes from Theorem 5.3.1(1). For the other inequality, we have the following theorem.

Theorem 5.3.5 *Let $T \supseteq \mathsf{PA}^*$ be a completion having a nonstandard prime model, and let $\alpha < \omega_1$. Then there is a set $\Phi(x)$ of formulas such that $h(\Phi(x)) = (\beth_{\alpha+1})^+$.*

Proof Corollary 5.3.2 covers finite α, so we assume that $\omega \leq \alpha < \omega_1$. Let M be the (nonstandard!) prime model of T. Considered as an ordered set, M embeds any countable, linearly ordered set (WHY?).

To get $\Phi(x)$, we will need an increasing bounded sequence $\langle c_\nu : \nu < \alpha \rangle$ from M that will be determined later on. For convenience we define a set $\Phi_\alpha(x, y)$ of 2-ary formulas instead of unary formulas (permissible due to the existence of a pairing function). This set consists of the following assertions:

(1) $x < y$;
(2) $c_\nu \leq y \rightarrow \forall u \leq c_\nu[\varphi(u, x) \leftrightarrow \varphi(u, y)]$, for each $\nu < \alpha$ and each 2-ary formula $\varphi(u, x)$;
(3) $\varphi(x) \leftrightarrow \varphi(y)$, for each unary formula $\varphi(x)$.

We show that $h(\Phi_\alpha(x, y)) = (\beth_{\alpha+1})^+$.

For the easier half of the proof, we show: if $N \models T$ and N omits $\Phi_\alpha(x, y)$, then $|N| \leq \beth_\alpha$. First, we show by induction on $\nu < \alpha$ that $|[0, c_\nu]| \leq \beth_{\nu+1}$. For $\nu > 0$, we consider three cases: $\nu = 0$, ν is a successor ordinal, and ν is a limit ordinal.

$\nu = 0$. The simple argument used here will be repeated. If $|[0, c_0]| > \beth_1$, then there would be a, b such that $a < b < c_0$ with a, b realizing the same type. Clearly, $\langle a, b \rangle$ realizes $\Phi_\alpha(x, y)$.

$\nu = \mu + 1 < \alpha$ *and* $|[0, c_\mu]| \leq \beth_{\mu+1}$. There are at most $\beth_{\nu+1}$ 1-types over $[0, c_\mu]$. If $|[0, c_\nu]| > \beth_{\nu+1}$, then $|[c_\mu, c_\nu]| > \beth_{\nu+1}$, so there would be a, b such that $c_\mu < a < b < c_\nu$, with a, b realizing the same type over $[0, c_\mu]$. Clearly $\langle a, b \rangle$ satisfies $\Phi_\alpha(x, y)$.

$\nu < \alpha$ *is a limit ordinal and* $|[0, c_\mu]| \leq \beth_{\mu+1}$ *for all* $\mu < \nu$. Let $C = \bigcup\{[0, c_\mu] : \mu < \nu\}$. Then $|C| \leq \beth_\nu$. If $|[0, c_\nu]| > \beth_{\nu+1}$, then there would be $a < b$ in $[0, c_\nu - 1] \setminus C$ realizing the same type over C. Clearly $\langle a, b \rangle$ satisfies $\Phi(x, y)$.

Now observe that the same simple argument that had been used in the three cases also shows that $|M \setminus \bigcap\{[c_\nu, \infty) : \nu < \alpha\}| \leq \beth_{\alpha+1}$, and therefore $|N| \leq \beth_{\alpha+1}$.

Now for the harder half of the proof: $h(\Phi_\alpha(x, y)) \geq \beth_{\alpha+1}$. Thus, it suffices to construct a model $N \models T$ which omits $\Phi_\alpha(x, y)$ such that $|N| = \beth_{\alpha+1}$. To do this, we need to take a look again at the AH/NR Theorem and its consequence, Theorem 5.1.3. We will take a short recess from the proof of Theorem 5.3.5 to do this.

Theorem 5.1.3 asserts the existence of a large collection $\mathcal{P}$ of complete types, among which are $2^{\aleph_0}$ distinct 1-types. Each 1-type $p(x) \in \mathcal{P}$ determines a proper cut K of the prime model M, where $K = \{a \in M : a < x \text{ is a formula in } \mathcal{P}\}$. However, all 1-types determine exactly the same cut since: if $p(x), q(x)$ are 1-types in $\mathcal{P}$, then $p(x) \cup q(y) \cup \{x < y\}$ is consistent by (5). We get a variation of Theorem 5.1.3 in which all the 1-types determine different cuts. For this we need a variation of the AH/NR Theorem.

Suppose that among the relation symbols in $\mathcal{L}$ are the unary relation symbols $P_0, P_1, \ldots, P_k$. For this discussion only, let us say that $\mathfrak{A} = (A, <, \ldots)$ is a *partitioned* $\mathcal{L}$-structure if $A = P_0 \cup P_1 \cup \cdots \cup P_k$ and $a < b$ whenever $i < j \leq k$ and $a \in P_i, b \in P_j$. If $\mathfrak{A}$ is a partitioned $\mathcal{L}$-structure, then there is an $\mathcal{L}$-structure $\mathfrak{B}$ that is not only as required by the AH/NR Theorem but is also partitioned. This fact, which we refer to as the Partitioned AH/NR Theorem, easily follows from the AH/NR Theorem (Do it!). By modifying the proof of Theorem 5.1.3 by using the Partitioned AH/NR Theorem instead of the AH/NR Theorem, we can prove the following variation of Theorem 5.1.3.

Theorem 5.3.6 *Let $T \supseteq \mathsf{PA}^*$ be a completion having a nonstandard prime model. Then there is a set $\mathcal{P}_0$ such that (1)–(4) of Theorem 5.1.3 hold and there is a bijection $r \mapsto q_r(x)$ from the set of rationals to the set of 1-types in $\mathcal{P}_0$ such that the following hold:*

(5) *If $r < s$ are rationals, then there is $a \in M$ such that the formulas $x < a$ and $a < x$ are in $q_r(x)$ and $q_s(x)$ respectively.*

(6) *Suppose that $3 \leq n < \omega$ and $p_I \in \mathcal{P}_0$ is an I-type whenever $I \subsetneq n$, and that $p_J \subseteq p_I$ whenever $J \subseteq I \subsetneq n$. Then there are $2^{\aleph_0}$ n-types $p \in \mathcal{P}_0$ such that $p \supseteq \bigcup\{p_I : I \subsetneq n\}$.*

(7) *Suppose that $r < s$ are rationals. Then there are $2^{\aleph_0}$ 2-types $p(x_0, x_1) \in \mathcal{P}_0$ such that $p(x_0, x_1) \supseteq q_r(x_0) \cup q_s(x_1)$.*

To prove this theorem, do a construction just like the one in the proof of Theorem 5.1.3 but with a small difference: the Partitioned AH/NR Theorem is used in place of the AH/NR Theorem.

We now resume the proof of Theorem 5.3.5 but with a rather sketchy presentation. What we have at this point is an ordinal $\alpha < \omega_1$ and the 2-type $\Phi_\alpha(x, y)$. Let $B = \beth_{\alpha+1}$ which, as usual, is the set of ordinals less than it. For each $\nu < \omega$, let $B_\nu = \{\nu\}$, and if $\mu \leq \alpha$ and $\nu = \omega + \mu \leq \omega + \alpha$, let $B_\nu = \{\beta \in B : \beth_\nu \leq \beta < \beth_{\nu+1}\}$. The B_ν's form a partition of B. It is an easy matter to get a compatible function $f : [B]^{<\omega} \longrightarrow \omega$ such that:

- $f(x) = f(y)$ iff there is $\nu \leq \omega + \alpha$ such that $x, y \in B_\nu$;
- if $x, y \in B_\nu$ are distinct, then there are $\mu < \nu$ and $z \in B_\mu$ such that $f(\{z, x\}) \neq f(\{z, y\})$.

As in the proof of Proposition 5.1.7, we get a compatible $g : [B]^{<\omega} \longrightarrow \mathcal{P}_0$ that is equivalent to f such that the following hold whenever $X \in [B]^k$:

- $g(X)$ is an $|X|$-type.
- Suppose $I \in [k]^s$, where $I = \{i_0, i_1, \ldots, i_{s-1}\}$ and $i_0 < i_1 < \cdots < i_{s-1} < k$. Then $g(X \circ I)$ is the type $p(x_0, x_1, \ldots, p_{s-1})$, where $p(x_{i_0}, x_{i_1}, \ldots, x_{i_{s-1}})$ is the I-subtype of $g(X)$.

Continuing as in the proof of Proposition 5.1.7, we get $N \succ_{\mathsf{cof}} M$ such that B is a solid basis for N and f and tp are equivalent functions on $[B]^{<\omega}$. Clearly $|N| = |B| = \beth_{\alpha+1}$.

It still has to be decided what the c_ν's in $\Phi_\alpha(x, y)$ are, and we do that now. For each $\nu < \alpha$, let $c_\nu \in M$ be such that $B_{\omega+\nu} < c_\nu < B_{\omega+\nu+1}$.

The bonuses (e) and (f) that the construction using the AH/NR Theorem yields have analogues in this situation. These help us to show that N omits $\Phi_\alpha(x, y)$. □

Theorem 5.3.5 leaves unanswered if for every (limit) ordinal $\alpha < \omega_1$ there is $\Phi(x)$ such that $h(\Phi(x)) = (\beth_\alpha)^+$. This appears to be open.

5.4 The automorphism group

The concern of this section is to investigate what are the possibilities for $\mathrm{Aut}(M)$, the group of automorphisms of an arbitrary model M of PA^*.

Definition 5.4.1 Let G be a group. A linear ordering $<$ of G is a right-ordering if, whenever $a, x, y \in G$ are such that $x < y$, then $xa < ya$. We say G is right-orderable if it has a right-ordering. □

The importance of right-orderable groups is that they characterize the possible automorphism groups of linearly ordered structures, putting a significant restriction on the possible automorphism groups of models of PA. We will see in Theorems 5.4.3 and 5.4.4 that there are no other restrictions.

Lemma 5.4.2 *Let G be a group. Then G is right-orderable iff there is a linearly ordered structure $\mathfrak{A} = (A, <, \ldots)$ such that $G \cong \mathrm{Aut}(\mathfrak{A})$.* □

Theorem 5.4.3 *Let G be a right-orderable group, and let M be a nonstandard prime model. Then there is $N \succ_{\mathsf{cof}} M$ such that $\mathrm{Aut}(N) \cong G$.*

Proof By Lemma 5.4.2, we can assume that $G = \mathrm{Aut}(\mathfrak{A})$, where $\mathfrak{A} = (A, <, \ldots)$. Let $\kappa = |A|$. If κ is finite, then G is trivial, so we can just let $N = M$. Assume for the rest of this proof that $\kappa \geq \aleph_0$. We get $N \succ_{\mathsf{cof}} M$ such that $\mathrm{Aut}(N) \cong G$ and $|N| = \kappa$. First, we give a proof for $\kappa \leq 2^{\aleph_0}$, and then, after that, give the proof for arbitrary κ.

Let $T = \mathrm{Th}(M)$. Assume $\kappa \leq 2^{\aleph_0}$. Recall Example 5.1.6 in which $orb : [A]^{<\omega} \longrightarrow \mathcal{O}$ is a compatible function. Clearly, $|\mathcal{O}| \leq \kappa \leq 2^{\aleph_0}$. Now apply Proposition 5.1.7, getting $N \succ_{\mathsf{cof}} M$ such that A is a solid basis for N and orb

and tp are equivalent. Let $\mathfrak{B} = (A, <, \langle \text{tp}^{-1}(p) : p \in S_n(T), n < \omega\rangle)$. It is clear that $G = \text{Aut}(\mathfrak{A}) = \text{Aut}(\mathfrak{B})$. Proposition 5.1.2 implies that $\text{Aut}(M) \cong G$. This completes the proof for $\kappa \leq 2^{\aleph_0}$.

Now let κ be arbitrary. Assume, without loss of generality, that κ and A are disjoint sets, thereby getting the linearly ordered set $(B, <)$, where $B = \kappa \cup A$, $(\kappa, <)$ and $(A, <)$ are both substructures of $(B, <)$, and $x < y$ whenever $x \in \kappa$ and $y \in A$.

We next define a compatible function $f : [B]^{<\omega} \longrightarrow \omega$. Let $|\mathcal{O}| = \lambda$. Clearly, $\lambda \leq \kappa$. Even though λ may be quite large, by a little coding trick, we will manage to get f to map into ω.

Let $\mathcal{O} = \{O_\alpha : \alpha < \lambda\}$. Let $f : [B]^{<\omega} \longrightarrow \omega$ be such that whenever $X = \{x_0, x_1, \ldots, x_{m-1}\} \in [B]^{<\omega}$ and $X = \{y_0, y_1, \ldots, y_{n-1}\} \in [B]^{<\omega}$, where $x_0 < x_1 < \cdots < x_{m-1}$, $y_0 < y_1 < \cdots < y_{n-1}$, then $f(X) = f(Y)$ iff the following hold:

(1) $m = n$;

(2) if $i < n$, then $x_i \in A$ iff $y_i \in A$;

(3) if $i < n$ and $I \subseteq n$ are such that $X \circ I \subseteq A$ and $x_i \in \kappa$, then $X \circ I \in O_{x_i}$ iff $Y \circ I \in O_{y_i}$.

Such a function f exists (Do it!).

Now apply Proposition 5.1.7, getting $N \succ_{\mathsf{cof}} M$ such that B is a solid basis for N and *orb* and tp are equivalent. Let $\mathfrak{B} = (B, <, \langle f^{-1}(m) \cap [B]^n : m, n < \omega\rangle)$. It is clear that $G = \text{Aut}(\mathfrak{A}) \cong \text{Aut}(\mathfrak{B})$ (Do it!). Proposition 5.1.2 implies that $\text{Aut}(M) \cong G$. □

A defect with the previous theorem is that it does not apply to prime models which are standard. For example, the standard model of TA has been left out. There are several ways that the theorem (and its proof) can be modified to yield a theorem covering TA. One possibility is Exercise 5.6.2. Another is by replacing cofinal extensions by end extensions as in the next theorem. (See Exercise 5.6.5 for an improved refinement of this theorem.)

Theorem 5.4.4 *Let G be a right-orderable group, and let M be a model of* PA^*. *Then there is $N \succ_{\mathsf{end}} M$ such that $G \cong \text{Aut}(N)$.* □

There are two main ingredients in the proof of this theorem. The first is Theorem 3.3.14 which is a result about the existence of very rigid models. The second is the AH-technology which was developed in Section 5.1 of this chapter. Some refinement of this technology is needed for this proof.

Let $T \supseteq \mathsf{PA}^*$ be a completion having the nonstandard prime model M, and let $p(\bar{x}) \in S_n(T)$. We defined in Section 5.1 what it means for $p(\bar{x})$ to be a solid type. We now jazz up the definition. If $c \in M$, then $p(\bar{x})$ is *solid over c* if it is solid and two more requirements are met:

(1) If N is a $p(\bar{x})$-extension of M, then N is a cofinal c-end extension.

Before stating the second requirement, we observe a consequence of (1). Let M_0 be a $p(\bar{x})$-extension of M. Thus, $M_0 = M(\bar{a})$ in which $\bar{a}$ realizes $p(\bar{x})$. Suppose that $M \prec_{\mathsf{cof}} M_1$ and that M_1 is generated by $[0,c]_{M_1}$. Then M_0 and M_1 have an amalgamation M_2 as discussed in Section 2.3. We denote this amalgamation by $M_1 \star M_0$. Then, $M_1 \star M_0 = M_0(\bar{a})$. Let $p'(\bar{x}) \in S_n(M_0)$ be the type realized by $\bar{a}$

(2) If $p'(\bar{x})$ is as just described, then $p'(\bar{x})$ is a solid type.

If $\mathcal{P}$ is an AH–set for T and, in addition, each type in $\mathcal{P}$ is solid over c, then we say that $\mathcal{P}$ is an AH-*set for T over c*. The proof of the next theorem is basically like the proof of Theorem 5.1.3. We leave the details of the proof to be worked out.

Theorem 5.4.5 *Let $T \supseteq \mathsf{PA}^*$ be a completion having a nonstandard prime model M, and let $c \in M$. Then there is an AH–set $\mathcal{P}$ for T over c.* □

Proof of Theorem 5.4.4 By Lemma 5.4.2, we can assume that $G = \mathrm{Aut}(\mathfrak{A})$, where $\mathfrak{A} = (A, <, \ldots)$. Let $\kappa = |A|$. The elementary end extension N that we obtain is not be very big; in fact $|N| = |M| + \kappa$.

By Theorem 3.3.14, we can assume that M is very rigid. Then let $p(x)$ be a minimal type for $\mathrm{Th}(M)$, and let $M(c)$ be a $p(x)$-extension of M. Therefore, $M(c) \succ_{\mathsf{end}} M$ and $|M(c)| = |M|$.

As in the second part of the proof of Theorem 5.4.3, assume κ and A are disjoint sets, and then get the same $(B, <)$ and the same compatible function $f : [B]^{<\omega} \longrightarrow \omega$ as in that proof.

Next, we want to apply Theorem 5.4.5. Consider the theory $T = \mathrm{Th}(M, c)$, and let (M_0, c) be its prime model. Let $\mathcal{P}$ be an AH-set for T over c. Let $f : [B]^{<\omega} \longrightarrow \omega$ be as in the proof of Theorem 5.4.3. Now let (N_0, c) be an extension of (M_0, c) generated by the solid basis B such that f and tp are equivalent on $[B]^{<\omega}$.

Let $(N, c) = (M(c), c) \star (N_0, c)$. Since $\mathcal{P}$ is an AH-set for T over c, it follows that B is a solid basis for $(M, c, i)_{i \in M_0}$ (DO IT!). Just as in the first part of the proof of Theorem 5.4.3, $G \cong \mathrm{Aut}((N, c, i)_{i \in M})$. Notice that c is in the last gap of N and realizes a minimal (hence, rare) type, so it cannot be moved by an automorphism of N. Thus $\mathrm{Aut}((N, c, i)_{i \in M}) = \mathrm{Aut}((N, i)_{i \in M})$. But $M \prec_{\mathsf{end}}$ and N is very rigid, so $\mathrm{Aut}((N, i)_{i \in M}) = \mathrm{Aut}(N)$. Therefore, $G \cong \mathrm{Aut}(N)$. □

5.5 Indiscernible generators

This section contains a proof of the theorem that every countable recursively saturated model of PA^* is generated by a set of indiscernibles. While technically the proof does not rely on AH-technology, its presence in this chapter can be justified because, at a crucial point in the proof, the AH/NR Theorem is called upon.

Theorem 5.5.1 *Every countable, recursively saturated model $M \models \mathsf{PA}^*$ is generated by a set of indiscernibles.*

Proof Let M be countable and recursively saturated. We will NOT find an indiscernible set $I \subseteq M$ such that $M = \mathrm{Scl}(I)$. What we will do is find an indiscernible set $I \subseteq M$ such that if $N = \mathrm{Scl}(I) \preccurlyeq M$, then (1) N is recursively saturated and (2) $\mathrm{SSy}(N) = \mathrm{SSy}(M)$. Then (1) and (2) imply that $N \cong M$. Since N is generated by a set of indiscernibles, so must M be.

For this proof we adopt the following notational convention: If $X \subseteq M$ and $n < \omega$, then $[X]^n$ is the set of increasing n-tuples from X.

We begin this proof with a discussion about "universal" functions.

There is a definable function $F : M \times M \longrightarrow M$ that is *universal* for all bounded, definable functions. This means: if $a \in M$ and $f : [0,a] \longrightarrow M$ is a definable function, then there is $u \in M$ (in fact, there is an unbounded set of u's) such that $F(x,u) = f(x)$ for every $x \leq a$. Such a function F can be found which is even 0-definable. Moreover, if $(M', F') \equiv (M, F)$, then F' is universal for all bounded, definable functions in M'.

Other types off universal functions can easily be derived from F. For example, if $n < \omega$, then let $F_n : [M]^{n+1} \longrightarrow M$ be defined by $F_n(x_0, x_1, \ldots, x_{n-1}, u) = F(\langle n, x_0, x_1, \ldots, x_{n-1}\rangle, u)$. Then the F_n's are universal in the following sense: if $a \in M$, $s < \omega$ and $f_n : [0,a]^n \longrightarrow M$ are definable functions for each $n \leq s$, then there is $u \in M$ (again, an unbounded set of them) such that $F_n(x_0, x_1, \ldots, n_{n-1}, u) = f_n(x_0, x_1, \ldots, x_{n-1})$ whenever $n \leq s$ and $x_0 < x_1 < \cdots < x_{n-1} \leq a$.

It is easy to get still other universal functions. Let $G_{i,n} : [M]^{n+1} \longrightarrow 2$ be such that $G_{i,n}(x_0, x_1, \ldots, x_{n-1}, u) = \min\{1, F(\langle 0, i, n, x_0, x_1, \ldots, x_{n-1}\rangle, u)\}$. Then the $G_{i,n}$'s are universal in a similar way that the F_n's are: if $a \in M$, $s < \omega$, and $g_{i,n} : [0,a]^n \longrightarrow 2$ are definable functions for each $i, n \leq s$, then there is $u \in M$ such that $G_{i,n}(x_0, x_1, \ldots, x_{n-1}, u) = g_{i,n}(x_0, x_1, \ldots, x_{n-1})$ whenever $i, n \leq s$ and $x_0 < x_1 < \cdots < x_{n-1} \leq a$. Moreover, there is a mutual universality; that is, if $a \in M$, $s < \omega$ and $f_n : [0,a]^n \longrightarrow M$ and $g_{i,n} : [0,a]^n \longrightarrow 2$ are definable functions for each $i, n \leq s$, then there is $u \in M$ such that $F_n(x_0, x_1, \ldots, x_{n-1}, u) = f_n(x_0, x_1, \ldots, x_{n-1})$ and $G_{i,n}(x_0, x_1, \ldots, x_{n-1}, u) = g_{i,n}(x_0, x_1, \ldots, x_{n-1})$ whenever $x_0 < x_1 < \cdots < x_{n-1} \leq a$ and $i, n \leq s$. We make use of the F_n's and the $G_{i,n}$'s and their mutual universality.

Let $\langle \varphi_j^n(\bar{x}, y) : j, n < \omega\rangle$ be a recursive, doubly indexed list of formulas such that each $\varphi_j^n(\bar{x}, y)$ is $(n+1)$-ary (that is, $\bar{x}$ is the n-tuple $x_0, x_1, \ldots, x_{n-1}$) and such that if $\langle \theta_j(\bar{x}, y) : j < \omega\rangle$ is a recursive sequence of $(m+1)$-ary formulas, then there is some $n < \omega$ such that $n \geq m$ and for each $k < \omega$ there is $l < \omega$ such that the sentence

$$\forall x_0, x_1, \ldots, x_{n-1} \forall y \left[\bigwedge_{j \leq k} \theta_j(x_0, x_1, \ldots, x_{m-1}, y) \leftrightarrow \bigwedge_{j \leq l} \varphi_j^n(x_0, x_1, \ldots, x_{n-1}, y) \right]$$

is in $\mathrm{Th}(M)$. (Notice that m and n may not be equal, but that little bit of inelegance is more than compensated for by the simplification in notation.) For $n, k < \omega$, let $\Phi_k^n(\bar{x}, y) = \bigwedge_{j \leq k} \varphi_j^n(\bar{x}, y)$.

We want to get an increasing sequence $b_0 < b_1 < b_2 < \cdots$ of elements in M such that whenever $k, n < \omega$ and $i_0 < i_1 < i_2 < \cdots < i_n < \omega$, then the sentence

$$\exists y \Phi^n_k(b_{i_0}, b_{i_1}, \ldots, b_{i_{n-1}}, y) \rightarrow \Phi^n_k\big(b_{i_0}, b_{i_1}, \ldots, b_{i_{n-1}}, F_n(b_{i_0}, b_{i_1}, \ldots, b_{i_{n-1}}, b_{i_n})\big)$$

is true in M. The sequence can be constructed inductively. Suppose that $m < \omega$ and that we already have $b_0 < b_1 < \cdots < b_{m-1}$ so that all of the above sentences involving $b_0, b_1, \ldots, b_{m-1}$ are true in M. Let $B = \{b_0, b_1, \ldots, b_{m-1}\}$. The b_m that we want should satisfy a certain recursive set $\Sigma(u)$ of formulas. A typical formula in $\Sigma(u)$ is obtained by taking one of the above sentences involving only $b_0, b_1, \ldots, b_m$ and then replacing each occurrence of b_m with the variable u. By the recursive saturation of M, there is such a b_m provided any finite subset of $\Sigma(u)$ is satisfiable. Let $\Sigma_0(u) \subseteq \Sigma(u)$ be a finite set, and suppose that $s < \omega$ is such that if Φ^n_k occurs somewhere in $\Sigma_0(u)$, then $n, k \leq s$. For each $n \leq s$, let $f_n : B^n \longrightarrow M$ be such that if $a_0, a_1, \ldots, a_{n-1} \in B$, then for each $k \leq s$ the sentence

$$\exists y \Phi^n_k(a_0, a_1, \ldots, a_{n-1}, y) \rightarrow \Phi^n_k\big(a_0, a_1, \ldots, a_{n-1}, f_n(a_0, a_1, \ldots, a_{n-1})\big)$$

is true in M. By the universal property of $F_0, F_1, \ldots, F_s$ there is $b \in M$ such that $b > b_{m-1}$ and $F_n(a_0, a_1, \ldots, a_{n-1}, b) = f_n(a_0, a_1, \ldots, a_{n-1})$ whenever $n \leq s$ and $a_0 < a_1 < \cdots < a_{n-1}$ are in B. Then b satisfies $\Sigma_0(u)$. Thus, let $b_m = b$.

We now have the increasing sequence $b_0 < b_1 < b_2 < \cdots$ of elements in M. Let $X = \{b_0, b_1, b_2, \ldots\}$. Thus $X \subseteq M$ is a nonempty subset with no largest element such that the expansion (M, X) satisfies the sentences $\sigma_{n,k}$ for each $n, k < \omega$, where $\sigma_{n,k}$ is

$$\forall x_0, x_1, \ldots, x_{n-1}, u \in X[x_0 < x_1 < \cdots < x_{n-1} < u \wedge$$

$$\exists y \Phi^n_k(\bar{x}, y) \rightarrow \Phi^n_k(\bar{x}, F_n(\bar{x}, u))] .$$

This set X has an interesting property: $\mathrm{Scl}(X)$ is recursively saturated. Moreover, if $(M, X) \models \sigma_{n,k}$ for all $n, k < \omega$ and $Y \subseteq X$ is nonempty and has no largest element, then $\mathrm{Scl}(Y)$ is recursively saturated. (DO IT!)

We are going to go back to the construction of the sequence $b_0 < b_1 < b_2 < \cdots$ and impose some additional conditions on it involving the $G_{i,n}$'s, thereby imposing further conditions on the set X.

Definition Suppose $r < \omega$ and $Y \subseteq M$. We say that Y is *r-free* if whenever $\emptyset \neq D \subseteq Y$ is finite, $s < \omega$, and $g_{i,n} : [D]^n \longrightarrow 2$ whenever $i \leq s$ and $r \leq n \leq |D|$, then there is $c \in Y$ such that $c > \max(D)$ and whenever $i \leq s$, $r \leq n \leq |D|$, and $\langle d_0, d_1, \ldots, d_{n-1}\rangle \in [Y]^n$, then $G_{i,n}(d_0, d_1, \ldots, d_{n-1}, c) = g_{i,n}(d_0, d_1, \ldots, d_{n-1})$.

The further condition that we want to impose on X is that it is 0-free. Remembering that the F_n's and the $G_{i,n}$'s have a mutual universality property, we have no difficulty building this condition into the construction of the sequence $b_0 < b_1 < b_2 < \cdots$.

The next definition should look familiar.

Definition Suppose $r < \omega$ and $Y \subseteq M$. We say that Y is *r-indiscernible* if whenever $n \leq r$, $\theta(\bar{x})$ is an n-ary formula, and $\langle c_0, c_1, \ldots c_{n-1}\rangle$, $\langle d_0, d_1, \ldots, d_{n-1}\rangle \in [Y]^n$, then $M \models \theta(\bar{c}) \leftrightarrow \theta(\bar{d})$.

Obviously, every set is 0-indiscernible. Therefore, we now have a set $X \subseteq M$ such that all but (1) of the following hold:

(0) $(M, X) \models \sigma_{n,k}$ for all $n, k < \omega$;
(1) (M, X) is recursively saturated;
(2) $X \neq \emptyset$ and X has no largest element;
(3) X is 0-indiscernible;
(4) X is 0-free.

However, by the chronic resplendence of M, by choosing another set X if needed, we can get all of (0)–(4) to hold.

An enumeration of $\mathrm{SSy}(M)$ is needed, so let $\mathrm{SSy}(M) = \{S_r : r < \omega\}$. Our goal is, by a careful thinning process, to get a set $I \subseteq M$ of indiscernibles such that $\mathrm{Scl}(I) \cong M$. We do this by constructing a decreasing sequence $X = X_0 \supseteq X_1 \supseteq X_2 \supseteq \cdots$ such that, for each $r < \omega$, we have:

(1_r) (M, X_r) is recursively saturated;
(2_r) $X_r \neq \emptyset$ and X_r has no largest element;
(3_r) X_r is r-indiscernible;
(4_r) X_r is r-free;
(5_r) if $r > 0$ and $i < \omega$, then $G_{i,r-1}([X_r]^r) = \{1\}$ iff $i \in S_{r-1}$.

Before we construct the sequence $X = X_0 \supseteq X_1 \supseteq X_2 \supseteq \cdots$, let us see how it is used to get the desired $I \subseteq M$. If $Y = \bigcap\{X_r : r < \omega\}$ has a nonempty subset I with no largest element, then I would do, for it would be indiscernible by (3_r), and $\mathrm{SSy}(M) \subseteq \mathrm{SSy}(\mathrm{Scl}(I))$ by (5_r). There is a problem since it might be that $Y = \emptyset$. However, the recursive saturation of M allows us to get around this problem.

Let $y_0 < y_1 < y_2 < \cdots$ be an increasing sequence from M so that the type of $\langle y_0, y_1, \ldots, y_{r-1}\rangle$ is the same as the type of the r-tuple of the first r elements of X_r. The recursive saturation of M allows us to obtain y_r having already chosen $y_0, y_1, \ldots, y_{r-1}$. Now $I = \{y_0, y_1, y_2, \ldots\}$ is an indiscernible set which will do the job.

We now construct the sequence $X = X_0 \supseteq X_1 \supseteq X_2 \supseteq \cdots$ so that each X_r satisfies (1_r)–(5_r). For starters, let $X_0 = X$. Now suppose we have X_r. For a preliminary move, we get $Z \subseteq X_r$ such that:

(1′) (M, Z) is recursively saturated;

(2′) $Z \neq \emptyset$ and Z has no largest element;

(3′) Z is r-indiscernible;

(4′) Z is $(r+1)$-free;

(5′) if $i < \omega$, then

(a) $G_{i,r}([Z]^{r+1}) = \{1\}$ iff $i \in S_r$;

(b) $G_{i,r}([Z]^{r+1}) = \{0\}$ iff $i \notin S_r$.

It suffices to get $Z \subseteq X_r$ satisfying just (2′),(4′), and (5′), condition (3′) being inherited from X_r, for then by the chronic resplendence of (M, X_r), we will be able to get such a Z also satisfying (1′).

The method for constructing Z is very similar to the method used for constructing X. We get an increasing sequence $c_0 < c_1 < c_2 < \cdots$ of elements in X_r. We want the sequence to be such that Z is $(r+1)$-free, and such that whenever $i_0 < i_1 < i_2 < \cdots < i_r < \omega$, then

$$G_{i,r}(c_{i_0}, c_{i_1}, \ldots, c_{i_{r-1}}, c_{i_r}) = 1 \Longleftrightarrow i \in S_r\,.$$

Since X_r is r-free, an inductive construction of this sequence is straightforward and not dependent on recursive saturation.

Now we have Z satisfying (1′)–(5′) and have arrived at the final step of the proof, which is to get $X_{r+1} \subseteq Z$ satisfying (1_{r+1})–(5_{r+1}). Condition (5_{r+1}) is not a problem as this will be inherited from (5′), and (1_{r+1}) can be handled by the chronic resplendence of (M, Z) as has been done twice before.

We are almost ready to apply the AH/NR Theorem. Let $\mathcal{L}$ be the relational language that comprises the binary relation symbol $<$ and the $(n+1)$-ary relation symbols R_{in} for $i < \omega$ and $r+1 \leq n < \omega$. If $s < \omega$, let $\mathcal{L}_s = \{<\} \cup \{R_{in} : i \leq s \text{ and } r+1 \leq n \leq s\}$. Let $\mathfrak{B} = (Z, <, \ldots)$ be the $\mathcal{L}$-structure such that if $z_0, z_1, \ldots, z_n \in Z$, then $\mathfrak{B} \models R_{ij}(z_0, z_1, \ldots, z_n)$ iff $z_0 < z_1 < \cdots < z_n$ and $G_{i,n}(z_0, z_1, \ldots, z_n) = 1$.

Because Z is $(r+1)$-free, there is a sequence $w_0 < w_1 < w_2 < \cdots$ such that if $W = \{w_0, w_1, w_2, \ldots\}$, then W is also $(r+1)$-free and $\mathfrak{B}|W$ is recursive. To be more precise, let $\mathfrak{C} = (\omega, <, \ldots) \cong \mathfrak{B}|W$, and then $\mathfrak{C}$ is recursive. For each $s < \omega$, let $\mathfrak{C}_s = (\mathfrak{C}|s){\restriction}\mathcal{L}_s$. Then, the sequence $\langle \mathfrak{C}_s : s < \omega \rangle$ is a recursive sequence.

By the AH/NR Theorem, for each $s < \omega$ there is an $\mathcal{L}_s$-structure $\mathfrak{B}_s = (B_s, <, \ldots)$ such that $\mathfrak{C}_s \cong \mathfrak{B}_s \subseteq \mathfrak{B}{\restriction}\mathcal{L}_s$, with the additional property that for each $i \leq s$, B_s is $(r+1)$-indiscernible for the first s $(r+1)$-ary formulas in some fixed recursive enumeration of these formulas. Observe that, if $s < t < \omega$ and

$B_t = \{d_0, d_1, \ldots d_{t-1}\}$, where $d_0 < d_1 < \cdots < d_{t-1}$, then

$$\mathfrak{B}_s \cong (\mathfrak{B}_t | \{d_0, d_1, \ldots, d_{s-1}\}) \restriction \mathcal{L}_s.$$

By the recursiveness of the sequence $\langle \mathfrak{C}_s : s < \omega \rangle$ and the resplendence of (M, Z) there is a sequence $d_0 < d_1 < d_2 < \cdots$ from Z such that (and now we can let $X_{r+1} = \{d_0, d_1, d_2, \ldots\}$) X_{r+1} is $(r+1)$-indiscernible and for each $s < \omega$,

$$\mathfrak{B}_s \cong (\mathfrak{B} | \{d_0, d_1, \ldots, d_{s-1}\}) \restriction \mathcal{L}_s.$$

Thus, (2_{r+1})–(5_{r+1}) hold. Finally, by the chronic resplendence of (M, Z) we can arrange that (1_{r+1}) also holds.

Recall how I is obtained from the sequence $X_0 \supseteq X_1 \supseteq X_2 \supseteq \cdots$ and that $N = \mathrm{Scl}(I)$ is recursively saturated. We see that $\mathrm{SSy}(N) = \mathrm{SSy}(M)$. Consider an arbitrary $S_r \in \mathrm{SSy}(M)$, and then let $\langle i_0, i_1, i_2, \ldots, i_r \rangle \in [I]^{r+1}$. Then, for any $i < \omega$, $i \in S_r$ iff $G_{i,r+1}(i_0, i_1, \ldots, i_r) = 1$, so $S_r \in \mathrm{SSy}(N)$. □

Once it has been determined that every countable, recursively saturated model of PA^* is generated by a set of indiscernibles, one can ask about the order types of these sets. Clearly, such generating sets are nonempty, countable and have no last element. The proof of the previous theorem shows that nothing else is needed.

Corollary 5.5.2 *Let η be the order type of a countable, nonempty linearly set having no last element. Then every countable, recursively saturated model $M \models$ PA^* is generated by a set of indiscernibles with order type η.*

Proof Let I be the set of indiscernibles constructed in the proof of Theorem 5.5.1. Consider a model N generated by a set J of indiscernibles of order type η, where increasing tuples from J and I realize the same types. Then N is recursively saturated (DO IT!) and $\mathrm{SSy}(N) = \mathrm{SSy}(M)$ (DO IT!). Therefore, $N \cong M$. □

5.6 Exercises

♣**5.6.1** Define *left-orderable* and prove: a group is left-orderable iff it is right-orderable.

♦**5.6.2** (Improving Theorem 5.4.3.) Let G be a right-orderable group, and let $T \supseteq \mathsf{PA}^*$ be a completion. Let M be a simple (nonstandard) extension of the prime model of T. Then there is $N \succ_{\mathsf{cof}} M$ such that $\mathrm{Aut}(N) \cong G$.

♦**5.6.3** Suppose $T \neq \mathsf{TA}$ is a completion of PA. If $n + 1 \leq k \leq 2^n$, then there is $M \models T$ such that $\mathrm{Lt}(M) \cong \mathbf{B}_n$ and M has exactly k nonisomorphic elementary substructures.

The next two problems concern Theorem 5.4.3.

♣**5.6.4** Show that the extension N constructed in the proof does not fill the standard cut of M.

♦**5.6.5** The conclusion can be improved so that N is a cofinal b-end extension of M, where $b \in M$ is arbitrary but given in advance.

♦**5.6.6** Let M be a countable, recursively saturated model, and let $G = \mathrm{Aut}(A, <, \ldots)$, where A is countable. Then M has an inductive satisfaction class S such that $\mathrm{Aut}(M, S) \cong G$.

♠**5.6.7** Does every (or even some recursively saturated) nonstandard countable model M have a rigid cofinal extension?

5.7 Remarks & References

The Ehrenfeucht–Mostowski paper mentioned in the first sentence of this chapter is [29].

The AH/NR Theorem was proved, independently by Nešetřil & Rödl [137] and by Abramson & Harrington [1]. In the latter paper, this theorem was proved just so it could be applied to models of Peano Arithmetic. Theorems 5.2.2, 5.3.4, and 5.2.3 are from [1]. Theorem 5.3.5 is a slight improvement of a theorem from [1] with a possibly slightly simpler proof. Theorems 5.3.4 answered questions raised by Julia Knight [87] who had proved in that paper that $H(\mathsf{TA}) \leq \beth_\omega$ and in [86] that each completion T of PA has a type $\Phi(x)$ for which $h(\Phi(x)) = \aleph_2$. Michael Morley's classic Theorem 5.3.1, originally proved in [134], was the first application of the Erdős–Rado Theorem in model theory. The Erdős–Rado Theorem and the Erdős–Rado–Hajnal Theorem were first proved in [38] and [37] respectively.

For any ordinal α, the Erdős cardinal $\kappa(\alpha)$ is the least cardinal κ such that $\kappa \longrightarrow (\alpha)_2^{<\omega}$. Theorem 5.2.3 concerns $\kappa(\omega)$. A good source of information on Erdős cardinals is the book [64].

The material in Section 5.4 on automorphism groups is based on Schmerl [174]. Theorem 5.4.3 is apparently new, not having appeared in [174]. There has been much written about right-orderable groups. The book [89] by Kopytov and Medvedev is a comprehensive account of the subject. Lemma 5.4.2 is due independently to P. M. Cohn [23] and P. Conrad [24].

The proof of Theorem 5.5.1 is from Schmerl [166] with only minor changes. Exercise 5.6.3 is from Schmerl [177].

6

GENERICS AND FORCING

Forcing is a standard technique in set theory. It can be used advantageously in the construction of models of Peano Arithmetic. This chapter is devoted to doing exactly that.

In this chapter it is sometimes convenient, although generally not necessary, to have the language $\mathcal{L}$ of PA^* be finite. Consequently, *throughout this chapter, the underlying assumption is that* PA^* *is in a finite language* $\mathcal{L}$.

6.1 Generics

If M is a model of PA^*, then a *notion of forcing* for M is a nonempty partially ordered set $\mathbb{P} = (P, \trianglelefteq)$ that is definable in M. The order relation has been written with the symbol $\trianglelefteq$, emphasizing that it is reflexive (i.e. $x \trianglelefteq x$ for all $x \in P$). It will be our convention that for a notion of forcing, the order relation is reflexive. It is convenient to identify $\mathbb{P}$ with its underlying set P, so we will often do that. There is often some ambiguity about how to interpret the ordering in a notion of forcing. For us, if $p \trianglelefteq q$, then q is an *extension* of p, so that q contains more information than p does. Elements of $\mathbb{P}$ are often referred to as *conditions*.

An example of a notion of forcing is the *full binary tree* as defined in M. This is the partially ordered set $\mathbb{B} = (M, \trianglelefteq)$ of all (codes of) finite sequences of 0's and 1's, as defined in M, where $p \trianglelefteq q$ if p is an initial segment of q. (Here we are assuming that every element of M codes a unique such sequence.) The empty sequence, which we always assume to be coded by 0, is the unique minimal condition in $\mathbb{B}$. A notion of forcing $\mathbb{P} = (P, \trianglelefteq)$ is a *binary tree* if $\mathbb{P} \subseteq \mathbb{B}$ such that $p \in P$ whenever $p \trianglelefteq q \in \mathbb{P}$. Every binary tree $\mathbb{P}$ has a unique minimal condition (namely 0), and each of its conditions has at most two immediate successors. If a condition $p \in \mathbb{P}$ has two immediate successors, namely $p\hat{\ }0$ and $p\hat{\ }1$, then we say that p *splits* in $\mathbb{P}$.

Some definitions concerning a notion of forcing $\mathbb{P}$ will be needed. For those already familiar with forcing, these are just the usual basic definitions. *We will omit many proofs since in most cases they are just repetitions of well-known arguments from set theory. The reader is encouraged to fill in the details.*

For a condition p, a subset $D \subseteq \mathbb{P}$ is *dense above* p if for each $q \trianglerighteq p$ there is $r \trianglerighteq q$ such that $r \in D$. The subset $D \subseteq \mathbb{P}$ is *dense* if it is dense above every $p \in \mathbb{P}$, and it is *open* if there is a condition p such that $\{q \in \mathbb{P} : p \trianglelefteq q\} \subseteq D$. Conditions $p, q \in \mathbb{P}$ are *compatible* if there is a condition that extends both p

and q and are *incompatible* otherwise. A subset $F \subseteq \mathbb{P}$ is a *filter* if (1) for any two conditions $p, q \in F$ there is $r \in F$ such that $p, q \trianglelefteq r$ and (2) whenever $p \trianglelefteq q \in F$, then $p \in F$. A filter $G \subseteq \mathbb{P}$ is *generic* relative to $\mathbb{P}$ if G has a nonempty intersection with every definable, dense subset of $\mathbb{P}$. For each $p \in \mathbb{P}$, the set $\{q \in \mathbb{P} : p \trianglelefteq q \text{ or } q \text{ is incompatible with } p\}$ is a definable, dense set, thereby showing that every generic filter is a maximal filter. If $\mathbb{P}$ is a binary tree, then the maximal filters are also known as *branches*. Generic filters will often be referred to simply as *generics*. Generics relative to the full binary tree $\mathbb{B}$ are *Cohen* generics. A subset $D \subseteq \mathbb{P}$ is *open* if whenever $q \trianglerighteq p \in D$, then $q \in D$. A filter is generic iff it has a nonempty intersection with every definable dense open set (Do IT!).

Lemma 6.1.1 *If M is countable and $\mathbb{P}$ is a notion of forcing, then each condition is in some generic filter.* □

It is generally desirable for generics not to be definable. In countable models, this is guaranteed if the notion of forcing is perfect, where $\mathbb{P}$ is defined to be *perfect* if each condition in $\mathbb{P}$ has two incompatible extensions. The full binary tree $\mathbb{B}$ is an obvious example of a perfect notion of forcing.

Lemma 6.1.2 *If M is countable and $\mathbb{P}$ is a perfect notion of forcing, then:*

(1) *each condition is in $2^{\aleph_0}$ generics;*

(2) *no generic is definable.* □

Conversely, if a notion of forcing for a countable model has no definable generics, then it is perfect (Do IT!). There is also a converse of (1) of the previous lemma: if a notion of forcing for a countable model is not perfect, then some condition is in exactly one (necessarily definable) generic (Do IT!).

Arbitrary notions of forcing are in some sense no better than just binary trees. The following sequence of three exercises is intended to make this a little more precise. A notion of forcing $\mathbb{P}$ is a *tree* if, for any $p \in \mathbb{P}$, the set $\{q \in \mathbb{P} : q \trianglelefteq p\}$ of the predecessors of p is linearly ordered by $\trianglelefteq$. A notion of forcing is *ranked* if (in M) there is a bound on the length of the linearly ordered subsets of the set of predecessors of any element $p \in \mathbb{P}$. The least of these bounds can be thought of as the rank of p. Every binary tree is a ranked tree.

Exercise 6.1.3 Suppose $\mathbb{P}$ and $\mathbb{Q}$ are notions of forcing for M and $f : \mathbb{P} \longrightarrow \mathbb{Q}$ is a definable embedding (i.e. if $p_1, p_2 \in \mathbb{P}$, then $p_1 \trianglelefteq p_2$ iff $f(p_1) \trianglelefteq f(p_2)$) onto a dense subset of $\mathbb{Q}$. Suppose Y is a filter of $\mathbb{Q}$. Then, Y is a generic relative to $\mathbb{Q}$ iff $f^{-1}(Y)$ is a generic relative to $\mathbb{P}$. □

Exercise 6.1.4 For every $\mathbb{P}$ there is a dense $\mathbb{Q} \subseteq \mathbb{P}$ which is a ranked tree. □

Exercise 6.1.5 If $\mathbb{Q}$ is a ranked tree, then there is a binary tree $\mathbb{P}$ and a definable embedding $f : \mathbb{P} \longrightarrow \mathbb{Q}$ onto a dense subset. □

Lemma 6.1.6 *Let $\mathbb{P}$ be a notion of forcing for M, and let X be generic relative to $\mathbb{P}$. Suppose that $(M, \mathbb{P}, X) \prec (N, \mathbb{Q}, Y)$. Then, $\mathbb{Q}$ is a notion of forcing for N, and Y is generic relative to $\mathbb{Q}$.* □

Exercise 6.1.7 If $\mathbb{P}$ is a notion of forcing for a model M of countable cofinality, then each condition is in a generic. If, moreover, $\mathbb{P}$ is perfect, then each condition is in an undefinable generic. (HINT: use Lemmas 6.1.1 and 6.1.2.)

6.2 Forcing

There are models (e.g. rather classless models) having no undefinable inductive subsets. On the other hand, all resplendent models have undefinable inductive subsets (DO IT!). The same is true of all countable models, but we do not know a proof of this fact that does not involve some version of forcing. In fact, every generic filter is inductive; thus, from the results of the previous section, we are able to get many undefinable, inductive subsets of a countable model of PA. The main ingredient in the proof that generics are inductive is forcing, which is the subject of this section. In this section we will develop this technique for constructing undefinable inductive sets.

6.2.1 *Definition*

We begin by defining the forcing relation $\Vdash$. For a model M, the forcing language $\mathcal{L}^F(M)$ is $\mathcal{L}(M)$ augmented with a unary predicate symbol U. It is quite convenient to assume that the only logical symbols are $\vee$, $\exists$, and $\neg$. The other logical symbols will be considered as operations on formulas: for example, we let

$$\varphi_1 \wedge \varphi_2 = \neg(\neg\varphi_1 \vee \neg\varphi_2)$$

and

$$\forall v \varphi(v) = \neg\exists v \neg\varphi(v)\,.$$

Fix a notion of forcing $(\mathbb{P}, \trianglelefteq)$ for M. We define a relation $p \Vdash \sigma$ (to be read: *p forces σ*), where $p \in \mathbb{P}$ and σ is a sentence of $\mathcal{L}^F(M)$, by induction on the complexity of σ.

Definition 6.2.1 Suppose σ is an $\mathcal{L}^F(M)$-sentence and p a condition. Then:

(1) if σ is an atomic $\mathcal{L}(M)$-sentence, then $p \Vdash \sigma$ iff $M \models \sigma$;
(2) if $\sigma = U(c)$ and $c \in M$, then $p \Vdash \sigma$ iff $c \trianglelefteq p$;
(3) if $\sigma = \sigma_1 \vee \sigma_2$, then $p \Vdash \sigma$ iff $p \Vdash \sigma_1$ or $p \Vdash \sigma_2$;
(4) if $\sigma = \exists v \psi(v)$, then $p \Vdash \sigma$ iff $p \Vdash \psi(c)$ for some $c \in M$;
(5) if $\sigma = \neg\sigma_1$, then $p \Vdash \sigma$ iff for no q extending p does $q \Vdash \sigma_1$.

We say that p *decides* σ if either $p \Vdash \sigma$ or $p \Vdash \neg\sigma$.

The forcing relation for conjunction and universal quantification is defined by double negation:

$$p \Vdash \sigma_1 \wedge \sigma_2 \quad \text{iff } p \Vdash \neg(\neg\sigma_1 \vee \neg\varphi_2);$$
$$p \Vdash \forall v \psi(v) \quad \text{iff } p \Vdash \neg\exists v \neg\psi(v).$$

The following proposition contains some, but not all, of the basic properties of forcing. It can be proved by induction on the complexity of σ (DO IT!).

Proposition 6.2.2 *Let $\mathbb{P}$ be a notion of forcing for M. For all conditions p and $\mathcal{L}^F(M)$-sentences σ, the following hold:*

(1) *If $p \Vdash \sigma$ and $p \trianglelefteq q$, then $q \Vdash \sigma$.*
(2) *If $p \Vdash \sigma$, then $p \nVdash \neg\sigma$.*
(3) *There is $q \trianglerighteq p$ such that q decides σ.* □

Proposition 6.2.3 *For every $n \geq 1$, there is a Σ_n formula $\mathsf{Forc}_n(x, y)$ in the language $\mathcal{L} \cup \{P, \trianglelefteq\}$ such that whenever $M \models \mathsf{PA}^*(\mathcal{L})$, $\mathbb{P} = (P, \trianglelefteq)$ is a notion of forcing for M, and σ is a Σ_n sentence in $\mathcal{L}^F(M)$, then $p \Vdash \sigma$ iff $(M, P, \trianglelefteq) \models \mathsf{Forc}_n(p, \sigma)$.* □

What the previous proposition is saying is that there is a Σ_n formula which uniformly defines forcing of Σ_n sentences. There is an analogous Π_n formula which uniformly defines forcing of Π_n sentences. This proposition really requires that the language $\mathcal{L}$ is finite. If $\mathcal{L}$ is infinite, then this proposition has a natural local version.

6.2.2 *n-Generics*

In this subsection we pay closer attention to the complexity of the forced sentences.

Definition 6.2.4 If $n < \omega$, then $X \subseteq \mathbb{P}$ is *n-generic* if X is a filter and every $\mathcal{L}^F(M)$-sentence that is Σ_n is decided by some $p \in X$.

The 0-generics are precisely the maximal filters (DO IT!). Consequently, we will usually tacitly assume that $n \geq 1$ when referring to n-generics.

Lemma 6.2.5 *For each $n < \omega$ and $p \in \mathbb{P}$, there is a definable n-generic $X \subseteq \mathbb{P}$ such that $p \in X$. In fact, there is such an X which is Δ_{n+1} definable in $(M, \mathbb{P})$, and if $\mathbb{P}$ is a binary tree, then there is one which is Δ_n in $(M, \mathbb{P})$.* □

Lemma 6.2.6 (Truth = Forcing) *Suppose that $X \subseteq \mathbb{P}$ is n-generic and that σ is a Σ_n $\mathcal{L}^F(M)$-sentence. Then $(M, X) \models \sigma$ iff there is $p \in X$ such that $p \Vdash \sigma$.* □

If X is a generic, then we write $X \Vdash \sigma$ to mean either $(M, X) \models \sigma$ or, equivalently, that there is $p \in X$ such that $p \Vdash \sigma$. The next proposition connects generics and forcing.

Proposition 6.2.7 *Let $\mathbb{P}$ be a notion of forcing for M, and let $X \subseteq \mathbb{P}$. Then X is generic iff X is n-generic for every $n < \omega$.* □

Corollary 6.2.8 *Let $\mathbb{P}$ be a notion of forcing for M and let X be generic relative to $\mathbb{P}$. Suppose that $(M, \mathbb{P}) \prec (N, \mathbb{Q})$, Y is generic relative to $\mathbb{Q}$ and $X = M \cap Y$. Then $(M, X) \prec (N, Y)$.*

Proof Consider an $\mathcal{L}^F(M)$-sentence σ that is Σ_n, and suppose that $M \models \sigma$. By Theorem 6.2.6, there is $p \in X$ such that $p \Vdash \sigma$. Then, from Proposition 6.2.3, $p \Vdash \sigma \implies (M, \mathbb{P}, \trianglelefteq) \models \mathsf{Forc}_n(p, \sigma) \implies (N, \mathbb{Q}, \trianglelefteq) \models \mathsf{Forc}_n(p, \sigma)$, and then by Theorem 6.2.6 again, $N \models \sigma$. □

Corollary 6.2.9 *Suppose that σ is an $\mathcal{L}^F(M)$-sentence and X is a generic such that $(M, X) \models \sigma$. Then there is a definable maximal filter $D \subseteq \mathbb{P}$ such that $(M, D) \models \sigma$.*

Proof Suppose σ is Σ_n and let $p \in X$ force σ. Then by Lemma 6.2.5 there is a definable n-generic containing p. Then $(M, D) \models \sigma$ by Lemma 6.2.6. □

Theorem 6.2.10 *Every generic is inductive.*

Proof Suppose that X is generic but not inductive. Then there is an $\mathcal{L}^F(M)$-sentence σ that is an instance of the induction scheme such that $(M, X) \models \neg\sigma$. Then, by Corollary 6.2.9 there is definable $D \subseteq \mathbb{P}$ such that $(M, D) \models \neg\sigma$. But this contradicts the fact that every definable set is inductive. □

6.2.3 *Prime expansions*

A countable model may be large in the sense that there are many nondefinable elements. However, it is always possible to adjoin a Cohen generic to create a prime model, as shown in the next theorem of Simpson.

Theorem 6.2.11 *Every countable model $M \models \mathsf{PA}$ has an inductive subset X such that (M, X) is a prime model.* □

Proof Let M be a countable model of PA. We will construct a subset $X \subseteq M$ which is a Cohen generic. Let $a_0, a_1, \ldots$ enumerate M, and let $\sigma_0, \sigma_1, \ldots$ be a list of all $\mathcal{L}^F(M)$-sentences, but being careful so that each σ_i involves only the constants $a_0, \ldots, a_{i-1}$.

We will inductively define a sequence $p_0 \triangleleft p_1 \triangleleft p_2 \triangleleft \cdots$ of conditions. Let p_0 be the empty condition, let p_{2n+1} be the extension of p_{2n} by appending a 1 followed by a_n 0's and then another 1, and then let p_{2n+2} be the first (in M) condition that extends p_{2n+1} and decides σ_n.

Let $X = \{p \in \mathbb{P} : p \trianglelefteq p_n \text{ for some } n < \omega\}$. Then X is generic, so it is inductive. Clearly, a_n is definable from p_{2n+2} in M. By induction on i one can show that each p_i is definable in (M, X). □

In the previous theorem, the class X was carefully constructed. For arbitrary M, not just any nondefinable class X will do. However, by contrast, Theorem 2.2.13 shows that there are many models M for which any nondefinable class X does yield a prime (M, X). See Exercise 6.6.9 for an exercise that goes in the opposite direction.

6.2.4 *The Low Basis Theorem*

The Low Basis Theorem is useful in conjunction with the Completeness Theorem, and the same for its formalized version and the Arithmetized Completeness Theorem. In this subsection, 1-generics are used to prove the Low Basis Theorem. But first, a few background remarks.

König's Lemma states that every infinite binary tree has an infinite branch. The Completenss Theorem can be viewed, in part, as a consequence of this. A refinement of König's Lemma is that every infinite recursive binary tree has a branch recursive in $\mathbf{0}'$ or, equivalently, a Δ_2 branch. The Low Basis Theorem has a further refinement: every infinite recursive binary tree has an infinite low branch B (where B is low iff $B' \leq_T \mathbf{0}'$). These refinements yield corresponding refinements to the Completeness Theorem. All these theorems have their formalized counterparts, the Completeness Theorem's being the Arithmetized Completeness Theorem and the Low Basis Theorem's being Theorem 6.2.14.

Borrowing some terms from Computability (né Recursion) Theory, we say that a class $X \subseteq M$ is *low* in M if whenever $Y \subseteq M$ is Δ_2 in (M, X), then it is Δ_2 in M.

We say that X is *generalized low* in M if whenever Y is Δ_2 in (M, X), then there is a set $K \subseteq M$ which is Σ_1 in M such that Y is Δ_1 in (M, X, K). Utilizing complete Σ_1 sets, this definition can be simplified. If K is the complete Σ_1 set of M, then X is generalized low iff whenever Y is Δ_2 in (M, X), then Y is Δ_1 in (M, X, K) (refer to Exercise 1.11.5 and (Do it!)). Neither of these definitions would be different if we only required Y to be Σ_1 in (M, X) (Do it!). Every low set is generalized low; in fact:

Exercise 6.2.12 If $X \subseteq M$, then X is low in M iff X is generalized low and Δ_2 in M.

Lemma 6.2.13 *Let $\mathbb{P}$ be a notion of forcing for a model M. Every 1-generic relative to $\mathbb{P}$ is generalized low in $(M, \mathbb{P})$.*

Proof Let G be 1-generic. To prove that G is generalized low, we need to consider an arbitrary Y that is Δ_2 in $(M, \mathbb{P}, G)$, but we need only consider one such Y, namely the complete Σ_1 subset $(M, \mathbb{P}, G)$. Let K be the Σ_1 complete subset in $(M, \mathbb{P})$. Both Y and K are sets of Σ_1 sentences in which constants denoting elements of M are allowed. Consider any such sentence σ in the language of $(M, \mathbb{P}, G)$. Then $\sigma \in Y$ iff there is $p \in G$ such that $p \Vdash \sigma$. Thus, $\sigma \in Y$ iff $[\exists p \in G(p \Vdash \sigma)] \in K$. Following Proposition 6.2.3, this yields that Y is Σ_1 in $(M, \mathbb{P}, G, K)$. □

Theorem 6.2.14 (Low Basis Theorem) *Let $\mathbb{P}$ be a definable unbounded binary tree in a model M. Then $\mathbb{P}$ has an unbounded branch B which is low in $(M, \mathbb{P})$.*

Proof By Exercise 6.2.12 and Lemma 6.2.13, it suffices to let B be an unbounded Δ_2 1-generic branch. That is exactly the kind of B we will construct. Notice that not just any old Δ_2 1-generic branch will do: the tree $\mathbb{P}$ may have (and, in interesting cases, does have) a maximal element a, and then $\{p \in \mathbb{P} : p \trianglelefteq a\}$ is bounded and 1-generic (in fact, it is generic) and is Δ_2 (in fact, it is Δ_0).

We work in the model $(M, \mathbb{P}, K)$, where K is the Σ_1 complete set of $(M, \mathbb{P})$. Note that a subset $X \subseteq M$ is Δ_2 in $(M, \mathbb{P})$ iff it is Δ_1 in $(M, \mathbb{P}, K)$. Let $\langle \sigma_i : i \in M \rangle$ be some standard enumeration of all Σ_1 sentences in the forcing language $\mathcal{L}^F(M, \mathbb{P})$.

Working in $(M, \mathbb{P}, K)$, we define an increasing Δ_1 sequence $\langle p_i : i \in M \rangle$ of conditions as follows. Let $p_0 = 0$. Having p_i, with $\{q \in \mathbb{P} : p_i \lhd q\}$ being unbounded, consider the set $F = \{q \in \mathbb{P} : p_i \lhd q \Vdash \sigma_i\}$. Clearly, F is unbounded iff the sentence

$$\forall x \exists y [y > x \wedge y \in F]$$

holds, but also iff the sentence

$$\exists x \forall y [y \trianglerighteq x \longrightarrow y \in F]$$

holds. If F is unbounded, let p_{i+1} be the first member of F for which $\{q \in \mathbb{P} : q \trianglerighteq p_{i+1}\} \subseteq F$; if F is bounded, let p_{i+1} be the first member of $\mathbb{P}$ for which $p_{i+1} \rhd p_i$ and $\{q \in \mathbb{P} : p_{i+1} \lhd q\}$ are unbounded. This sequence determines a branch $B = \{x \in \mathbb{P} : x \lhd p_i \text{ for some } i \in M\}$, which is unbounded and is Δ_2 in $(M, \mathbb{P})$ since it is Δ_1 in $(M, \mathbb{P}, K)$.

To complete the proof, we need to see that B is 1-generic. Clearly, either $p_{i+1} \Vdash \sigma_i$ or there is some $j > i$ for which $p_j \Vdash \neg\sigma_i$. □

The branch constructed in the previous proof was 1-generic. It is too much to expect to be able to get an unbounded 2-generic, because: *An unbounded definable binary tree $\mathbb{P}$ has an unbounded 2-generic branch iff the set of maximal conditions is not dense.*

The Low Basis Theorem implies a correpsonding improvement to the Arithmetized Completeness Theorem in which the Δ_2 in that theorem is replaced by low. This improved theorem is called the Low Arithmetized Completeness Theorem.

6.3 Product forcing

This section discusses product forcing, our main uses of which appear in the next section on how the MacDowell–Specker Theorem reacts to an uncountable language.

Definition 6.3.1 Let $\mathbb{P}_1 = (P_1, \trianglelefteq_1)$ and $\mathbb{P}_2 = (P_2, \trianglelefteq_2)$ be two notions of forcing for a model M. Define their *product* $\mathbb{P}_1 \times \mathbb{P}_2$ to be $(P_1 \times P_2, \trianglelefteq)$, where $(p_1, p_2) \trianglelefteq (q_1, q_2)$ iff both $p_1 \trianglelefteq q_1$ and $p_2 \trianglelefteq q_2$.

There are some routine facts about products which should be checked. If $\mathbb{P}_1$ and $\mathbb{P}_2$ are notions of forcing, then their product $\mathbb{P}_1 \times \mathbb{P}_2$ is also a notion of forcing. The two notions of forcing $\mathbb{P}_1 \times \mathbb{P}_2$ and $\mathbb{P}_2 \times \mathbb{P}_1$ are isomorphic in the obvious sense. If $F_1 \subseteq \mathbb{P}_1$ and $F_2 \subseteq \mathbb{P}_2$, then $F_1 \times F_2$ is a filter of $\mathbb{P}_1 \times \mathbb{P}_2$ iff F_1, F_2 are filters of $\mathbb{P}_1, \mathbb{P}_2$ respectively (DO IT!). If $G = G_1 \times G_2$ is a generic relative to $\mathbb{P}_1 \times \mathbb{P}_2$, then G_1 and G_2 are generics. But more is true.

Lemma 6.3.2 (The Product Lemma) *Suppose $\mathbb{P}_1$ and $\mathbb{P}_2$ are notions of forcing for M and $G \subseteq \mathbb{P}_1 \times \mathbb{P}_2$. Then G is a generic iff there are generics G_1 and G_2, relative to $\mathbb{P}_1$ and $\mathbb{P}_2$ respectively, such that $G = G_1 \times G_2$.* □

Lemma 6.3.3 *If $G_1 \times G_2$ is generic relative to $P_1 \times P_2$, then* $\mathrm{Def}(M) = \mathrm{Def}(M, G_1) \cap \mathrm{Def}(M, G_2)$.

Proof Suppose $D \in \mathrm{Def}(M, G_1) \cap \mathrm{Def}(M, G_2)$, and let $\varphi_1(G_1, v)$ and $\varphi_2(G_2, v)$ be defining formulas in (M, G_1) and (M, G_2), respectively. Then there is a condition $(p_1, p_2) \in G_1 \times G_2$ such that $(p_1, p_2) \Vdash \varphi_1(G_1, v) \leftrightarrow \varphi_2(G_2, v)$. Since forcing is definable (Proposition 6.2.3), it suffices to show that $d \in D$ iff, for $i = 1, 2$, the set $\{q_i \in \mathbb{P}_i : p_i \trianglerighteq q_i \Vdash \varphi_i(G_i, d)\}$ is dense above p_i. The one direction is obvious (DO IT!), so assume that there is $d \in D$ and, without loss of generality, a condition $q_1 \trianglerighteq p_1$ such that $q_1 \Vdash \neg\varphi_1(G_1, d)$. Then let $q_2 \trianglerighteq p_2$ be such that $q_2 \in G_2$ and $q_2 \Vdash \varphi_2(G_2, d)$. But then $(p_1, p_2) \trianglelefteq (q_1, q_2) \Vdash \neg\varphi_1(G_1, d) \wedge \varphi_2(G_2, d)$, which is a contradiction. □

This theorem can be used to show that Cohen generics are not minimal with respect to definability (Corollary 6.3.4). This will contrast them with perfect generics which are discussed in the next section.

Consider the full binary tree $\mathbb{B}$. Then $\mathbb{B} \times \mathbb{B}$ is not much different from $\mathbb{B}$ since there is a definable embedding $f : \mathbb{B} \longrightarrow \mathbb{B}^2$ onto a dense subset of $\mathbb{B} \times \mathbb{B}$. (See Exercise 6.1.3 for the relevance of this.) We will describe one such embedding.

If $p \in \mathbb{B}$, then $p^{(e)} \in \mathbb{B}$ is its *even* part if $2 \cdot \ell(p^{(e)}) \leq \ell(p) \leq 2 \cdot \ell(p^{(e)}) + 1$ and $p_i^{(e)} = p_{2i}$ for each $i < \ell(p^{(e)})$. Similarly, $p^{(o)}$ is the *odd* part, where $2 \cdot \ell(p^{(o)}) \leq \ell(p) + 1 \leq 2 \cdot \ell(p^{(o)}) + 1$ and $p_i^{(o)} = p_{2i+1}$ for each $i < \ell(p^{(o)})$. For example, if $p = \langle 0, 0, 1, 0, 1, 1, 0, 1, 1, 1, 0 \rangle$, then $p^{(e)} = \langle 0, 1, 1, 0, 1, 0 \rangle$ and $p^{(o)} = \langle 0, 0, 1, 1, 1 \rangle$. Then, let f be such that $f(p) = \langle p^{(e)}, p^{(o)} \rangle$. This f is a definable embedding onto a dense subset (DO IT!).

If X is a Cohen generic, then let its *even* part be $X^{(e)} = \{p^{(e)} : p \in X\}$, and its *odd* part be $X^{(o)} = \{p^{(o)} : p \in X\}$. Clearly, $X \in \mathrm{Def}(M, X^{(e)}, X^{(o)})$ and $X^{(e)}, X^{(o)} \in \mathrm{Def}(M, X)$ (Do IT!).

Corollary 6.3.4 *Let X be a Cohen generic in M. Then there are $Y, Z \in \mathrm{Def}(M, X)$ such that Y, Z are Cohen generic and $\mathrm{Def}(M) = \mathrm{Def}(M, Y) \cap \mathrm{Def}(M, Z)$.*

Proof You probably have already figured out to let $Y = X^{(o)}$ and $Z = X^{(e)}$. Then, since f is a definable embedding of $\mathbb{B}$ onto a dense subset of $\mathbb{B}^2$, the set $\{\langle p_1, p_2\rangle \in \mathbb{B}^2 : \langle p_1, p_2\rangle \trianglelefteq f(p)$ for some $p \in X\}$ is a generic relative to $\mathbb{B}^2$. But this set is just $Y \times Z$, so by the Product Lemma, both Y and Z are Cohen generics. Lemma 6.3.3 implies that $\mathrm{Def}(M) = \mathrm{Def}(M, Y) \cap \mathrm{Def}(M, Z)$. □

Given three notions of forcing $\mathbb{P}_1, \mathbb{P}_2, \mathbb{P}_3$ we can form their products $(\mathbb{P}_1 \times \mathbb{P}_2) \times \mathbb{P}_3$ and $\mathbb{P}_1 \times (\mathbb{P}_2 \times \mathbb{P}_3)$, which are canonically isomorphic, so we can unambiguously write $\mathbb{P}_1 \times \mathbb{P}_2 \times \mathbb{P}_3$. More generally, we can form $\mathbb{P}_1 \times \mathbb{P}_2 \times \cdots \times \mathbb{P}_n$. It is intended here that n is a standard integer. However, this makes good sense even if n is a nonstandard element in M as long as the sequence $\langle \mathbb{P}_1, \mathbb{P}_2, \ldots, \mathbb{P}_n\rangle$ is definable in M. If all of the factors $\mathbb{P}_i$ are identical, say $\mathbb{P}_i = \mathbb{P}$ for each i, then we will write $\mathbb{P}^n$ instead of $\mathbb{P}_1 \times \mathbb{P}_2 \times \cdots \times \mathbb{P}_n$. The Product Lemma 6.3.2 easily implies its own generalization to cover n-fold products.

Definition 6.3.5 If $\mathbb{P}$ is a notion of forcing and $\mathcal{G}$ is a set of generics, then we say that $\mathcal{G}$ is a set of *mutual generics* if whenever $G_1, G_2, \ldots, G_n$ are finitely many distinct generics in $\mathcal{G}$, then $G_1 \times G_2 \times \cdots \times G_n$ is a generic relative to $\mathbb{P}^n$.

Equivalently, $\mathcal{G}$ is a set of mutual generics if whenever $G_0, G_1, \ldots, G_n$ are finitely many distinct generics in $\mathcal{G}$, then G_n is a generic relative to $\mathbb{P}$ in $(M, G_0, G_1, \ldots, G_{n-1})$ (Do IT!). For perfect notions of forcing, there can be large sets of mutual generics as we will see in Theorem 6.3.7. First, the case of countable models.

Lemma 6.3.6 *If M is countable and $\mathbb{P}$ is a perfect notion of forcing, then there is a set $\mathcal{G}$ of $2^{\aleph_0}$ mutual generics.* □

Proof For $n < \omega$, let $S_n = \{0,1\}^n$ be the set of all sequences of 0's and 1's having length n, and then let $\mathbb{P}_n = \mathbb{P}^{S_{n+1}}$ (which is the set of all funtcions $p : S_{n+1} \longrightarrow \mathbb{P}$ ordered pointwise: $p \trianglelefteq q$ iff $p(s) \trianglelefteq q(s)$ for each $s \in S_{n+1}$). Let $D_0, D_1, D_2, \ldots$ be a list of definable subsets of M, where each D_n is a dense open subset of $\mathbb{P}_n$, and the following holds: whenever $1 \leq k < \omega$ and $D \subseteq \mathbb{P}^k$ is a dense open subset, then there are arbitrarily large $n < \omega$ such that whenever $e : k \longrightarrow S_{n+1}$ is one-to-one, then $D_n \subseteq \{p \in \mathbb{P}_n : \langle p_{e(0)}, p_{e(1)}, \ldots, p_{e(k-1)}\rangle \in D\}$.

Inductively on $n < \omega$, we obtain $p_s \in \mathbb{P}$ for each $s \in S_n$. For $n = 0$, let $p_\emptyset \in \mathbb{P}$ be arbitrary. Now suppose we have p_s for each $s \in S_n$. Let $q = \langle q_t : t \in S_{n+1}\rangle \in \mathbb{P}_n$ be such that $q_{s^\frown 0}$ and $q_{s^\frown 1}$ are incompatible extensions of p_s for each $s \in S_n$. Then, since D_n is dense, we can get $\langle p_t : t \in S_{n+1}\rangle \in D_n$ extending q. For each

$s \in \{0,1\}^\omega$, let $G_s = \{p \in \mathbb{P} : p \lhd p_{s|n}$ for some $n < \omega\}$. Then distinct s's yield distinct G_s's. For, if $s \neq t$, then let $n < \omega$ be the least for which $s_n \neq t_n$. Then $p_{s|n} \in G_s$ and $p_{t|n} \in G_t$ are incompatible. Thus, the set $\mathcal{G}$, consisting of all the G_s's, has cardinality $2^{\aleph_0}$. It is clear that each G_s is generic with respect to $\mathbb{P}$. In fact $\mathcal{G}$ is a set of mutual generics (DO IT!). □

Theorem 6.3.7 *Suppose M has countable cofinality and that $|M| = \kappa$. If $\mathbb{P}$ is a perfect notion of forcing, then there is a set $\mathcal{G}$ of mutual generics such that each condition $p \in \mathbb{P}$ is in $\kappa^{\aleph_0}$ of these generics.*

Proof The proof will be like the proof of the previous lemma, but more of the work will be done inside the model. Since $\mathbb{P}$ is perfect, it can be proved using formal induction in M that for each $a \in M$ and each condition p there is a set of a pairwise incompatible conditions extending p.

Since M has countable cofinality, we can let $b_0 < b_1 < b_2 < \cdots$ be a cofinal sequence. Inductively, we will define another cofinal sequence $a_0 < a_1 < a_2 < \cdots$ and definable bounded subsets $A_0 \subseteq A_1 \subseteq \cdots$. For each $n < \omega$, we will let $A_n = \{p \in \mathbb{P} : p \leq a_n\}$. First, choose $a_0 > b_0$ large enough so that $A_0 \neq \emptyset$. Having a_n and also A_n, let $a_{n+1} > b_{n+1}$ large enough so that for each $p \in A_n$, there is a definable set of $Q \subseteq A_{n+1}$ of pairwise incompatible extensions of p such that $M \models |Q| \geq b_n$.

Let $S_n = \{s \in A_0 \times A_1 \times \cdots \times A_n : s_0 \lhd s_1 \lhd \cdots \lhd s_n\}$, and then let $\mathbb{P}_n = \mathbb{P}^{S_n}$. Each of A_n, S_n, and $\mathbb{P}_n$ is definable in M.

Let $D_0, D_1, D_2, \ldots$ be much like the sequence in the proof of the previous lemma; specifically, each D_n is a dense open subset of $\mathbb{P}_{n+1}$, and the following holds: whenever $1 \leq k < \omega$ and $D \subseteq \mathbb{P}^{a_k}$ is a dense open subset, then there are arbitrarily large $n < \omega$ such that whenever $e : a_k \longrightarrow S_{n+1}$ is a definable bijection, then $D_n \subseteq \{p \in \mathbb{P}_{a_k} : \langle p_{e(i)} : i < a_k\rangle \in D\}$.

Inductively on $n < \omega$, we obtain $p_n = \langle p_{n,s} : s \in S_n\rangle \in \mathbb{P}_n$. For $n = 0$, choose $p_0 \in \mathbb{P}_0$ arbitrarily. Now suppose we have $p_n = \langle p_s : s \in S_n\rangle \in \mathbb{P}_n$.

Let $q = \langle q_t : t \in S_{n+1}\rangle \in \mathbb{P}_{n+1}$ be such that whenever $s, t \in S_{n+1}$ are distinct $r = s{\restriction}n = t{\restriction}n$, then q_s and q_t are incompatible extensions of $p_{n,r}$. Then, since D_n is a dense subset of $\mathbb{P}_{n+1}$, we can get $p_{n+1} = \langle p_{n+1,t} : t \in S_{n+1}\rangle \in D_n$ extending q.

Let $I = \{s \in \mathbb{P}^\omega : s{\restriction}n \in S_n$ for each $n < \omega\}$, and for each $s \in I$, let $G_s = \{p \in \mathbb{P} : p \lhd p_{\#}$ for some $n < \omega\}$. As in the previous lemma, distinct s's yield distinct G_s's, and the set $\mathcal{G}$ of all the G_s's is mutually generic.

It remains to check the cardinality of $|\mathcal{G}| = |I|$. Let $\kappa_n = |\{x \in M : x \leq b_n\}|$. Then, $\kappa = \bigcup_{n<\omega} \kappa_n$. Therefore, $|I| \geq \prod_{n<\omega} \kappa_n = \kappa^{\aleph_0}$. □

6.4 MacDowell–Specker vs the uncountable

The proof of the MacDowell–Specker Theorem proceeds by a double induction. One is an induction on parameters, which is done internally, allowing the proof to work for uncountable models. The other takes place in the real world and seems to require the countability of the language. Thus, the proof of the MacDowell–Specker Theorem works well for models of $\mathsf{PA}^*(\mathcal{L})$ when $\mathcal{L}$ is countable but is dubious for uncountable $\mathcal{L}$. The goal of this section is to investigate what the situation is when the MacDowell–Specker Theorem meets an uncountable language. In the first subsection, the theorem of Mills puts a damper on the possibility of completely generalizing the MacDowell–Specker Theorem to uncountable languages. In the second section, Theorem 6.4.3 shows that all is not lost and that in certain cases MacDowell–Specker can overcome the uncountability of the language and produce elementary end extensions. In the third subsection, Theorem 6.4.3 is applied to construct models having many classes.

6.4.1 *No end extension*

The MacDowell–Specker may fail when the language is uncountable, as the following theorem shows.

Theorem 6.4.1 *Let M be a nonstandard countable model. There are X_α for $\alpha < \omega_1$ such that*

$$(M, X_\alpha)_{\alpha<\omega_1} \models \mathsf{PA}^*$$

and $(M, X_\alpha)_{\alpha<\omega_1}$ has no elementary end extension.

Proof Let us fix a nonstandard countable model M and a nonstandard element $e \in M$. Let $\mathbb{P}$ be the notion of forcing consisting of (codes of) definable functions $p : [0, c] \longrightarrow [0, e]$, for some $c \in M$, ordered by extension of functions. Different conditions in $\mathbb{P}$ may have different domains. For $n < \omega$ and $c \in M$, let $\mathbb{P}_c^{n+1}$ be the subset of $\mathbb{P}^{n+1}$ consisting of all $\langle p_0, p_1, \ldots, p_n\rangle$ such that whenever $i < j \leq n$ and $c < x < \ell(p_i), \ell(p_j)$, then $p_i(x) \neq p_j(x)$. With these definitions the following lemma, which is a cousin of Lemma 6.3.6, is the key to proving the theorem. Notice that the generics which the lemma asserts to exist are not mutual generics, but they do have a very similar sort of mutual relation. Its proof is so similar to the proof of Lemma 6.3.6 that we will omit it.

Lemma 6.4.2 *There is a set $\mathcal{G}$ of $2^{\aleph_0}$ generics relative to $\mathbb{P}$ with the added property that whenever $n < \omega$ and $G_0, G_1, G_2, \ldots, G_n \in \mathcal{G}$ are distinct, then there is $c \in M$ such that $G_0 \times G_1 \times \cdots \times G_n$ is generic relative to $\mathbb{P}_c^{n+1}$.* □

It is worth noting that if we had chosen e to be standard, then for $n > e$, there would no condition $\langle p_0, p_1, \ldots, p_n\rangle \in \mathbb{P}_c^{n+1}$ such that $\ell(p_0), \ell(p_1), \ldots, \ell(p_n) \geq c + 2$, and then the lemma would no longer be true.

Let $\mathcal{G}$ be the set of generics from the lemma above. Then $(M, X)_{X\in\mathcal{G}} \models \mathsf{PA}^*$ since, (by the previous lemma) whenever $X_0, X_1, \ldots, X_{n-1} \in \mathcal{G}$ are distinct, there is $c \in M$ such that $X_0 \times X_1 \times \cdots \times X_{n-1}$ is generic relative to $\mathbb{P}_c^n$. Now let X_α, for $\alpha < \omega_1$, be distinct generics in $\mathcal{G}$.

To obtain a contradiction, assume that $(M, X_\alpha)_{\alpha<\omega_1} \prec_{\mathsf{end}} (N, Y_\alpha)_{\alpha<\omega_1}$. Notice that each X_α determines a function $f_\alpha : M \longrightarrow [0, e]_M$ and each Y_α determines a function $g_\alpha : N \longrightarrow [0, e]_N$. Clearly, $(M, f_\alpha) \prec_{\mathsf{end}} (N, g_\alpha)$ so that $f_\alpha = g_\alpha | M$. Consider some $a \in N \backslash M$. We claim: if $\alpha < \beta < \omega_1$, then $g_\alpha(a) \neq g_\beta(a)$. For, let $c \in M$ be such that $X_\alpha \times X_\beta$ is generic relative to $\mathbb{P}^2_c$. Then $(M, X_\alpha, X_\beta) \models \forall x[x > c \rightarrow f_\alpha(x) \neq f_\beta(x)]$, and then also $(N, Y_\alpha, Y_\beta) \models \forall x[x > c \rightarrow g_\alpha(x) \neq g_\beta(x)]$. So, in particular, $g_\alpha(c) \neq g_\beta(c)$.

Therefore, the set $Y = \{g_\alpha(c) : \alpha < \omega_1\}$ is uncountable, yet $Y \subseteq [0, e]_N = [0, e]_M \subseteq M$, which is a contradiction. □

6.4.2 *Extensions with mutual generics*

The significance of the next theorem, generalizing the MacDowell–Specker Theorem, becomes apparent with the realization that there may be uncountably many mutual Cohen generics.

Theorem 6.4.3 *Suppose M is a model and $\mathcal{G}$ is a set of mutual Cohen generics. Then the model $(M, X)_{x\in\mathcal{G}}$ has a conservative elementary end extension.* □

The proof of this theorem makes use of four lemmas (Lemmas 6.4.4–6.4.7) and will be completed after them.

If the set $\mathcal{G}$ in the theorem consists of just one generic, then there is nothing new here. For, if G is its unique member, then the expansion (M, G) is a model of PA^* for a countable language and the MacDowell–Specker Theorem applies. Despite this, we will first consider the case of one generic, but in a broader context.

Let $\mathbb{P}$ be a notion of forcing for M (which may or may not be the full binary tree $\mathbb{B}$). If $X \subseteq \mathbb{P}$ and $C \subseteq M$ are definable and if $p \in \mathbb{P}$, then X is *C-homogeneous above p* if X is dense above p and whenever $p \trianglelefteq q, q' \in X$, then $q \in C \Longleftrightarrow q' \in C$. Informally, if X is dense above p, then X is C-homogeneous above p iff either C is a subset of X above p or C is disjoint from X above p. We say that X is *eventually C-homogeneous* if there is a dense set of condtions $p \in \mathbb{P}$ for which X is C-homogeneous above p. If X is eventually C-homogeneous, then it is dense (Do it!). A sequence of definable sets $X_0, X_1, X_2, \ldots$ is *complete* for $\mathbb{P}$ if $\mathbb{P} \supseteq X_0 \supseteq X_1 \supseteq X_2 \supseteq \cdots$ and whenever $\langle C_i : i \in M\rangle$ is a definable sequence of subsets of $\mathbb{P}$ that is definable (meaning that $\{\langle i, x\rangle : x \in C_i\}$ is definable), then there is some $k < \omega$ such that for each $i \in M$, X_k is eventually C_i-homogeneous. Given a generic G relative to $\mathbb{P}$ and a complete sequence $X_0 \supseteq X_1 \supseteq \cdots$, we define the set $\Phi_G(v)$ of unary $\mathcal{L}(M)$-formulas so that a formula $\varphi(v)$ is in $\Phi_G(v)$ iff $G \cap X_m \cap \varphi(v)^M$ is unbounded for all $m < \omega$.

Lemma 6.4.4 *Suppose that G is generic relative to $\mathbb{P}$ and $X_0 \supseteq X_1 \supseteq X_2 \supseteq \cdots$ is complete for $\mathbb{P}$. Then:*

(1) *$\Phi_G(v)$ is a complete type over M.*

(2) *Let $M(c)$ be a $\Phi_G(v)$-extension of M, and let $\mathbb{P}'$ be such that $(M,\mathbb{P}) \prec (M(c),\mathbb{P}')$. Then $c \rhd s$ for every $s \in G$, and if $G' \subseteq M(c)$ is generic with respect to $\mathbb{P}'$ such that $c \in G'$, then $(M(c),G')$ is a conservative extension of (M,G).*

Proof To prove that $\Phi_G(v)$ is a complete type, it suffices to prove the following:

(a) for any $\varphi(v)$, either $\varphi(v) \in \Phi_G(v)$ or $\neg\varphi(v) \in \Phi_G(v)$.
(b) if $\varphi_0(v), \varphi_1(v) \in \Phi_G(v)$, then $\varphi_0(v) \wedge \varphi_1(v) \in \Phi_G(v)$.

For (a), suppose $\varphi(v) \notin \Phi_G(v)$. Let $\varphi(v)$ define C, and let $k < \omega$ be such that $G \cap X_k \cap C$ is bounded. Then, if $m > k$, $G \cap X_m \cap C$ is bounded, so $(G \cap X_m)\backslash C$ is unbounded. Hence $\neg\varphi(v) \in \Phi_G(v)$.

For (b), let C_0, C_1 be the sets defined by $\varphi_0(v)$ and $\varphi_1(v)$. Let $k < \omega$ be large enough so that X_k is eventually C_0-homogeneous and eventually C_1-homogeneous. Then there is $p \in G$ such that whenever $p \trianglelefteq q \in X_k$, then $p \in C_0$ iff $q \in C_0$ and $p \in C_1$ iff $q \in C_1$. Therefore, if $p \trianglerighteq q \in G$, then $q \in C_0 \cap C_1$, so that $\varphi_0(v) \wedge \varphi_1(v) \in \Phi_G(v)$, proving (b).

We now prove (2). For $p \in G$, let C_p be the set defined by the formula $v \rhd p$. It is clear that each $X_m \cap C_p$ is dense above p, so that $G \cap X_m \cap C_p \neq \emptyset$. Since this is so for all $p \in G$, we that $G \cap X_m \cap C_p$ is unbounded. Therefore, $v \rhd p$ is in $\Phi_G(v)$, so that $c \rhd p$.

Next, consider an $\mathcal{L}^F(M)$-formula $\varphi(x,v)$ intending to show that if $D \subseteq M(c)$ is defined by $\varphi(c,v)$ in $(M(c),G')$, then $D \cap M \in \mathrm{Def}(M,G)$. Even though the formula $\varphi(x,v)$ is an $\mathcal{L}^F(M)$-formula, we can assume, without loss, that U does not occur in it. For every $q, i \in M$, there is $p \in G$ which decides $\varphi(q,i)$, so, in $M(c)$, c decides all $\varphi(q,i)$ for $q, i \in M$. Since forcing is definable, there is d such that $c \lhd d \in G'$ which forces all the sentences "c decides $\varphi(q,i)$." But $d = g(c)$ for some M-definable function $g : M(c) \longrightarrow M(c)$. Thus, instead of $\varphi(x,v)$ we use the formula $q(x) \Vdash \varphi(x,v)$ in which U does not occur.

Thus, let $\varphi(x,v)$ be an $\mathcal{L}(M)$-formula. Let $C_i \subseteq M$ be defined by $\varphi(x,i)$. Let $k < \omega$ be such that X_k is eventually C_i-homogenous for each i. Then $i \in D$ iff $G \cap X_k \cap C_i$ is unbounded. This gives a definition of D in (M,G). □

Some notation concerning $\mathbb{B}$ is needed. The lexicographical relation $<_{\mathsf{lex}}$ is defined on $\mathbb{B}$ by letting $s <_{\mathsf{lex}} t$ iff there is $n < \ell(s), \ell(t)$ such that $s{\restriction}n = t{\restriction}n$, $s(n) = 0$ and $t(n) = 1$. Notice that $\mathbb{B}$ is not linearly ordered by $<_{\mathsf{lex}}$ but any set of pairwise incompatible conditions is. This relation can be extended to Cohen generics by letting $G_0 <_{\mathsf{lex}} G_1$ iff there are $s \in G_0$ and $t \in G_1$ such that $s <_{\mathsf{lex}} t$. Then $<_{\mathsf{lex}}$ linearly orders the set of Cohen generics (DO IT!).

If $X \subseteq \mathbb{B}$ and $1 \leq n < \omega$, then let $\langle X\rangle^n$ be the set of $\langle s_0, s_1, \ldots, s_{n-1}\rangle \in X^n$ such that $\ell(s_0) = \ell(s_1) = \cdots = \ell(s_{n-1})$ and $s_0 <_{\mathsf{lex}} s_1 <_{\mathsf{lex}} \cdots <_{\mathsf{lex}} s_{n-1}$. This notation routinely extends to nonstandard $n \in M$. With no risk of confusion, we let $\langle\mathbb{B}\rangle^n$ be ordered by $\trianglelefteq$ in the natural way: if $s,t \in \langle\mathbb{B}\rangle^n$, then $s \trianglelefteq t$ iff $s_i \leq t_i$ for each $i < n$. The two notions of forcing, $\mathbb{B}^n$ and $\langle\mathbb{B}\rangle^n$, are not a whole lot different from each other: if $G = G_0 \times G_1 \times \cdots \times G_{n-1}$ is generic relative to $\mathbb{B}^n$ and $G_0 <_{\mathsf{lex}} G_1 <_{\mathsf{lex}} \cdots <_{\mathsf{lex}} G_{n-1}$, then $G \cap \langle\mathbb{B}\rangle^n$ is generic relative to $\langle\mathbb{B}\rangle^n$. (There is a converse to this. EXERCISE: what is it?)

Now for an important definition: a subset $X \subseteq \mathbb{B}$ is *large* if $\langle X\rangle^n$ is a dense subset of $\langle\mathbb{B}\rangle^n$ (or, equivalently, of $\mathbb{B}^n$) whenever $1 \le n \in M$.

Lemma 6.4.5 *Let M be a model and $1 \le n < \omega$. Let $X \subseteq \mathbb{B}$ be a definable large set, and let $\langle C_i : i \in M\rangle$ be a definable sequence of subsets of $\langle\mathbb{B}\rangle^n$. Then there is a definable large $Y \subseteq X$ such that $\langle Y\rangle^n$ is eventually C_i-homogeneous for each i.* □

The main ingredient the proof of this lemma is the Halpern–Laüchli–Laver–Pincus Theorem, which is a weaker version of Lemma 6.4.5 for the standard model. Recall that $\{0,1\}^{<\omega}$ is just the standard version of $\mathbb{B}$. The following statement of the HLLP Theorem is not exactly the traditional one, but it can either be derived from the traditional one or be proved by a straightforward modification of the "manipulative" proof (as given, for example, in [130]) of the traditional one.

> THE HLLP THEOREM: *Let $1 \le n < \omega$, and let $X, C \subseteq \{0,1\}^{<\omega}$ be such that X is large. Then there is a large $Y \subseteq X$ such that $\langle Y\rangle^n$ is eventually C-homogeneous.*

There are two related observations to be made about this theorem. One is that the large set Y can be chosen to be recursive in C and X. The second is that the theorem, even in this stronger form, can be formalized and proved as a scheme in PA^*. This gives us the variation of the Lemma 6.4.5 in which there is only one C_i, (i.e. all C_i are the same C) and with the additional feature that Y is Δ_1 in (M, X, C). From this, Lemma 6.4.5 can be derived (an exercise), even with the strengthened conclusion that Y is Δ_1 in (M, X, C).

Using Lemma 6.4.5, we fix a decreasing sequence $\mathbb{B} = X_0 \supseteq X_1 \supseteq X_2 \supseteq \cdots$ of large subsets such that whenever $n < \omega$ and $\langle C_i : i \in M\rangle$ is a definable sequence of definable subsets of $\langle\mathbb{B}\rangle^n$, then there is some $k < \omega$ such that $\langle X_k\rangle^n$ is eventually C_i-homogeneous for each i. Not only is this sequence complete for $\mathbb{B}$, but the sequence $\langle X_0\rangle^n \supseteq \langle X_1\rangle^n \supseteq \cdots$ is complete for $\langle\mathbb{B}\rangle^n$. Thus, associated with each generic F relative to $\langle\mathbb{B}\rangle^n$ is a well-defined 1-type $\Phi_F(v)$ whose nice properties are stated in Lemma 6.4.4. We will, however, be more interested in a related n-type than in the 1-type $\Phi_F(v)$.

Suppose that $G = G_0 \times G_1 \times \cdots \times G_{n-1}$ is generic relative to $\mathbb{B}^n$, where $G_0 <_{\mathsf{lex}} G_1 <_{\mathsf{lex}} \cdots <_{\mathsf{lex}} G_{n-1}$. Then $F = G \cap \langle\mathbb{B}\rangle^n$ is generic relative to $\langle\mathbb{B}\rangle^n$. Let $\Psi_G(v_0, v_1, \ldots, v_{n-1})$ be the set of n-ary $\mathcal{L}(M)$-formulas $\psi(v_0, v_1, \ldots, v_{n-1})$ such that $\psi(\langle (v)_0, (v)_1, \ldots, (v)_{n-1}\rangle)$ is in $\Phi_F(v)$. The nice properties of $\Phi_F(v)$ from Lemma 6.4.4 are easily transferred to nice properties of $\Psi_G(\bar{v})$.

The next lemma follows directlly from Lemma 6.4.4 (DO IT!).

Lemma 6.4.6 *Assume $1 \leq n < \omega$ and G is as described above. Then:*

(1) *$\Psi_G(\bar{v})$ is a complete n-type over M.*
(2) *Let $M(\bar{c})$ be a $\Psi_G(\bar{v})$-extension of M, and let $\mathbb{P}'$ be such that $(M, \mathbb{P}) \prec (M(\bar{c}), \mathbb{P}')$. Then $\bar{c} \rhd \bar{s}$ for every $\bar{s} \in G$ and if $G' \subseteq M(\bar{c})$ is generic with respect to $\mathbb{B}'$ such that $\bar{c} \in G'$, then $(M(\bar{c}), G')$ is a conservative extension of (M, G).* □

There is a compatibilty condition that connects up all the types $\Psi_G(\bar{v})$ that we have constructed.

Lemma 6.4.7 (Compatibility Condition) *Let $G = G_0 \times G_1 \times \cdots \times G_{n-1}$, where $G_0 <_{\mathsf{lex}} G_1 <_{\mathsf{lex}} \cdots <_{\mathsf{lex}} G_{n-1}$, is generic relative to $\mathbb{B}^n$. Let $i_0 < i_1 < \cdots < i_{m-1} < n$ and $H = G_{i_0} \times G_{i_1} \times \cdots \times G_{i_{m-1}}$. Then H is generic relative to $\mathbb{B}^m$ and $\Psi_H(v_{i_0}, v_{i_1}, \ldots, v_{i_{m-1}}) \subseteq \Psi_G(v_0, v_1, \ldots, v_{n-1})$.*

Proof Just unravel the definitions (DO IT!). □

Proof of Theorem 6.4.3 The set $\mathcal{G}$ is linearly ordered by $<_{\mathsf{lex}}$. Let $(I, <)$ be a linearly ordered set having the same order type as $\mathcal{G}$, and then let $\mathcal{G} = \{G_i : i \in I\}$, where $i < j \iff G_i <_{\mathsf{lex}} G_j$. Let I^* be the set of increasing finite sequences from I, and for each $j \in I^*$ of length n, let $G_j = G_{j_0} \times G_{j_1} \times \cdots \times G_{j_{n-1}}$, which is generic relative to $\mathbb{P}^n$. We form the giant I-type

$$\Psi(\bar{v}) = \bigcup \{\Psi_{G_j}(v_{j_0}, v_{j_1}, \ldots, v_{j_{n-1}}) : j \in I^*\}.$$

Lemmas 6.4.6 and 6.4.7 guarantee that this is a complete I-type over M which generates an extension $M(\bar{c})$. It is clear that $M(\bar{c})$ has countable cofinality (either (DO IT!) or replace $M(\bar{c})$ with a simple elementary end extension of itself), so Theorem 6.3.7 applies, and we can find a set $\{G_i' : i \in I\}$ of mutual Cohen generics in $M(\bar{c})$ such that $c_i \in G_i$ for each $i \in I$. Then Lemma 6.4.6 shows that $(M(\bar{c}), G_i')_{i \in I}$ is a conservative extension of $(M, G_i)_{i \in I}$. □

6.4.3 *Getting many classes*

Theorem 6.4.3 is critical for showing the existence of models having many classes. Every model M of cardinality κ has at least κ classes. It could be that it has more. For example, if $\mathrm{cf}(M) = \omega$, then M has at least $\kappa^{\aleph_0}$ classes which are quite easy to construct once you realize that a class can have order type ω. Theorem 6.3.7 improves upon this by asserting the existence of $\kappa^{\aleph_0}$ mutual Cohen generics. Of course, if we are unlucky, it might be that $\kappa^{\aleph_0} = \kappa$ and then we still have the minimum number of classes. A saturated model of cardinality κ has 2^κ classes (DO IT!), but the existence of saturated models requires some additional set-theoretic assumptions. Another approach to getting models with many classes would be to use a two-cardinal theorem from model theory. There are two candidate theorems: Chang's, which would also require some additional set-theoretic hypothesis such as the GCH, and Shelah's, which would require some additional work to implement. The next theorem shows that for any infinite cardinal κ, there are models having more than κ classes.

Theorem 6.4.8 *Suppose that M is a model and $|M| \geq \kappa$. Then M has an elementary end extension N of cardinality κ having a set of κ^+ mutual Cohen generics.*

Proof The construction of the model N in the theorem will make use of trees. A *tree* is a partially ordered set $(T, <)$ such that for any $x \in T$, the set $\{y \in T : y < x\}$ of its predecessors is well-ordered by $<$. The order type of this set is the *rank* of x, denoted by $rk(x)$. For each ordinal α, we let T_α be the set of elements having rank α. The *height* $ht(T)$ of T is the least ordinal α for which $T_\alpha = \emptyset$. If T is a tree of height γ, then a *branch* of T is a linearly ordered subset $B \subseteq T$ such that $B \cap T_\alpha \neq \emptyset$ whenever $\alpha < \gamma$. We let $[T]$ be the set of branches of T. A tree T is *normal* if three conditions are met:

(1) whenever $x \in T_\alpha$, then $|\{y \in T_{\alpha+1} : x < y\}| \leq 2$;
(2) if α is not a successor ordinal and $x, y \in T_\alpha$ have exactly the same predecessors, then $x = y$;
(3) $ht(T)$ is a limit ordinal.

We begin the proof with a fact about the existence of certain kinds of trees.

> *For each κ there is a normal tree T such that $|T| = \kappa$, $|[T]| > \kappa$, and $ht(T) < \kappa$.*

To get such a tree, we first consider the tree S of height $\kappa + 1$ consisting of all functions f, where for some ordinal $\alpha \leq \kappa$, $f : \alpha \longrightarrow \{0, 1\}$. Let $<$ on S be extension, so that $rk(f)$ is the domain of f. Thus, $f < g$ iff there is $\alpha < rk(g)$ such that $f = g{\restriction}\alpha$. Clearly, $|S_\kappa| = 2^\kappa > \kappa$. Let $\gamma \leq \kappa$ be the least ordinal such that $|S_\gamma| > \kappa$. Some simple cardinal arithmetic shows that γ is a limit ordinal and that $T = \{f \in S : rk(f) < \gamma\}$ is a normal tree such that $|T| = \kappa$ and $|[T]| = |S_\gamma| > \kappa$. This proves the fact.

We next describe a method for constructing elementary end extensions along a normal tree. (Of course, we will have in mind one of the trees whose existence was just proved.) Let M be any model and let T be a normal tree of height $\gamma < \kappa$. We will construct an elementary chain $\langle M_\alpha : \alpha \leq \gamma \rangle$ such that:

(0) $M \prec_{\mathsf{end}} M_0$;
(1) if $\alpha < \beta \leq \gamma$, then $M_\alpha \prec_{\mathsf{end}} M_\beta$;
(2) if $\alpha < \gamma$ is not a limit ordinal, then $\mathrm{cf}(M_\alpha) = \omega$;
(3) if $\alpha \leq \gamma$ is a limit ordinal, then $M_\alpha = \bigcup\{M_\beta : \beta < \alpha\}$;
(4) if $\alpha \leq \gamma$, then $|M_\alpha| = \kappa$.

As we construct this chain, we will also construct $\langle G_x : x \in T\rangle$ such that:

(5) if $\alpha < \gamma$, then $\{G_x : x \in T_\alpha\}$ is a set of mutual Cohen generics in M_α;
(6) if $\alpha < \gamma$ and $x, y \in T_\alpha$ are distinct, then $G_x \neq G_y$;
(7) if $\alpha < \beta < \gamma$, $x \in T_\alpha$, $y \in T_\beta$, then $G_x = G_y \cap M_\beta$.

To get started, let M' be an elementary end extension of M having cardinality κ, and then let M_0 be a simple elementary end extension of M', so that $|M_0| = \kappa$ and $\mathrm{cf}(M_0) = \omega$. Let $G_x \subseteq M_0$ be a Cohen generic for the unique $x \in T_0$. For ordinals $\alpha > 0$, we consider separately the cases of successor ordinal and of limit ordinal.

α is a successor ordinal. Let $\alpha = \beta + 1$. By Theorem 6.4.3 and the remark following it, the model $(M_\beta, G_x)_{x\in T_\beta}$ has an elementary end extension $(M_\alpha, G'_x)_{x\in T_\beta}$, where $|M_\alpha| = \kappa$ and $\mathrm{cf}(M_\alpha) = \omega$. Use Theorem 6.3.7 to get a set $\mathcal{H}$ of mutual Cohen generics such that each condition in $\mathbb{B}^{M_\alpha}$ is in some generic in $\mathcal{H}$. Let $c \in M_\alpha \backslash M_\beta$, and then for each $x \in T_\beta$, let $a_x \in G'_x$ be such that $\ell(a_x) = c$. Let $f : T_\alpha \longrightarrow M_\alpha$ be a one-to-one function such that whenever $x < y$, where $x \in T_\beta$ and $y \in T_\alpha$, then $f(y) \in \{a_x\hat{\ }0, a_x\hat{\ }1\}$, and then let $G_y \in \mathcal{H}$ be such that $f(y) \in G_y$.

α is a limit ordinal. We have no choice. By (3), we must let $M_\alpha = \bigcup\{M_\beta : \beta < \alpha\}$, and by (7), (only for $\alpha < \gamma$) if $x \in T_\alpha$, then $G_x = \bigcup\{G_y : y < x\}$.

It is clear (Do it!) that (0)–(7) hold, except possibly for (5) in the case that α is a limit ordinal. For this case, consider distinct $x_0, x_1, \ldots, x_n \in T_\alpha$ to show that $G_{x_0}, G_{x_1}, \ldots, G_{x_n}$ are mutual Cohen generics. For each $\beta < \alpha$, let $x_i|\beta$ be the unique $y \in T_\beta$ such that $y < x_i$. Let $\beta < \alpha$ be large enough so that $x_i|\beta \neq x_j|\beta$ whenever $i < j \leq n$. Then, by induction on ordinals and Corollary 6.2.8, it follows that whenever $\beta < \nu \leq \alpha$, then $(M_\beta, G_{x_0|\beta}, G_{x_1|\beta}, \ldots, G_{x_n|\beta}) \prec (M_\nu, G_{x_0|\nu}, G_{x_1|\nu}, \ldots, G_{x_n|\nu})$. We then can conclude (5) by letting $\nu = \alpha$.

Now let $N = M_\gamma$. For each branch $B \in [T]$, let $G_B = \bigcup\{G_x : x \in B\}$, and then let $\mathcal{G} = \{G_B : B \in [T]\}$. Distinct B's yield distinct G_B's, so $|\mathcal{G}| \geq \kappa^+$. Finally, that $\mathcal{G}$ is a set of mutual Cohen generics is the same as showing (5) for limit ordinals. □

Remarks 1. The construction in the proof of Theorem 6.4.8 started with a normal tree T and then produced a model N and a set $\mathcal{G} = \{G_B : B \in [T]\}$ of mutual generics. If $ht(T)$ has uncountable cofinality, then the model $(N, G)_{G\in\mathcal{G}}$ will be rather classless. (See Theorem 6.3.7 for comparison.)

2. Let κ be an uncountable regular cardinal. A tree T is a *κ-tree* if $ht(T) = \kappa$ and $|T_\alpha| < \kappa$ whenever $\alpha < \kappa$. By slightly modifying the notion of a normal tree so as to better apply to κ-trees, we can use the construction of the theorem to get: corresponding to each normal κ-tree T is a κ-like model N having a set $\mathcal{G} = \{G_B : B \in [T]\}$ of mutual Cohen generics such that

$(N, G)_{G \in \mathcal{G}}$ is rather classless. Some additional properties of the tree T may be reflected in the model N.

6.5 Perfect generics

The method of forcing that has just been introduced is very useful for getting inductive subsets. But not all can be obtained in this way. The following exercise is aimed at showing this. It should be compared with Theorem 6.5.8.

Exercise 6.5.1 Let $X \subseteq M$ be an undefinable generic relative to $\mathbb{P}$. Then there are $Y, Z \in \mathrm{Def}(M, X)$ such that $X \in \mathrm{Def}(M, Y, Z)$ and $\mathrm{Def}(M) = \mathrm{Def}(M, Y) \cap \mathrm{Def}(M, Z)$. (HINT: consult Corollary 6.3.4.)

Suppose $\mathbb{P} \subseteq M$ is a definable perfect binary tree; that is $\mathbb{P}$ is a subtree of $\mathbb{B}$, so $\mathbb{P}$ also can be thought of as a notion of forcing. To define the forcing relation for $\mathbb{P}$, denoted by $\Vdash_{\mathbb{P}}$, restrict the range of conditions in Definition 6.2.1 to $\mathbb{P}$. For $n < \omega$, we say that the subtree $\mathbb{Q} \subseteq \mathbb{P}$ is an *n-deciding* subtree of $\mathbb{P}$ if $\mathbb{Q}$ is a definable perfect binary tree and for each Σ_n sentence σ of $\mathcal{L}^F(M)$ there is $c \in M$ such that whenever $c < p \in \mathbb{Q}$, then either $p \Vdash_{\mathbb{P}} \sigma$ or $p \Vdash_{\mathbb{P}} \neg\sigma$.

Lemma 6.5.2 *Let M be a model and $n < \omega$, and suppose that $\mathbb{P}$ is a definable perfect binary tree. Then there is an n-deciding subtree $\mathbb{Q} \subseteq \mathbb{P}$.*

Proof Let $\langle \sigma_i : i \in M \rangle$ be a definable list of all Σ_n sentences in the language $\mathcal{L}^F(M)$. By induction on $\mathbb{B}$ in M, we define a function $f : \mathbb{B} \longrightarrow \mathbb{P}$. Let $f(0)$ be the shortest p which splits in $\mathbb{P}$. If $e \in \{0, 1\}$, $s \in \mathbb{B}$, and $\ell(s) = i$, then let $f(s\hat{\ }e) = p\hat{\ }e$, where $p \trianglerighteq f(s)$ and p is the shortest such condition which splits and decides σ_i. There is such a definable f since forcing is definable (Proposition 6.2.3). Then $\mathbb{Q} = \{p \in \mathbb{P} : p \triangleleft f(s) \text{ for some } s \in \mathbb{B}\}$ is as required. □

Definition 6.5.3 In a model M, a branch G of $\mathbb{B}$ is a *perfect generic* if there is a sequence $\mathbb{B} = \mathbb{P}_0 \supseteq \mathbb{P}_1 \supseteq \mathbb{P}_2 \supseteq \cdots$ such that, for each $n < \omega$, $\mathbb{P}_{n+1}$ is an n-deciding subtree of $\mathbb{P}_n$ and $G \subseteq \mathbb{P}_n$.

Why do perfect generics exist? After all, there may be no branch in the intersection of the decreasing sequence of trees.

Exercise 6.5.4 In a model M, let $\mathbb{P}_0 \supseteq \mathbb{P}_1 \supseteq \mathbb{P}_2 \supseteq \cdots$ be a sequence of definable binary trees such that for each $a \in M$ there is $n < \omega$ such that no $p < a$ splits in $\mathbb{P}_n$. Then $B = \bigcap_{n<\omega} \mathbb{P}_n$ is a branch of $\mathbb{B}$.

Proposition 6.5.5 *Suppose G is a perfect generic in a model M. Then:*

(1) *if σ is an $\mathcal{L}^F(M)$-sentence and $(M, G) \models \sigma$, then there is a definable $D \subseteq M$ such that $(M, D) \models \sigma$;*

(2) $(M, G) \models \mathsf{PA}^*$;
(3) $G \notin \mathrm{Def}(M)$.

Proof Let $G \subseteq \bigcap_{n<\omega} \mathbb{P}_n$, where $\mathbb{B} = \mathbb{P}_0 \supseteq \mathbb{P}_1 \supseteq \mathbb{P}_2 \supseteq \cdots$ and each $\mathbb{P}_{n+1}$ is a definable n-deciding subtree of $\mathbb{P}_n$.

(1) Suppose σ is Σ_n. Then there is $p \in G$ such that $p \Vdash_{\mathbb{P}_n} \sigma$. By Corollary 6.2.9, there is a definable $D \subseteq \mathbb{P}_n$ such that $(M, D) \models \sigma$.
(2) Just like the proof of Theorem 6.2.10.
(3) Suppose that $G \in \mathrm{Def}(M)$ is definable by a formula $\varphi(x)$, and let σ be the sentence $\forall x[U(x) \longleftrightarrow \varphi(x)]$. Then there is $n < \omega$ such that $p \Vdash_{\mathbb{P}_n} \sigma$ for all $p \in \mathbb{P}_n$ (Do it!). Let $p \in \mathbb{P}_n \backslash G$. Then $p \Vdash_{\mathbb{P}_n} \sigma \wedge U(p)$, so by (1) there is a definable $D \subseteq \mathbb{P}_n$ such that $(M, D) \models \sigma \wedge U(p)$. But then $\varphi(x)$ also defines D in M, so $D \neq G$. However, $p \in D \backslash G$, which is a contradiction. □

The next theorem improves Simpson's Theorem 6.2.11.

Theorem 6.5.6 *Suppose M is a countable model and $R_0, R_1, R_2, \ldots$ are countably many subsets of M such that $M^* = (M, R_0, R_1, R_2, \ldots) \models \mathsf{PA}^*$. Then there is $G \subseteq M$ such that $(M, G) \models \mathsf{PA}^*$ and $R_n \in \mathrm{Def}(M, G)$ for each $n < \omega$.*

Proof The G that works will be a carefully constructed perfect generic of M^*. For this we need to define a decreasing sequence $\mathbb{B} = \mathbb{P}_0 \supseteq \mathbb{P}_1 \supseteq \mathbb{P}_2 \supseteq \cdots$ so that, for each $n < \omega$, $\mathbb{P}_{n+1}$ is an n-deciding subtree of P_n that is definable in M^*. At the same time we will define another decreasing sequence $\mathbb{Q}_0 \supseteq \mathbb{Q}_1 \supseteq \mathbb{Q}_2 \supseteq \cdots$ of perfect binary trees definable in M^*. These sequences of trees will be dovetailed; that is

$$\mathbb{B} = \mathbb{P}_0 \supseteq \mathbb{Q}_0 \supseteq \mathbb{P}_1 \supseteq \mathbb{Q}_1 \supseteq \mathbb{P}_2 \supseteq \cdots .$$

We need a cofinal sequence $a_0 < a_1 < a_2 < \cdots$ in M. Given $\mathbb{P}_n$, we refer to Lemma 6.5.2 to get

$$\mathbb{Q}_n \in \mathrm{Def}(M, \mathbb{P}_n) \tag{1}$$

that is an n-deciding subtree of $\mathbb{P}_n$. We want that $\mathbb{Q}_n$ has no elements $q < a_n$ that split. If this is not the case, then let $p \in \mathbb{Q}_n$ be such that if $p \trianglelefteq q \in \mathbb{Q}_n$, then $q \geq a_n$, and then replace $\mathbb{Q}_n$ with $\{q \in \mathbb{Q}_n : q \trianglelefteq p \text{ or } p \trianglelefteq q\}$. We then get $\mathbb{P}_{n+1}$ as a definable perfect binary subtree of $\mathbb{Q}_n$ as follows: if $q \in \mathbb{Q}_n$, then $q \in \mathbb{P}_{n+1}$ iff whenever $i \in M$ and $p \triangleleft q$ are such that p splits in $\mathbb{Q}_n$ and there are exactly

$2i$ conditions $r \lhd p$ that split in $\mathbb{Q}_n$, then $p\hat{\ }0 \unlhd q$ iff $i \in R_n$. Therefore,

$$\mathbb{P}_{n+1} \in \mathrm{Def}(M, \mathbb{Q}_n, R_n). \tag{2}$$

The point of this definition is that if G is any branch of $\mathbb{P}_{n+1}$, then

$$R_n \in \mathrm{Def}(M, \mathbb{Q}_n, G). \tag{3}$$

Since $\mathbb{Q}_n$ is an n-deciding subtree of $\mathbb{P}_n$, then so is $\mathbb{P}_{n+1}$.

By Exercise 6.5.4, $G = \bigcap_{n<\omega} \mathbb{P}_n$ a branch of $\mathbb{B}$. Clearly, it is perfect generic. It follows from (1), (2), (3) that each $R_n \in \mathrm{Def}(M, G)$. □

The following technical lemma, which is in the spirit of Lemma 6.5.2, will be needed in the theorem that comes right after it.

Lemma 6.5.7 *Let M be a model and $\varphi(x)$ an $\mathcal{L}^F(M)$-formula. Suppose that $\mathbb{P}$ is a definable perfect binary tree. Then there is a definable binary subtree $\mathbb{Q}$ such that either:*

(1) *there is $c \in M$ such that whenever $a \in M$, then either $p \Vdash_{\mathbb{P}} \varphi(a)$ whenever $c < p \in \mathbb{Q}$, or else $p \Vdash_{\mathbb{P}} \neg\varphi(a)$ whenever $c < p \in \mathbb{Q}$;*
(2) *for each $i \in M$ there are $j, c \in M$ such that $j > i$ and whenever $p, q \in \mathbb{Q}$ are such that $\ell(p) = \ell(q) = j$ and $p{\restriction}i \neq q{\restriction}i$, then there is $a \in M$ such that either $p \Vdash_{\mathbb{P}} \varphi(a)$ and $q \Vdash_{\mathbb{P}} \neg\varphi(a)$ or else $p \Vdash_{\mathbb{P}} \neg\varphi(a)$ and $q \Vdash_{\mathbb{P}} \varphi(a)$.*

Proof Suppose $\varphi(x)$ is a Σ_n formula. For the first move, let $\mathbb{P}_0$ be an n-deciding subtree of $\mathbb{P}$. For the second move, define $\mathbb{P}_1$ to be the perfect subtree of $\mathbb{P}$ such that letting $f : \mathbb{B} \longrightarrow \mathbb{P}_1$ be the unique definable order-preserving function onto the set of those points in $\mathbb{P}_1$ which split, whenever $s \in \mathbb{B}$ and $\ell(s) = i$, then $f(s)$ decides $\varphi(i)$. There are two possiblities:

(a) *There is $s \in \mathbb{B}$ such that whenever $s \lhd t_1, t_2$ and $\ell(t_1) = \ell(t_2)$, then $f(t_1) \Vdash_{\mathbb{P}} \varphi(i)$ iff $f(t_1) \Vdash_{\mathbb{P}} \varphi(i)$.* Then we let $\mathbb{Q} = \{p \in \mathbb{P}_1 : p \unrhd f(s)\}$, and (1) holds.
(b) *Possibility* (a) *fails.* By induction in M, we define a function $g : \mathbb{B} \longrightarrow \mathbb{P}_1$. Let $g(0) = f(0)$. If $s \in \mathbb{B}$ let $g(s\hat{\ }0) = p_0$ and $g(s\hat{\ }1) = p_1$, where p_0, p_1 are the first incompatible pair in $\mathbb{B}_1$ such that for some $i \in M$, $g(s) \lhd p_0, p_1$, $\ell(p_0) = \ell(p_1) = i$, $p_0 \Vdash_{\mathbb{P}} \varphi(i)$ and $p_1 \Vdash_{\mathbb{P}} \neg\varphi(i)$. Then we let $\mathbb{Q} = \{p \in \mathbb{P} : p \lhd g(s)$ for some $s \in \mathbb{B}\}$, and (2) holds. □

Cohen generics are inductive nondefinable classes. Corollary 6.3.4 says that Cohen generics not "minimally nondefinable." The next theorem shows that perfect generics can be different in that they may be "minimally nondefinable."

Theorem 6.5.8 *Let M be countable. Then there is an inductive $A \subseteq M$ such that $A \notin \mathrm{Def}(M)$ and whenever $B \in \mathrm{Def}(M, A)$, then either $B \in \mathrm{Def}(M)$ or $A \in \mathrm{Def}(M, B)$.*

Proof The set A will be a perfect generic with an additional feature that Lemma 6.5.7 provides. Let $a_0 < a_1 < a_2 < \cdots$ be a cofinal sequence in M. Let $\varphi_0(x), \varphi_1(x), \varphi_2(x), \ldots$ enumerate all the $\mathcal{L}^F(M)$-formulas with one free variable. Construct a sequence

$$\mathbb{B} = \mathbb{P}_0 \supseteq \mathbb{Q}_0 \supseteq \mathbb{P}_1 \supseteq \mathbb{Q}_1 \supseteq \mathbb{P}_2 \supseteq \cdots$$

of definable perfect binary trees such that each $\mathbb{Q}_n$ is an n-deciding subtree of $\mathbb{P}_n$, and each $\mathbb{P}_{n+1}$ is a subtree of $\mathbb{Q}_n$ as described in Lemma 6.5.7 using the formula $\varphi_n(x)$. In order to guarantee that the sequence $\mathbb{B} = \mathbb{P}_0 \supseteq \mathbb{P}_1 \supseteq \mathbb{P}_2 \supseteq \cdots$ produces a perfect generic (as in Definition 6.5.3), we will require that $\mathbb{Q}_n$ has no elements $q < a_n$ that split. This is easily accomplished (DO IT!) as in the proof of Theorem 6.5.6. Then $A = \bigcap_{n<\omega} \mathbb{P}_n$ is a perfect generic according to Exercise 6.5.4, and then by Proposition 6.5.5(2), A is inductive.

We show that A has the required property. Consider $B \in \mathrm{Def}(M, A)$. There is a formula $\varphi_n(x)$ which defines B. The subtree $\mathbb{P}_{n+1} \subseteq \mathbb{Q}_n$ is obtained using Lemma 6.5.7, so one of the two possibilities from that lemma hold.

Suppose (1) holds. Let $p \in \mathbb{P}_{n+1}$ be large enough to decide all instances of $\varphi_n(a)$. Thus $B = \{a \in M : p \Vdash_{\mathbb{Q}_n} \varphi_n(a)\}$, so it is definable by Proposition 6.2.3.

Suppose (2) holds. In this case $A \in \mathrm{Def}(M, B)$. In fact,

$$\forall c \exists q \rhd p \forall b \Big[q \in \mathbb{P}_{n+1} \wedge q > c \wedge \big[\big(q \Vdash_{\mathbb{Q}_n} \varphi_n(b) \big) \longrightarrow b \in B \big] \\ \wedge \big[\big(q \Vdash_{\mathbb{Q}_n} \neg\varphi_n(b) \big) \longrightarrow b \notin B \big] \Big]$$

holds iff $p \in A$. □

Corollary 6.5.9 *Every countable model has an inductive undefinable subset which is not generic.*

Proof See Exercise 6.5.1. □

6.6 Exercises

♣**6.6.1** Every notion of forcing $\mathbb{P}$ in M has a maximal filter $F \in \Sigma_1((M, \mathbb{P}))$.

♣**6.6.2** Every binary tree $\mathbb{P}$ in M has a branch $B \in \Delta_1((M, \mathbb{P}))$.

♣**6.6.3** (**Weak König's Lemma**) Every unbounded binary tree $\mathbb{P}$ in M has an unbounded branch $B \in \Delta_2((M, \mathbb{P}))$.

♦**6.6.4** Can Corollary 6.3.4 be modified by replacing each "Def" with "Def$_0$," to be interpreted as definability without parameters?

♦**6.6.5** If M is countable, then there is an (undefinable!) perfect subtree $\mathbb{P} \subseteq \mathbb{B}$ such that every branch of $\mathbb{P}$ is a perfect generic.

♦**6.6.6** Let G be a Cohen generic in M. There is a function $f \in \mathrm{Def}(M,G)$, such that for every function $g \in \mathrm{Def}(M)$, $(M,G) \models \exists w \forall x > w[f(x) > g(x)]$.

♦**6.6.7** Let G be a perfect generic in M. For every function $f \in \mathrm{Def}(M,G)$ there is a function $g \in \mathrm{Def}(M)$, such that $(M,G) \models \exists w \forall x > w[f(x) < g(x)]$.

♥**6.6.8** Every countable model M has a strongly minimal extension N which has a Cohen generic G such that (N,G) is not a prime model.

♠**6.6.9** Characterize those countable models M that have a generic X such that any element definable in (M,X) is already definable in M.

6.7 Remarks & References

Forcing was invented by Paul Cohen in order to prove independence results in set theory. The definition of generic that is given in Section 6.1 is attributed to Solovay. Subsection 6.2.2 was influenced by Odifreddi [140]. Theorem 6.2.11 is due to Steve Simpson [186] who expanded on an idea of Felgner [40]. The Low Basis Theorem was first proved, in its usual recursion-theoretic form, by Jockusch & Soare [63]. The formalized version was proved implicitly by Paris [147] whose use was adapted by Kossak [99] and others. Product forcing was in Cohen's original exposition of forcing. The set-theoretic version of the results in Section 6.3 are quite standard. Theorem 6.4.1 was proved by Mills [131] and answered in the negative the strongest (and therby all) of the five question asked by Gaifman [43]. Theorem 6.4.3 was first correctly proved by Schmerl [163] following an earlier incorrect attempt that did not invoke the Pincus–Laver Theorem.

The PL Theorem has a somewhat mysterious history. This theorem is a slight improvement of a theorem first proved by Halpern & Läuchli in [52]. Then, according to what is written in [132], Laver proved the PL Theorem in 1969 but did not publish the results, and then indepndently and later in 1974, Pincus proved it. The proof that our proof is based on is from Milliken [132], who adapted the proof of the HL Theorem in [52]. The proof of the HL Theorem is ingenious, but idiosyncratic. There have been attempts to give a more conventional proof, one of the most recent being by [2].

The results in Section 6.4, and others along these lines, have appeared in Schmerl [166] and Keisler & Schmerl [79]. These results have applications to the construction of some real-closed fields. In the first paragraph of Subsection 6.4.3 reference is made to two two-cardinal theorems, the first due to C.C. Chang [20] and the second to Saharon Shelah [183].

Perfect generics are an adaptation of Sacks reals to models of Peano Arithmetic. Theorem 6.5.6 appears here for the first time, although one of the authors discussed the possibility of such a theorem with Simpson around 1978 give or take a half decade.

7

CUTS

Strong cuts play an important role in the model theory of arithmetic. They were introduced by Kirby and Paris in their work that lead to the discovery of combinatorial principles independent of PA. For us, the important connection to the material of the subsequent chapters is that a recursively saturated model is arithmetically saturated iff the standard cut is strong in it.

The main result of this chapter is Theorem 7.3.4 which explains the importance of strongness. To prove it, we need to discuss other combinatorial properties of cuts first. This preliminary discussion takes us a bit longer than necessary because of a detour to see how bad nonconservative extensions can be. If N is a nonconservative elementary end extension of M, then N codes some undefinable class X of M. Can this class be noninductive? Can we get an X as above for which (M, X) is not a model of $I\Sigma_1$? It turns out that the answer depends on the model M.

We conclude the chapter with a section titled *Why* PA*?*. It contains two interesting theorems illustrating the special status that PA has among extensions of $I\Sigma_0$. These results are of independent interest. We include them in this chapter because of the techniques used in their proofs.

Let I be a cut of a model M. The combinatorial properties of I that we consider here are definable in terms of properties of the subsets of I which are coded in M. All these properties are first-order definable in (M, I), but it will also be convenient to formalize them as properties of the second-order structure $(I, \text{Cod}(M/I))$, where $\text{Cod}(M/I)$ is the collection of subsets of I which are coded in M. In particular, we show that I is strong in M iff $(I, \text{Cod}(M/I)) \models \mathsf{ACA}_0$.

The other two properties of cuts that we discuss here are semiregularity and regularity. These notions can be characterized in terms of the induction and collection schemes. For this discussion, we will work with the first-order structures $(I, A_0, A_1, \dots)$, where $A_0, A_1, \dots$ is a list of all the sets in $\text{Cod}(M/I)$. Let Γ be either the collection of Σ_n or Π_n formulas of $\mathcal{L} \cup \{A_0, A_1, \dots\}$, for some $n < \omega$. Since it is usually clear what M is, to simplify notation, we just write $I^* \models I\Gamma$ if $(I, A_0, A_1, \dots, A_{k-1}) \models I\Gamma$, for all $A_0, A_1, \dots, A_{k-1} \in \text{Cod}(M/I)$ and similarly for $B\Gamma$. Then for all $I \subseteq_{\mathsf{end}} M$ and all $n < \omega$,

$$I^* \models I\Sigma_0 + B\Sigma_{n+1} \Rightarrow I^* \models I\Sigma_n \Rightarrow I^* \models B\Sigma_n,$$

and all other well-known results concerning induction, collection, and the least number scheme for the arithmetical hierarchy hold. We will show that if $I \subseteq_{\mathsf{end}} M$ is a cut of a model M, then I is semiregular iff $I^* \models I\Sigma_1$, and I is regular iff $I^* \models B\Sigma_2$.

Let us note that all coding machinery we use can be formalized in $I\Sigma_1$, and, with some extra effort, in $I\Sigma_0 + \mathsf{exp}$. In particular, we will use the fact that if $I \subseteq_{\mathsf{end}} M$ is an exponentially closed cut in M and $A \subseteq I$ is M-finite, then A has a code in I.

All Σ_0 formulas are absolute with respect to cuts which are closed under multiplication. The next proposition generalizes this.

Proposition 7.0.1 *Let $I \subseteq_{\mathsf{end}} M$ be a cut closed under multiplication. Then for every $B \in \mathrm{Cod}(M/I)$, there is $b \in M$ such that for every Σ_0 formula $\varphi(x, Y)$ with a set variable Y, there is an arithmetic formula $\varphi'(x, y)$ such that for all $a \in I$,*

$$(I, B) \models \varphi(a, B) \quad \text{iff } M \models \varphi'(a, b).$$

Proof Let $b \in M$ be a code of a bounded definable $X \subseteq M$ such that $B = X \cap I$. Obtain φ' by replacing each occurrence of a subformula of the form $v \in B$, by $v \in b$. □

7.1 Semiregular cuts

The notion of semiregularity, like many other combinatorial properties, is an analogue of the corresponding property of cardinal numbers. This particular analogy is not exact, as semiregularity in arithmetic corresponds to regularity in set theory. Regularity is reserved for a stricter property which is discussed in the next section.

Definition 7.1.1 Let $I \subseteq_{\mathsf{end}} M$ be a cut of a model M. Then I is semiregular if for every function $f \in M$ whose domain is bounded in I, $\mathrm{rg}(f) \cap I$ is bounded in I.

Semiregularity can be also defined via the notion of cofinality. For $I \subseteq_{\mathsf{end}} M$, the *cofinality* of I in M, $\mathrm{cf}^M(I)$, is the intersection of those $J \subseteq_{\mathsf{end}} M$ for which there is a function $f \in M$ such that $J \subseteq \mathrm{dom}(f)$ and $\sup(f(J)) = I$.

Exercise 7.1.2 Let I be a cut of M. Then I is semiregular in M iff $\mathrm{cf}^M(I) = I$.

Clearly, $\mathrm{cf}^M(\mathbb{N}) = \mathbb{N}$ in any nonstandard model M. If $a \in M$ codes an infinite increasing sequence and $I = \sup_{n \in \mathbb{N}}(a)_n$, then $\mathrm{cf}^M(I) = \mathbb{N}$ (DO IT!). More generally, if J is a cut of M, $a \in M$ codes an increasing sequence, $\ell(a) > J$, and $I = \sup_{i \in J}(a)_i$, then $\mathrm{cf}^M(I) = \mathrm{cf}^M(J)$ (DO IT!). If I is downward ω coded, that is, there is $a \in M$ which codes an infinite decreasing sequence, and $I = \inf_{n\in\mathbb{N}}(a)_n$, then $\mathrm{cf}^M(I) = \mathbb{N}$. However, for each $b \in M$, coding an increasing ω-sequence, $I \neq \sup_{n\in\mathbb{N}}(b)_n$ (JUST DO IT!).

Let $I \subseteq_{\mathsf{end}} M$ be semiregular. Our first goal is to determine how much induction or collection holds in $(I, A_0, \ldots, A_{k-1})$ for all $A_0, \ldots, A_{k-1}$ coded in M. Let us begin with a simple observation, the proof of which is left to the reader. (HINT: use Proposition 7.0.1.)

Proposition 7.1.3 *Let I be a cut of a model M. Then $I^* \models I\Sigma_0 + B\Sigma_1$.* □

For semiregular cuts one can prove more.

Proposition 7.1.4 *Let I be a semiregular cut of a model M. Then I is closed under addition and multiplication.* □

Proof For $a \in I$, let $f : (a+1)_M \longrightarrow M$ be defined by $f(i) = a + i$. We will show that $\mathrm{rg}(f) \subseteq I$, so, in particular, $f(a) \in I$. If not, then, by semiregularity, $\mathrm{rg}(f) \cap I$ is bounded in I. Hence, there is $c < a$ such that $f(c) = \max(\mathrm{rg}(f) \cap I)$. Then $f(c+1) = a + c + 1 \notin I$, which is a contradiction, since I, being a cut, is closed under successor.

Now, to prove that I is closed under multiplication, take $a \in I$ and prove in a similar manner that the range of $g : (a+1)_M \longrightarrow M$, defined by $g(i) = ai$, is bounded in I. □

We could continue strengthening the above proposition by showing that semiregular cuts are closed under exponentiation, superexponentiation, etc.. Instead, we now prove a result which implies that semiregular cuts are closed under all primitive recursive functions.

Theorem 7.1.5 *Let I be a cut of a model M. Then I is semiregular in M iff $I^* \models I\Sigma_1$.*

Proof First suppose I is semiregular. Let $\varphi(x, y, Y)$ be a Σ_0 formula, let $B \in \mathrm{Cod}(M/I)$, and let $\varphi'(x, y, b)$ be the first-order translation of $\varphi(x, y, B)$ given by Proposition 7.0.1.

Suppose that

$$(I, B) \models \exists y\varphi(0, y, B) \wedge \forall x[\exists y\varphi(x, y, B) \longrightarrow \exists y\varphi(x+1, y, B)]. \qquad (*)$$

We will show that

$$(I, B) \models \forall x\exists y\varphi(x, y, B).$$

Consider the function

$$f(x) = \begin{cases} \min\{z : M \models \forall i < x\exists y < z\varphi'(i, y, b)\} & \text{if such } z \text{ exists,} \\ b & \text{otherwise.} \end{cases}$$

The conclusion will follow if we prove that for each $i \in I$, $f(i) \in I$. Suppose not, and let $c \in I$ be such that $f(c) > I$. By semiregularity, there is $d \in I$ such that $\mathrm{rg}(f \restriction c_M) \cap I \subseteq d_M$. Let $c' = \min\{x : x \leq c \text{ and } f(x) \geq d\}$. By $(*)$, $c' > 0$ and $f(c'-1) < d$. Then $f(c') = \max\{f(c'-1), e\}$, where $e = \min\{y : M \models \varphi'(c', y, b)\}$. By $(*)$, $e \in I$, so $f(c') \in I$, a contradiction.

For the converse, assume that $I^* \models I\Sigma_1$ and a function $f \in M$ be given. Let $F = f \cap I^2$. We define another function g using $I\Sigma_1$ in (I, F):

$$g(0) = \min\{y : \exists x < a(\langle x, y\rangle \in F)\},$$

and for $i \in I$,

$$g(i) = \begin{cases} \min\{y : \exists x < a(\langle x, y\rangle \in F) \wedge \forall j < i[y \neq g(j)]\} & \text{if such } y \text{ exists,} \\ g(i) & \text{otherwise.} \end{cases}$$

By Σ_1 induction, for each $i < a$, $g(i)$ is defined, and $g(i) \leq g(a-1)$. Hence $\mathrm{rg}(f \restriction a_M) \cap I$ is bounded, proving that I is semiregular in M. □

By Theorem 7.1.5, semiregular cuts are models of $I\Sigma_1$. The converse is not true. For example, every nonstandard model M has cuts I which are models of PA such that $\mathrm{cf}^M(I) = \mathbb{N}$. (See Exercise 7.5.2). However, not all countable models of PA have end extensions in which they are semiregular. If a countable I is not semiregular in a model M, then I has continuum many automorphisms. (See Exercise 7.5.3.) In particular, all rigid models are semiregular in all of their end extensions.

7.1.1 *Semiregularity and* WKL_0

If I is an $\mathcal{L}_{\mathsf{PA}}$ structure and $\mathfrak{X}$ is a collection of subsets of I, then $(I, \mathfrak{X})$ is a model of RCA_0 if;

(1) $(I, A_0, \ldots, A_{n-1}) \models I\Sigma_1$ for all $A_0, \ldots, A_{n-1} \in \mathfrak{X}$;
(2) $\mathfrak{X}$ is closed under Δ_1 definability.

Let I be a model of $I\Sigma_1$ and let $2^{<I}$ be the set of 0–1 sequences coded in I. Coding here is the same coding we use for models of PA. All its properties we need are formalizable and provable in $I\Sigma_1$. As usual, we identify coded objects with their codes. We call $T \subseteq 2^{<I}$ a *binary tree* if for every $\sigma \in T$ and every $i < \ell(\sigma)$, the restriction of σ to i, $\sigma \restriction i = \langle(\sigma)_0, \ldots, (\sigma)_{i-1}\rangle$ is in T. A tree T is of *unbounded* if for each $i \in I$ there is a $\sigma \in T$ such that $\ell(\sigma) = i$. An $f : I \longrightarrow \{0, 1\}$ is an *unbounded path* in T if for each $i \in I$, $f \restriction i \in T$.

> WEAK KÖNIG'S LEMMA: Every unbounded binary tree has an unbounded path.

Let $(I, \mathfrak{X})$ be a model of RCA_0. Then $(I, \mathfrak{X})$ is a model of WKL_0 if it satisfies Weak König's Lemma, that is every unbounded binary tree $T \in \mathfrak{X}$ has an unbounded path in $\mathfrak{X}$.

If I is a cut of a model M, then, even if I is not a model of $I\Sigma_1$, we can still consider $2^{<I}$ defined as the set of M-finite 0–1 sequences σ such that $\mathrm{dom}(\sigma)$ is a bounded subset of I. With this adjustment, we can show the following:

Proposition 7.1.6 *Let I be a cut of a model M. Then $(I, \mathrm{Cod}(M/I))$ satisfies Weak König's Lemma.*

Proof Let T be a binary tree in $\mathrm{Cod}(M/I)$ such that for each $i \in I$, there is $\sigma \in T$ such that $\ell(\sigma) = i$. Let $t \in M$ be a code of T. For each $i \in I$,

$$M \models \exists \sigma \in t\ [\ell(\sigma) = i \wedge \forall j < i(\sigma \restriction j \in t)]. \qquad (*)$$

By overspill, we get a $\sigma \in t$ satisfying $(*)$ for some $i > I$. This σ codes an unbounded path of T. □

If a cut $I \subseteq_{\mathsf{end}} M$ is not semiregular, then $I^* \not\models I\Sigma_1$, hence $(I, \mathrm{Cod}(M/I))$ is not a model of WKL_0. In fact, we can now rephrase Theorem 7.1.5 as follows:

Theorem 7.1.7 *Let I be a cut of a model M. Then I is semiregular in M iff $(I, \mathrm{Cod}(M/I)) \models \mathsf{WKL}_0$.* □

7.2 Regular cuts

Let M be a countable model, and let $I \subseteq_{\mathsf{end}} M$ be a cut of M. Is there a model K such that $M \prec K$, $I = \mathrm{GCIS}(M, K)$, and $I < c < (M \setminus I)$, for some $c \in K$? It is not difficult to see that this is impossible if I is not semiregular in M (Do it!). To prove a positive result, we need to replace semiregularity with a stricter property.

Definition 7.2.1 A cut $I \subseteq_{\mathsf{end}} M$ is *regular* in M if for every function $f \in M$, if $I \subseteq \mathrm{dom}(f)$ and $\mathrm{rg}(f \restriction I)$ is bounded in I, then there is $i \in I$ such that $f^{-1}(i) \cap I$ is unbounded in I.

The standard cut is regular in every nonstandard model. It is also not difficult to show that regular cuts are semiregular (Do it!). The next theorem shows that the converse does not hold. The proof of the next theorem is rather involved. An easier construction can be given to show that there are semiregular cuts which are not regular (see Propositional 2 in [83]). The extra level of difficulty in the theorem below is due to the fact that we are dealing with elementary cuts.

Theorem 7.2.2 *Every countable model M has an elementary end extension N such that M is semiregular but not regular in N.*

Proof Let M be a countable model. First we will construct an elementary end extension of M in which M is not regular, and then we will refine the construction to make sure that M is semiregular in the extension. Symbols $f, g, h, \dots$ will denote functions coded in M. We identify coded sets, sequences, and functions with their codes. For $f \in M$ let

$$\langle f \rangle = \{g \in M : f \subseteq g\}.$$

We will now define a decreasing sequence of definable subsets of M with the following largeness property: for a $\beta \in M$ and an M-finite set X, we say that the set Z (of coded functions) is (β, X)-*large*, if every f such that $\text{dom}(f) > \beta - 1$ and $\text{rg}(f) \subseteq X$, has an extension in Z.

Claim: Suppose that Z is (β, X)-large. If $Z = \bigcup_{j<b} Z_j$ is a definable partition of Z for some $b \in M$, then there are $j < b$, $\delta > \beta$, and $f : [\beta, \delta) \longrightarrow X$ such that $Z_j \cap \langle f \rangle$ is (δ, X)-large.

Proof of the claim: Assume to the contrary that there are no j, δ and f as in the claim. We will obtain a contradiction by defining a function $g \in M$ such that $\text{dom}(g) > \beta - 1$, $\text{rg}(g) \subseteq X$, and g has no extension in Z. The function g will be defined as $\bigcup_{i<b} g_i$, where g_i's are constructed inductively as follows. First notice that Z_0 is not (β, X)-large, because if it were, then this would prove the claim with $j = 0$, any $\delta > \beta$, and the empty f. Hence there is g_0 such that $\text{dom}(g_0) > \beta - 1$, $\text{rg}(g_0) \subseteq X$, and g_0 has no extension in Z_0. Let $\beta_0 = \beta$. We can assume that $\text{dom}(g_0) = [\beta_0, \beta_1)$ for some $\beta_1 > \beta_0$. Similarly, since $Z_1 \cap \langle g_0 \rangle$ is not (β_1, X)-large, there is g_1 such that $\text{dom}(g_1) = [\beta_1, \beta_2)$ for some $\beta_2 > \beta_1$, $\text{rg}(g_1) \subseteq X$, and g_1 has no extension in $Z_1 \cap \langle g_0 \rangle$. Continuing in this fashion, we define an increasing sequence $\langle \beta_i : i < b \rangle$ and a sequence of functions $g_i : [\beta_i, \beta_{i+1}) \longrightarrow X$, $i < b$ such that for each i, g_i has no extension in $Z_i \cap \langle g_{i-1} \rangle$. But then, $g = \bigcup_{i<b} g_i$ has no extension in Z. Contradiction.

Continuing the proof of the theorem, let $a \in M$ be nonstandard and let $\langle a_i : i < \omega \rangle$ be an enumeration of $[0, a]$. Let $X_0 = [0, a]$, and let $X_{n+1} = X_n \setminus \{a_n\}$, and let $\langle t_n : n < \omega \rangle$ be an enumeration of all Skolem terms with parameters from M.

We will construct a descending sequence $\langle Z_n : n < \omega \rangle$ of definable subsets of M such that $Z_0 = M$ and

(1) Z_n is (β_n, X_n)-large for some β_n in M;
(2) There exists $g_n \in M$ such that $\text{dom}(g_n) \subseteq [0, \beta_n - 1]$ and $Z_n \subseteq \langle g_n \rangle$;
(3) For $n > 0$, if for every $z \in Z_{n-1}$, $t_n(z) < b$ for some $b \in M$, then for some $c < b$, $Z_n \subseteq \{z \in M : t_n(z) = c\}$.

Suppose we have found Z_n with the above properties and for every $z \in Z_n$, $t_{n+1}(z) < b$. Then $Z_n \subseteq \bigcup \{t_{n+1}^{-1}(c) : c < b\}$. By (1), Z_n is (β_n, X_n)-large, hence it is (β_n, X_{n+1})-large as well. By the claim, there are $\delta > \beta_n$, $X \subseteq Z_n$, and

$f : [\beta, \delta) \longrightarrow X_{n+1}$ such that for some $c < b$, $X \subseteq t_{n+1}^{-1}(c)$ and $X \cap \langle f \rangle$ is (δ, X_{n+1})-large. Then we set $Z_{n+1} = X \cap \langle f \rangle$, $\beta_{n+1} = \delta$, and $g_{n+1} = g_n \cup f$.

Now let $p(x)$ be the type $\{\varphi(x) : \exists n < \omega[Z_n \subseteq \varphi(M)]\}$, let N be the $p(x)$-extension of N, and let $\pi \in N$ be such that $N = M(\pi)$ and π realizes $p(x)$. Then $M \prec_{\mathsf{end}} N$ and π codes a function $\bigcup_{n<\omega} g_n : M \longrightarrow [0, a]$. By the construction, each $e \leq a$ is in only finitely many Z_n's, hence, for each $e \leq a$, $\pi^{-1}(e)$ is bounded in M. Because each Z_n is unbounded, $\pi^{-1}([0, a])$ is unbounded in M, and it follows that M is not regular in M.

Now we will modify the construction to make sure that M is semiregular in N. Enumerate all definable maps $x \mapsto h_x$ such that for all x, h_x codes an increasing function $h_x : [0, e] \longrightarrow M$ for some $e \in M$. In the nth step we first define Z_{2n} from Z_{2n-1} as before.

Suppose Z_{2n} satisfies (1), (2), and (3) above. Take the nth map $x \mapsto h_x$ in the enumeration and assume that for all x, $\mathrm{dom}(h_x) = [0, e]$. Let $Z = Z_{2n}, X = X_{2n}$, $\beta = \beta_{2n}$, and let

$$R = \{f \in M : \mathrm{dom}(f) > \beta - 1 \wedge \mathrm{rg}(f) \subseteq X\}.$$

Suppose that there is $i \leq e$ such that

$$\forall f \in R \exists g \in Z \cap \langle f \rangle \, [h_g(i) > f].$$

Then let i_0 be the least such i (notice that (2) implies that $i_0 \geq \beta$). If there is no such i, let $i_0 = e + 1$. Then there is $f_0 \in M$, $f_0 : [\beta, \beta') \longrightarrow X$ such that

$$\forall g[g \in Z \cap \langle f_0 \rangle \longrightarrow h_g(i_0 - 1) \leq f_0].$$

For each $f \in R \cap \langle f_0 \rangle$, let g_f be the least $g \in Z \cap \langle f \rangle$ such that $h_{g_f}(i_0) > f$ and let $Z_{2n+1} = \{g_f : f \in R \cap \langle f_0 \rangle\}$. Then Z_{2n+1} satisfies (1) and (2).

As before, let $p(x)$ be the complete type determined by $\langle Z_n : n < \omega \rangle$ and let N be the $p(x)$-extension of M. Then for any $\hat{h} \in N$, which corresponds to some definable $x \mapsto h_x$, there are $i_0, f_0 \in M$ such that $\hat{h}(i_0 - 1) \leq f_0$, while $\hat{h}(i_0) > M$. Hence M is semiregular in N. □

Theorem 7.2.2 generalizes to higher levels of the arithmetic hierarchy but not in a straightforward way and not (so far) in full generality. We say more about it in Remarks and references at the end this chapter.

The following proposition can be proved directly form the definition, using the fact that regular cuts are semiregular (Do IT!).

Proposition 7.2.3 *Let $I \subseteq_{\mathsf{end}} M$ be a regular cut. Then for every unbounded $X \in \mathrm{Cod}(M/I)$ and every function $f \in M$ such that $X \subseteq \mathrm{dom}(f)$ and $\mathrm{rg}(f \restriction I)$ is bounded in I, there is $i \in I$ such that $f^{-1}(i) \cap X$ is unbounded in I.* □

For $I \subseteq K$ and $M \prec K$, let $K_{M\setminus I} = \inf_K(M \setminus I)$. We write $M \prec_I K$ if I is a cut both in M and in K and there is $c \in K$ such that $I < c < K_{M\setminus I}$.

Theorem 7.2.4 *Let M be a countable model and let I be a cut of M. Then the following are equivalent:*

(1) *I is regular in M;*
(2) *There is a K such that $M \prec_I K$;*
(3) *$I^* \models B\Sigma_2$.*

Proof To prove (1) $\Longrightarrow$ (2) we will define a descending sequence $\langle X_n : n < \omega\rangle$ of sets in $\mathrm{Cod}(M/I)$ such that:

(i) each X_n is unbounded in I;
(ii) if $f \in M$ is a function, $I \subseteq \mathrm{dom}(f)$, and $\mathrm{rg}(f)$ is bounded in I, then there are $i \in I$ and $n < \omega$ such that $X_n \subseteq f^{-1}(i)$.

Let $X_0 = I$, and suppose X_n has been defined. Let f be the nth function in a fixed enumeration of all coded functions whose domains include I. If $\mathrm{rg}(f \restriction (I \cap X_n))$ is bounded in I, then, by Proposition 7.2.3, there is $i \in I$ such that $f_n^{-1}(i) \cap X_n$ is unbounded in I. Let $X_{n+1} = f^{-1}(i) \cap X_n$ for this i.

Let U be a filter in $\mathrm{Cod}(M/I)$ generated by $\langle X_n : n < \omega\rangle$. For every $X \in \mathrm{Cod}(M/I)$, exactly one of X and $I \setminus X$ is in U, hence U is an ultrafilter.

Let K be the model whose universe is the set of equivalence classes of the coded functions $f : I \longrightarrow M$ under the equivalence relation

$$f \sim g \Longleftrightarrow \{i \in I : f(i) = g(i)\} \in U.$$

Then, after identifying each $a \in M$ with the equivalence class of the constant function $f(x) = a$, one can verify that Łoś's Theorem holds (DO IT!); hence $M \prec K$. It is also easy to verify that $M \prec_I K$ (DO IT!).

To prove of (2) $\Longrightarrow$ (3), suppose that for $a \in I$ and $B \in \mathrm{Cod}(M/I)$,

$$(I, B) \models \forall x < a \exists y \forall z\ \varphi(x, y, z, B),$$

for some Σ_0 formula $\varphi(x, y, z, X)$. Let $\varphi'(x, y, z, b)$ be the first-order translation of $\varphi(x, y, z, B)$ given by Proposition 7.0.1. For each $x_0 < a$ there is $y_0 \in I$ such that

$$\forall z \in I\ M \models \varphi'(x_0, y_0, z, b).$$

Fix such x_0 and y_0. By overspill, there is $z_0 > I$ such that

$$M \models \forall z < z_0 \varphi'(x_0, y_0, z, b).$$

Since $M \prec K$, the same is true in K. Let $c \in K$ be such that $I < c < (M \setminus I)$. We have proved that

$$\forall x < a \exists y \in I\ K \models \forall z < c\ \varphi'(x, y, z, b). \tag{$*$}$$

Let $e \in K$ be the smallest such that

$$K \models \forall x < a \exists y < e \forall z < c\ \varphi'(x, y, z, b). \tag{$**$}$$

Clearly $e \in I$, since otherwise the failure of $(**)$ for $e - 1$ would contradict $(*)$. This proves that

$$(I, B) \models \forall x < a \exists y < e \forall z\ \varphi(x, y, z, B).$$

It remains to prove (3) $\Longrightarrow$ (1). Suppose that $f \in M$ is such that $I \subseteq \text{dom}(f)$, $\text{rg}(f \restriction I)$ is bounded in I, and $f^{-1}(i) \cap I$ is bounded in I, for all $i \in I$. Let $F = f \cap I^2$ and let $a \in I$ be such that $\text{rg}(f \restriction I) \subseteq a_M$. Then, since, for each $i \in I$, $f^{-1}(i)$ is bounded in I,

$$(I, F) \models \forall x < a \exists y \forall z [z > y \longrightarrow \langle z, x \rangle \notin F].$$

By $B\Sigma_2$, there is $b \in I$ such that

$$(I, F) \models \forall x < a \exists y < b \forall z [z > y \longrightarrow \langle z, x \rangle \notin F].$$

Hence, for each $i \in I$, if $i \geq b$, then $f(i) \geq a$, which is a contradiction. □

Ramsey Theorem plays an important role in the theory of strong cuts. Let M be a model, and let $I \subseteq_{\text{end}} M$ be a cut of M. For $n < \omega$, $[I]^n$ is the set on increasing n-tuples $(x_0, \ldots, x_{n-1})$ of elements of I. For $n < \omega$ and $m \in I$, let RT^n_m be the sentence of the second-order arithmetic expressing that for every partition $f : [I]^n \longrightarrow [0, m-1]$, there are an $i < m$ and an unbounded X such that $f(a_0, \ldots, a_{n-1}) = i$, for all $(a_0, \ldots, a_{n-1}) \in [X]^n$. By RT^n_∞ we denote the statement $\forall x \mathsf{RT}^n_x$.

Proposition 7.2.5 *Let I be a semiregular cut of a model M, and assume that $(I, \text{Cod}(M/I)) \models \mathsf{RT}^2_2$. Then I is regular in M.*

Proof Suppose $f : I \longrightarrow M$ is coded in M and such that $\text{rg}(f \restriction I)$ is bounded in I. Define a partition of $g : [I]^2 \longrightarrow \{0, 1\}$ by

$$g(b, c) = \begin{cases} 0 & \text{if f(b)=f(c)}, \\ 1 & \text{otherwise}. \end{cases}$$

Suppose $C \subseteq I$ is such that $\forall x, y \in C$ if $x < y$, then $g(x,y) = 1$. In this case f is one-to-one on C; hence C is bounded in I. By RT^2_2, there is $C \in \mathrm{Cod}(M/I)$, which is unbounded in I and such that for all x, y in C, if $x < y$, then $f(x) = f(y)$, which finishes the proof. □

By the above proposition and Theorems 7.1.7 and 7.2.2 we have

Corollary 7.2.6 (1) $\mathsf{WKL}_0 \nvdash \mathsf{RT}^1_\infty$.
(2) $\mathsf{WKL}_0 \nvdash \mathsf{RT}^2_2$. □

7.3 Many faces of strongness

The definition below is one of many equivalent ways in which strongness can be expressed. The main result of this section, Theorem 7.3.4, is a long list of other characterizations.

Definition 7.3.1 A cut $I \subseteq_{\mathsf{end}} M$ is *strong* in M if for each $c \in M$ there exists $d \in M$ such that $d > I$ and for all $i \in I$, $(c)_i > I$ iff $(c)_i > d$.

We begin with a short summary of some results concerning Ramsey's Theorem and König's Lemma.

In the subsection on WKL_0 we discussed binary trees. All conventions introduced there apply also to trees. Let I be a model of $I\Sigma_1$ and let $I^{<I}$ be the set of sequences of elements of I coded in I. $T \subseteq I^{<I}$ is a *tree* if for every $\sigma \in T$, and every $i < \ell(\sigma)$, $\sigma \restriction i = \langle (\sigma)_0, \ldots, (\sigma)_{i-1} \rangle \in T$. A tree T is *unbounded* if for every $i \in I$ there is a $\sigma \in T$ such that $\ell(\sigma) = i$. A tree T is *finitely branching*, if for each $\sigma \in T$ $\mathrm{card}\,\{i \in I : \sigma\hat{\ }i \in T\} \in I$.

> KÖNIG'S LEMMA, KL: Every unbounded finitely branching tree has an unbounded path.

In Chapter 2 we used the MacDowell-Specker Theorem to show that $\mathsf{ACA}_0 \vdash \mathrm{RT}^n_\infty$ for all $n < \omega$ (Theorem 2.2.8). This also can be proved directly. In fact, the following theorem is well-known.

Theorem 7.3.2 *Let* $(I, \mathfrak{X})$ *be a model of* RCA_0. *Then the following are equivalent:*

(1) $(I, \mathfrak{X}) \models \mathsf{RT}^3_2$;
(2) $(I, \mathfrak{X}) \models \mathsf{RT}^n_\infty$ *for all* $n < \omega$;
(3) $(I, \mathfrak{X}) \models \mathsf{KL}$;
(4) $(I, \mathfrak{X}) \models \mathsf{ACA}_0$.

Since RCA_0 includes the induction schema for Σ_1 formulas, Theorem 7.3.2 does not apply to all models of the form $(I, \mathrm{Cod}(M/I))$. Every model M has a cut $I \subseteq_{\mathsf{end}} M$ such that for all $m, n < \omega$, $(I, \mathrm{Cod}(M/I)) \models \mathsf{RT}^n_m + \neg\mathsf{ACA}_0$.

(See Exercise 7.5.5.) For semiregular cuts, by Theorem 7.1.7, we can reformulate Theorem 7.3.2 as follows:

Corollary 7.3.3 *Let I be a semiregular cut of a model M. Then the conditions (1)–(4) of Theorem 7.3.2 are equivalent for $\mathfrak{X} = \mathrm{Cod}(M/I)$.*

Proof By Theorem 7.1.5, if I is a semiregular cut of M, then $(I, \mathrm{Cod}(M/I)) \models \mathsf{RCA}_0$; hence the result follows from Theorem 7.3.2. □

Recall that if $M \prec K$, then $K_{M \setminus I} = \inf_K(M \setminus I)$, and we write $M \prec_I K$ if $M \prec K$, $I \subseteq_{\mathsf{end}} M$, $I \subseteq_{\mathsf{end}} K$, and there is $c \in K$ such that $I < c < K_{M\setminus I}$.

Theorem 7.3.4 *Let I be a semiregular cut of a countable model M. Then the following are equivalent:*

(1) $(I, \mathrm{Cod}(M/I)) \models \mathsf{RT}^3_2$.
(2) *There is a model K such that $M \prec_I K$ and $\mathrm{Cod}(M/I) = \mathrm{Cod}(K/I)$.*
(3) $(I, \mathrm{Cod}(M/I)) \models \mathsf{KL}$.
(4) *For any infinite cardinal λ, there is a model K such that $M \prec_I K$, $K_{M\setminus I} = \{a \in K : |a_K| \leq \lambda\}$ and $|K_{M\setminus I}| = \lambda^+$.*
(5) *There is a model K such that $M \prec_I K$ and $K_{M\setminus I}$ is semiregular in K.*
(6) *I is strong in M.*
(7) *If $A \in \mathrm{Cod}(M/I)$ and B is Σ_1 definable in (I, A), then $B \in \mathrm{Cod}(M/I)$.*
(8) *If $A \in \mathrm{Cod}(M/I)$ and B is definable in (I, A), then $B \in \mathrm{Cod}(M/I)$.*

Proof (1) $\Longrightarrow$ (2) Suppose $(I, \mathrm{Cod}(M/I)) \models \mathsf{RT}^3_2$. By Proposition 7.2.5, I is regular in M. As in the proof of Theorem 7.2.4, we define a descending sequence $\langle X_n : n < \omega \rangle$ of elements of $\mathrm{Cod}(M/I)$ such that:

(i) each X_n is unbounded in I;
(ii) if $f \in M$ is a function, $I \subseteq \mathrm{dom}(f)$, and $\mathrm{rg}(f)$ is bounded in I, then there are $i \in I$ and $n < \omega$ such that $X_n \subseteq f^{-1}(i)$.

Since $(I, \mathrm{Cod}(M/I)) \models \mathsf{RT}^3_2$, we also can impose that:

(iii) if $g : [I]^3 \longrightarrow \{0, 1\}$ is coded in M, then there are $n < \omega$ and $i < 2$ such that for all $(a, b, c) \in [X_n]^3$, $g(a, b, c) = i$.

As in the proof of Theorem 7.2.4, let U be the ultrafilter

$$\{X \in \mathrm{Cod}(M/I) : \exists n\ X_n \subseteq X\},$$

and let K be the corresponding ultrapower, whose elements are equivalence classes modulo U of coded functions $f : I \longrightarrow M$. Conditions (i) and (ii) imply that $M \prec_I K$.

Suppose that $A \in \mathrm{Cod}(K/I)$ is coded by the equivalence class of a function $f : I \longrightarrow M$ coded in M. For $i \in I$, let A_i be the subset of I coded by $f(i)$. Let

$$g(a,b,c) = \begin{cases} 0 & \text{if } a_I \cap A_b = a_I \cap A_c, \\ 1 & \text{otherwise.} \end{cases}$$

Let $X \in U$ be an homogeneous set for g. Suppose that for all $(a,b,c) \in [X]^3$, $g(a,b,c) = 1$. Fix $a, b \in X$ such that $a < b$. Then for all $c \in X$ such that $b < c$, $a_I \cap A_b \neq a_I \cap A_c$. Since there are only 2^a coded subsets of a_I, by semiregularity of I, X is bounded in I. This contradicts the fact that $X \in U$. Hence, for all $(a,b,c) \in [X]^3$, $g(a,b,c) = 0$.

Let d be an element of I. For all $a, b, c \in X$ such that $d < a < b < c$ we have $d \in A_b$ iff $d \in A_c$. Let

$$b_d = \min\{x \in X : \exists y \in X \ d < y < x\}.$$

Then for every d there is $j \in I$ such that

$$d \in A_{b_d} \Longleftrightarrow \forall i > j[i \in X \longrightarrow d \in A_i]. \qquad (*)$$

By Łoś's Theorem

$$K \models d \in A \Longleftrightarrow \{i \in I : d \in A_i\} \in U.$$

Since $X \in U$ and U is an ultrafilter, $(*)$ implies that

$$K \models d \in A \Longleftrightarrow d \in A_{b_d}.$$

Since the last condition is expressible in M (using the codes for f and X), this shows that $A \in \mathrm{Cod}(M/I)$. Thus we proved that $\mathrm{Cod}(K/I) \subseteq \mathrm{Cod}(M/I)$, and since the other inclusion is obvious, this finishes the proof of (1) $\Longrightarrow$ (2).

(2) $\Longrightarrow$ (3) Let $T \in \mathrm{Cod}(M/I)$ be a finitely branching tree. The code for T also codes a tree T^K in K. Since $M \prec_I K$, for every $a \in I$, the sequences of length a in the two trees are the same. Pick an element $b \in K$ such that $I < b < (M \setminus I)$. Then, since $\ell(b) > I$,

$$B = \left\{x \in T^K : x <_{T^K} b\right\} \cap I$$

is a branch of T coded in K and is unbounded in I. By the assumption, B is coded in M.

(3) $\Longrightarrow$ (4) Notice that by Corollary 7.3.3 we have now proved that the statements (1), (2), and (3) are equivalent and are also equivalent to the statement: for all $n < \omega$, $(I, \mathrm{Cod}(M/I)) \models \mathsf{RT}^n_\infty$.

Let $\langle \varphi_i : i \in \omega \rangle$ be an enumeration of all formulas of $\mathcal{L}(M)$. Using RT_2^n we will inductively define a sequence $X_0 \supseteq X_1 \supseteq \cdots$ of unbounded sets in $\mathrm{Cod}(M/I)$ such that for all $\bar{a}, \bar{b} \in [X_i]^n$

$$M \models [\varphi_i(\bar{a}) \longleftrightarrow \varphi_i(\bar{b})].$$

We add to $\mathcal{L}(M)$ a set of new constants $\{c_\nu : \nu < \lambda^+\}$, and we define a theory S as

$$\left\{\varphi(c_{\nu_0}, \ldots, c_{\nu_{n-1}}) : \langle \nu_0, \ldots, \nu_{n-1} \rangle \in [\lambda^+]^n \wedge \exists i \forall \bar{a} \in [X_i]^n \; M \models \varphi(\bar{a})\right\}.$$

The choice if the X_i's guarantees that S is consistent and $\mathrm{Th}(M, a)_{a \in M} \subseteq S$. Let K be a model of S generated by the set C of (interpretations of) c_ν's in K.

For each $c \in C$, and all $a \in I$ and $b \in M \setminus I$, the sentence $a < c < b$ is in S. Hence $I < c < M \setminus I$.

Suppose $K \models t(\bar{c}) < a$ for some Skolem term t of $\mathcal{L}(M)$, $\bar{c} \in [C]^n$, and $a \in I$. Then there is i such that

$$M \models t(\bar{a}) < a \quad \text{for all } \bar{a} \in [X_i]^n. \tag{$*$}$$

Consider the formula

$$\varphi = [t(x_0, \ldots, x_{n-1}) = t(x_n, \ldots, t_{2n-1})].$$

Let X_j be the homogeneous set for φ. Since $X_i \cap X_j$ is unbounded (and I is semiregular), there are $\bar{a} \in [X_j]^{2n}$ such that

$$M \models t(a_0, \ldots, a_{n-1}) = t(a_n, \ldots, a_{2n-1}) = d,$$

for some $d \in I$. Then $t(\bar{a}) = d$ for all $\bar{a} \in [X_j]^n$. Hence $K \models t(\bar{c}) = d$, and this proves that $M \prec_I K$.

It remains to show that $|\{a \in K : |a_K| \leq \lambda\}| = \lambda^+$. To this end we will show that for all infinite ν, $|(c_\nu)_M| = |\nu|$. Let t be a Skolem term of $\mathcal{L}(M)$. We will show that if

$$K \models t(c_{\nu_0}, \ldots, c_{\nu_{n-1}}, c_{\mu_0}, \ldots, c_{\mu_{m-1}}) \leq c_{\nu_{n-1}},$$

where $\nu_0 < \cdots < \nu_{n-1} < \mu_0 < \cdots < \mu_{m-1}$, then

$$t(c_{\nu_0}, \ldots, c_{\nu_{n-1}}, c_{\mu_0}, \ldots, c_{\mu_{m-1}}) = t(c_{\nu_0}, \ldots, c_{\nu_{n-1}}, c_{\mu'_0}, \ldots, c_{\mu'_{m-1}}),$$

where $\nu_0 < \cdots < \nu_{n-1} < \mu'_0 < \cdots < \mu'_{m-1}$. The proof is similar to the one showing that $M \prec_I K$, and is left to the reader. HINT: consider the formula

$$t(x_0, \ldots, x_{n-1}, y_0, \ldots, y_{m-1}) = t(x_0, \ldots, x_{n-1}, z_0, \ldots, z_{m-1}).$$

The last claim implies that $|(c_\nu)_K| = \max(\aleph_0, |\nu|)$, finishing the proof.

(4) $\Longrightarrow$ (5) This implication is immediate. Let K be a model given by (4). Then the cut $\{a \in K : |a_K| \le \lambda\}$ is λ^+-like, hence semiregular.

(5) $\Longrightarrow$ (6) Let $f \in M$ be a function such that $I \subseteq \operatorname{dom}(f)$. Let K be a model given by (5), and let f^K be the function coded by f in K. Since $K_{M\setminus I} \subseteq \operatorname{dom}(f^K)$ and $K_{M\setminus I}$ is semiregular, for each b, if $I < b \in K_{M\setminus I}$, then $f^K(b_K) \cap K_{M\setminus I}$ is bounded in $K_{M\setminus I}$. Hence, for each b, if $I < b \in K_{M\setminus I}$, then there is $c \in K_{M\setminus I}$ such that for each $d \in K_{M\setminus I}$,

$$K \models \forall x < b\ [f(x) > c \longrightarrow f(x) > d]. \qquad (*)$$

By overspill, there is $d > K_{M\setminus I}$ such that $(*)$ holds. By taking a smaller d if necessary, we can assume that $d \in M$. It follows that for all $i \in I$, if $f(i) > I$, then $f(i) > d$.

(6) $\Longrightarrow$ (7) For $A \in \operatorname{Cod}(M/I)$, let

$$B = \{x \in I : (I, A) \models \exists y\ \varphi(x, y, A)\},$$

where $\varphi(x, y, X)$ is a Σ_0 formula with parameters from I.

For a fixed $c > I$, let

$$f(x) = \begin{cases} \min\{y : \varphi(x, y, A)\} & \text{if such } y \text{ exists,} \\ c & \text{otherwise.} \end{cases}$$

Let $d \in M \setminus I$ be such that for all $x \in I$, if $f(x) > I$, then $f(x) > d$. Then

$$B = \{x \in I : M \models \exists y < d\ \varphi'(x, y, a)\},$$

where $\varphi'(x, y, a)$ is the translation of $\varphi(x, y, A)$ given by Proposition 7.0.1. Hence, $B \in \operatorname{Cod}(M/I)$.

(7) $\Longrightarrow$ (8) This is proved by standard induction on the quantifier complexity of formulas.

Since (1), (2), and (3) are equivalent, to finish the proof of the theorem we show now that (8) $\Longrightarrow$ (3). This already follows from Corollary 7.3.3. The argument is just the proof of König's Lemma formalized in PA^*. First notice that (8) implies that $(I, \operatorname{Cod}(M/I)) \models \mathsf{PA}^*$. Indeed, if the set defined by $\varphi(x, A_0, \ldots, A_n)$ in $(I, A_0, \ldots, A_n)$ is nonempty, then, since it is coded in M, it has a least element.

Let $T \in \mathrm{Cod}(M/I)$ be a finitely branching tree. Define a branch B in T by induction in (I, T):

$$f(0) = \min\{x \in T : \ell(x) = 1 \wedge \forall i > 0 \exists y \in T(\ell(y) = i \wedge x \subseteq y)\},$$

$$f(i+1) = \min\{x \in T : \ell(x) = i+2 \wedge f(i) \subseteq x \wedge [\forall j > i \exists y \in T(\ell(y) = j \wedge x \subseteq y)]\}.$$

Since I is semiregular, if $i \in I$ and $f(i)$ is defined (i.e. the corresponding set is nonempty), then $f(i+1)$ is defined.

Let $B = f(I)$. Then B is an unbounded branch of T in I and, because B is definable from $<_T$, $B \in \mathrm{Cod}(M/I)$. □

Notice that while proving Theorem 7.3.4 we gave an almost complete proof of Theorem 7.3.2. The missing link is the implication $\mathsf{RCA}_0 + \mathsf{KL} \vdash \mathsf{RT}^n_\infty$, for all $n < \omega$. This is a formalization of the classical result of Ramsey Theory using the so called Erdős–Rado tree. See [83] or [187].

The assumption of semiregularity of I was used several times in the proof of Theorem 7.3.4. For an argument that this assumption is necessary, see Exercise 7.5.5.

7.4 Why PA?

There are several lines of argument one could take to explain why Peano Arithmetic deserves a special place in foundational studies. Still, instead of an argument, one prefers a theorem which shows PA is not only a convenient formal system, but is, in a sense, necessary. In this section we present two theorems in this direction. Of course, it is debatable whether they fulfill the objective. The first result, Theorem 7.4.2, reverses two theorems in the model theory of PA. The objective is to prove that if a theory T in $\mathcal{L}_{\mathsf{PA}}$ has such and such model theoretic properties, then T is at least as strong as PA. Now we describe the assumptions on T and the relevant properties.

For the rest of this section, we assume that T is a consistent theory in $\mathcal{L}_{\mathsf{PA}}$ extending $I\Sigma_0$.

It is not difficult to see that PA is "necessary" for the MacDowell-Specker Theorem. In fact, it is not even the MacDowell-Specker Theorembut its weaker form restricted to countable models. This is the content of the following exercise.

Exercise 7.4.1 Suppose that every countable model of T has an elementary end extension. Then $T \vdash \mathsf{PA}$.

The next theorem shows that PA is also necessary for Gaifman's splitting theorem suitably modified.

Theorem 7.4.2 *Let T be an $\mathcal{L}_{\mathsf{PA}}$ theory extending $I\Sigma_0 + \mathsf{exp}$, and suppose that every complete extension of T has a countable model K such that:*

(1) *K has no proper elementary substructures.*

(2) *Whenever $K \prec L$, and L is countable, there exist K' and L' such that $L \prec L'$ and $K \prec_{\mathsf{cof}} K' \subseteq_{\mathsf{end}} L'$.*

Then $T \vdash \mathsf{PA}$. □

Theorem 7.4.2 follows from the next two theorems (Do it!). The proof of the first one is not very difficult and is left as an exercise. The rest of this section is devoted to an outline of the proof of the second theorem. For details consult [70].

Theorem 7.4.3 *Let $K \models I\Sigma_n$ for some $n < \omega$, and suppose*

$$K \prec_{\mathsf{cof}} N \subseteq_{\mathsf{end}} L \models I\Sigma_n,$$

$N \neq L$, and $K \prec L$. Then $K \models B\Sigma_{n+1}$. □

Theorem 7.4.4 *Let $K \models B\Sigma_n + \mathsf{exp} + \neg I\Sigma_n$ for some $0 < n < \omega$. Then K has a proper elementary submodel. Moreover, if K is countable, then it is isomorphic to a proper elementary submodel of itself and it has continuum many automorphisms.* □

Notice that by Theorem 7.4.4, condition (1) in Theorem 7.4.2 can be replaced by

(1′) K is countable and has fewer than continuum many automorphisms;

The following lemma is a refinement of Theorem 7.1.5.

Lemma 7.4.5 *Suppose that $K \subseteq_{\mathsf{end}} N \models I\Sigma_0$, $K \neq N$, and $K \models B\Sigma_n + \mathsf{exp} + \neg I\Sigma_n$, where $0 < n < \omega$. For $n > 1$, assume also that $K \prec_{\Sigma_n} N$. Then K is not semiregular in N. Moreover, there are $a, f \in N$ and $b \in K$ such that $f \leq a$, f is a function from b_K into N such that $\mathrm{rg}(f) \cap K$ cofinal in K, and a^{a^a} exists in N.* □

Proof Consider the case of $n > 1$. The assumption $K \prec_{\Sigma_n} N$ implies that $N \models I\Sigma_{n-2} + \mathsf{exp}$. Since K is not a model of $I\Sigma_n$, there is $\psi(x, y) \in \Pi_{n-1}$, possibly with parameters from K, such that $K \models \exists x, y\psi(x, y)$ and there is no least x in K such that $\exists y\psi(x, y)$. Let $\psi(x, y)$ be $\forall z\varphi(x, y, z)$ with $\varphi(x, y, z) \in \Sigma_{n-2}$.

The formula

$$\exists x' \leq x \exists y' \leq y\ \forall z\varphi(x', y', z)$$

is equivalent to

$$\forall w[\exists x' \leq x \exists y' \leq y\ \forall z \leq w\varphi(x', y', z)].$$

Let $\theta(x, y, w)$ be the formula in the square brackets above. Since K and N are models of $B\Sigma_{n-2}$, $\theta(x, y, w)$ is equivalent both in K and N to a Σ_{n-2} formula. Notice that $K \models \exists x, y \forall w \theta(x, y, w)$ and there is no least $x \in K$ for which $\exists y \forall w \theta(x, y, w)$ holds.

Now, let $r, s \in K$ be such that $K \models \forall w \theta(r, s, w)$. Then for all $k \in K$, $K \models \forall w < k \theta(r, s, w)$. By overspill, there is $\beta \in N \setminus K$ such that

$$N \models \forall w < \beta \theta(r, s, w).$$

Now define $f : r_K \longrightarrow N$ by

$$f(x) = \begin{cases} \min\{u : \forall w < \beta \exists y < u \varphi(x, y, z)\} & \text{if such } u \text{ exists,} \\ \beta & \text{otherwise.} \end{cases}$$

As in the proof of Theorem 7.1.5, one can prove that the range of f must be unbounded in K (Do it!); hence f has the required properties.

Since N is closed under exponentiation, there is an $a \in N$ such that $a > f$ and a^{a^a} exists in N. In the case of $n = 1$ the proof is similar, but since N might not be closed under exponentiation, extra care must be taken when using overspill to select a β as above. See the details in [70]. □

The techniques used in the proof of Theorem 7.4.4 are slightly outside the scope of this book. We omit the details and refer the reader to [70]. Here is the idea of the proof. Let a, b, and f be as in Lemma 7.4.5. Let $x_0, x_1, \ldots$ be an enumeration of K. Using the fact that $(a_N, +, \times)$, considered as a relational structure, is recursively saturated, one can construct a sequence $\alpha_0, \alpha_1, \ldots$ of automorphisms of a_N such that α_i fixes $b_n \cup \{x_0, \ldots, x_{i-1}, f\}$ pointwise. Then the required embedding β is defined by

$$\beta(x_i) = \alpha_0^{-1} \alpha_1^{-1} \ldots \alpha_i^{-1}(x_i).$$

One can easily show that β is an embedding of K into itself. The main difficulty is to arrange the α_i's so that $\beta(K)$ is a proper substructure of K and the extension $\beta(K) \subseteq K$ is elementary. This takes some work.

7.4.1 *Schemes axiomatizing arithmetic*

The induction schema of PA is obtained by schematizing the second-order induction axiom IA:

$$\forall X[0 \in X \wedge \forall x(x \in X \longrightarrow x + 1 \in X) \longrightarrow \forall x\, x \in X].$$

The *schematization* of a second-order sentence $\forall X\ \Phi(X)$, with no other second-order quantifiers, is the scheme consisting of the universal closures of all instances of $\Phi(\varphi)$, where φ is a first-order formula, and $\Phi(\varphi)$ is obtained from $\Phi(X)$ by replacing each occurrence of $x_i \in X$ by $\varphi(x_i)$, where $\varphi(x_i)$ is first-order and it can have other variables. The schema obtained this way is denoted by $\Phi(\text{Def})$.

Notice that IA is equivalent to $\forall X\ \text{LA}(X)$, where $\text{LA}(X)$ declares that if X is nonempty, then it has a least element. Formally, $\text{LA}(X)$ can be written as follows (notice the use of x_0, it is not a typo):

$$\forall x_0 \in X \exists x_1 \in X \forall x_2 \in X (x_1 \leq x_2).$$

Thus, PA can be considered a schematization of $\forall X\ \text{LA}(X)$.

The slightly unusual form of $\text{LA}(X)$ is needed to conform to the following definition. We say that a second-order formula is *restricted* if it is of the form

$$Q_0 x_0 \in X \ldots Q_{n-1} x_{n-1} \in X \varphi(x_0, \ldots, x_{n-1}),$$

where each Q_i is either $\exists$ or $\forall$, and φ is in $\mathcal{L}_{\mathsf{PA}}$. So, $\text{LA}(X)$ is restricted.

The standard model is the only model IA and the basic semi-ring axioms of PA. In other words, PA is the schematization of a second-order axiom which is categorical for $\mathbb{N}$ over a finite subset of $\text{Th}(\mathbb{N})$. We will show that PA is, in a sense, the weakest such theory. More precisely:

Theorem 7.4.6 *Suppose $\Phi(X)$ is a restricted formula of second-order arithmetic such that $\forall X \Phi(X)$ is categorical for $\mathbb{N}$ over some finite theory T in $\mathcal{L}_{\mathsf{PA}}$. Then there is a finite set T_1 of $\mathcal{L}_{\mathsf{PA}}$ sentences such that $\mathbb{N} \models T_1$ and $T_1 + \Phi(\text{Def}) \vdash \mathsf{PA}$.* □

Before we begin the proof we need some preparation. Let $\mathcal{L}$ be a relational language and let $\mathcal{I} = (I, <)$ be a linearly ordered set. An *$\mathcal{I}$-structure* is a sequence $\mathcal{A} = \langle A_i : i \in I \rangle$ of $\mathcal{L}$-structures such that for $i < j$, $A_i \subseteq A_j$. Let A be the $\mathcal{L}$-structure $\bigcup_{i \in I} A_i$. If φ is a sentence of $\mathcal{L}(A)$ in prenex normal form, then the relation $\mathcal{A} \Vdash \varphi$ is defined by induction on φ as follows:

$\mathcal{A} \Vdash \varphi$ iff $A \models \varphi$ if φ is quantifier-free;
$\mathcal{A} \Vdash \forall x \varphi(x)$ iff for all $a \in A$, $\mathcal{A} \Vdash \varphi(a)$;
$\mathcal{A} \Vdash \exists x \varphi(x)$ iff for all i such that the parameters of φ are in A_i, and for all $j > i$, there is $b \in A_j$ such that $\mathcal{A} \Vdash \varphi(b)$.

The following lemma is easy and it is left as an exercise.

Lemma 7.4.7 (1) *If $\mathcal{I}$ has no last element and $\mathcal{A} \Vdash \varphi$, then $A \models \varphi$.*

(2) *Suppose $I' \subseteq I$. Let $\mathcal{A}' = \langle A_i : i \in I' \rangle$, and let $A' = \bigcup_{i \in I'} A_i$. If $\varphi \in \mathcal{L}(A')$ and $\mathcal{A} \Vdash \varphi$, then $\mathcal{A}' \Vdash \varphi$.* □

Let U be the set of formulas of $\mathcal{L}(A)$ which are in the prenex normal form, in which all negations are applied to atomic formulas. For every formula φ of $\mathcal{L}(A)$, let φ^* be a formula in U logically equivalent to $\neg\varphi$. A subset S of U is *closed* if it is closed under subformulas and if φ^* is in S, then φ is in S. Notice that any finite $S \subseteq U$ is contained in a finite closed subset of U.

If $\varphi(\bar{x})$ is in U, then we say that $\mathcal{A}$ *determines* $\varphi(\bar{x})$ if for all $\bar{a}$ in $\mathcal{A}$ either $\mathcal{A} \Vdash \varphi(\bar{a})$ or $\mathcal{A} \Vdash \varphi^*(\bar{a})$.

Lemma 7.4.8 *Suppose that $\mathcal{I} = (\omega, <)$ and $\mathcal{A}$ is an $\mathcal{I}$-structure with each A_i finite. Let S be a finite closed subset of U. Then there is an infinite increasing sequence $\langle i_j : j < \omega\rangle$ such that the $\mathcal{I}$-structure $\langle A_{i_j} : j < \omega\rangle$ determines S.*

Proof A relatively straightforward proof by induction on the size of S is left to the reader (Do IT!). HINTS: The result is obvious if S contains only quantifier-free formulas. Otherwise let $\varphi \in S$ be of maximal length. Then the inductive hypothesis applies to $S' = S \setminus \{\varphi, \varphi^*\}$. Without loss of generality, we can assume that φ is of the form $\exists y \psi(y, \bar{x})$. Now inductively define a sequence $i_0 < i_1 < \cdots$ such that for all $k < \omega$ and all $\bar{b}$ in $A_{i_{k-1}}$, either $\mathcal{A}^{(k)} \Vdash \varphi(\bar{b})$ or $\mathcal{A}^{(k)} \Vdash \varphi(\bar{b})^*$, where

$$\mathcal{A}^{(k)} = \langle A_{i_0}, \ldots, A_{i_k}, A_{i_{k+1}}, A_{i_{k+2}}, \ldots\rangle. \qquad \square$$

Proof of Theorem 7.4.6 Let

$$\Phi(X) = Q_0 x_0 \in X, \ldots, Q_{k-1} x_{k-1} \in X \varphi(x_0, \ldots, x_{k-1})$$

be a restricted formula, where $\varphi(x_0, \ldots, x_{k-1})$ is in $\mathcal{L}_{\mathsf{PA}}$. Suppose that $\mathbb{N}$ is the only model $\forall X \Phi(X) + T$ for some finite theory T in $\mathcal{L}_{\mathsf{PA}}$. Let $\mathcal{L}$ be the language containing just one k-ary relation symbol R, and let

$$\sigma = Q_0 x_0 \ldots Q_{k-1} x_{k-1} R(x_0, \ldots, x_{k-1}).$$

First we claim that for each $n < \omega$, there is an $(n+1, <)$-structure $\mathcal{A} = \langle A_i : i \leq n\rangle$ for the language $\mathcal{L}$ such that for each $i \leq n$, $A_i \subseteq \omega$ is finite,

$$\forall n_0, \ldots, n_{k-1} \in A_i \; [A_i \models R(n_0, \ldots, n_{k-1}) \text{ iff } \mathbb{N} \models \varphi(n_0, \ldots, n_{k-1})],$$

and $\mathcal{A} \Vdash \sigma^*$.

To this end, let $M \models \mathsf{TA}$ be countable and nonstandard. Since $\mathbb{N} \not\cong M$, there is some $X \subseteq M$ such that $M \models \neg\Phi(X)$. Notice that for all $n < \omega$,

$$\forall Y \subseteq \mathbb{N} \left[|Y| = n \longrightarrow (\mathbb{N}, Y) \models \Phi(Y)\right]$$

is a true statement which can be written as an equivalent sentence of $\mathcal{L}_{\mathsf{PA}}$. Since $M \models \mathsf{TA}$, the statement is also true in M, from which it follows that X must be infinite.

Let $B_0 \subseteq B_1, \dots$ be finite sets such that $X = \bigcup_{i<\omega} B_i$. Let

$$V = \{(a_0, \dots, a_{k-1}) \in M^k : M \models \varphi(a_0, \dots, a_{k-1})\}.$$

Finally, let $\mathcal{B}$ be the $(\omega, <)$-structure $\langle (B_i, V \cap B_i^k) : i < \omega \rangle$. By Lemma 7.4.8 we can assume that $\mathcal{B}$ determines σ. If $\mathcal{B} \Vdash \sigma$, then, by Lemma 7.4.7, $\bigcup \mathcal{B} \models \sigma$. Now, since $\bigcup \mathcal{B} = (X, V \cap B^k)$, this implies that $M \models \Phi(X)$, which is a contradiction. Hence, $\mathcal{B} \Vdash \sigma^*$ and, again by Lemma 7.4.7, for any $n < \omega$, $\langle (B_i, V \cap B_i^k) : i \leq n \rangle \Vdash \sigma^*$. Now let $G(x)$ be an $\mathcal{L}_{\mathsf{PA}}$ formula expressing the following:

$$\exists y \{ y \text{ is a (code of a) } (x+1, <)\text{-structure such that } y \Vdash \sigma^* \text{ and}$$
$$\forall a_0, \dots, a_{k-1} \in \bigcup y \; [(y \Vdash R(a_0, \dots, a_{k-1})) \longleftrightarrow \varphi(a_0, \dots, a_{k-1})]\}.$$

We have shown that for every $n < \omega$, $M \models G(n)$. Since $\mathbb{N} \prec M$, this finishes the proof of the claim.

To finish the proof of the theorem, let the base theory T_1 be $I\Sigma_1 + \forall x G(x)$. Since $I\Sigma_1$ is finitely axiomatizable, T_1 is finite. It remains to show that $T_1 + \Phi(\mathrm{Def}) \vdash \mathsf{PA}$. Suppose not. Then there is a model $M \models T_1 + \Phi(\mathrm{Def})$ such that for some $\mathcal{L}_{\mathsf{PA}}$ formula $\psi(x)$

$$M \models \psi(0) \wedge \forall x(\psi(x) \longrightarrow \psi(x+1)) \wedge \exists x \neg \psi(x).$$

Let J be the cut $\{b \in M : \forall x \leq b \; \psi(x)\}$. Let $a \in M$ be such that $J < a$. Since $M \models G(a)$, there is a coded sequence $\mathcal{A} = \langle A_i : i \leq a \rangle$ of $\mathcal{L}$-structures such that

$$M \models \forall i \leq j \leq a \{ A_i \subseteq A_j \wedge \mathcal{A} \Vdash \sigma^* \wedge$$

$$\forall a_0, \dots, a_{k-1} \in A_i \; [A_i \models R(a_0, \dots, a_{k-1}) \longleftrightarrow \psi(a_0, \dots, a_{k-1})]\}. \qquad (*)$$

Using $I\Sigma_1$, one can verify that $\mathcal{A}$ is an $(a_M, <)$-structure and $\mathcal{A} \Vdash \sigma^*$. It follows that $\langle A_i : i \in J \rangle \Vdash \sigma^*$. The formula

$$\chi(x) = \exists y [\forall z \leq y (\psi(z) \wedge x \in A_y)]$$

defines $A = \bigcup_{i \in J} A_i$ in M. By $(*)$, the interpretation of R is exactly $A^k \cap \psi^(M)$. Since $A \models \sigma^*$, this implies that M satisfies the negation of the sentence

$$Q_0 x_0 \dots Q_{n-1} x_{n-1} [\bigwedge_{i<n} \chi(x_i) \wedge \psi(x_0, \dots, x_{n-1})]$$

in violation of $M \models \Phi(\mathrm{Def})$. This contradiction finishes the proof. $\square$

7.5 Exercises

♣**7.5.1** There are nonstandard $M \prec_{\mathsf{end}} N$ such that all $X \in \mathrm{Cod}(N/M)$ are inductive and $(M, \mathrm{Cod}(N/M))$ is not a model of ACA_0.

♦**7.5.2** Every nonstandard model M has cuts I such that $I \models \mathsf{PA}$ and $\mathrm{cf}^M(I) = \mathbb{N}$. (HINT: every countable recursively saturated model M has elementary cuts I such that $\mathrm{cf}^M(I) = \mathbb{N}$, and every nonstandard model has a cut which is a recursively saturated model of PA.)

♦**7.5.3** Let M be a countable model, and let N be such that $M \subseteq_{\mathsf{end}} N$ and $\mathrm{cf}^N(M) = \mathbb{N}$. Then $|(\mathrm{Aut}(M))| = 2^{\aleph_0}$. More generally, if $M \subseteq_{\mathsf{end}} N$ and M is not semiregular in N, then $|(\mathrm{Aut}(M))| = 2^{\aleph_0}$.

♦**7.5.4** Every countable model has an elementary extension in which $\mathbb{N}$ is strong and has elementary extension in which $\mathbb{N}$ is not strong.

♣**7.5.5** Let I be a strong cut of a model M. For $c > I$, let $J = \{x : \exists i \in I \ (x < c^i)\}$. Then for all $n < \omega$ and all $a \in I$, $(J, \mathrm{Cod}(M/J)) \models \mathsf{RT}^n_a$.

♦**7.5.6** If I is a strong cut of countable model M, then there is K such that $M \prec_I K$ and I is strong in K.

♦**7.5.7** If I is a cut of a countable model M and $(I, \mathrm{Cod}(M/I)) \models \mathsf{RT}^2_\infty$, then there is K such that $M \prec_I K$ and I is regular in K.

♥**7.5.8** If I is a strong cut of a countable model M and K is the model constructed in the proof of (1) $\Longrightarrow$ (2) in Theorem 7.2.4, then $K_{M\setminus I}$ is strong in K.

♥**7.5.9** If I is a strong cut of a countable model M and $M \prec_I K$, then $I \prec K_{M\setminus I}$.

Superstrong cuts: for $I \subseteq_{\mathsf{end}} M$ and $c \in I$, $[I]^c$ is the collection of coded increasing sequences σ of elements of I such that $\ell(\sigma) = c$. A cut I of M is *superstrong* if it is semiregular, and there is a nonstandard c such that every coded partition of $[I]^c$ into two sets has a coded homogeneous set which unbounded in I.

♣**7.5.10** If I is strong in M and (M, I) is recursively saturated, then I is superstrong.

♦**7.5.11** If I is a nonstandard superstrong cut of M, then $\mathrm{Th}(I) \in \mathrm{SSy}(M)$.

♣**7.5.12** There is a cut which is strong but not superstrong. (HINT: use previous exercise.)

♦**7.5.13** Let $\Psi(X)$ be the formula expressing the following: "Either $(X)_0 = \emptyset$ or $(X)_0$ has a least element or $(X)_1$ is not a full satisfaction class." Then:

(1) $\forall X \ \psi(X)$ is categorical for $\mathbb{N}$ (over some finite $\mathcal{L}_{\mathsf{PA}}$-theory).
(2) For a sufficiently strong finite fragment T of TA, $T \vdash \Psi(\mathrm{Def})$.

7.6 Remarks & References

The study of the hierarchy of cuts in models of arithmetic was initiated by Paris and developed by Kirby & Paris in [80], [83], and [146], by Mills and Paris in [148], and in many papers that followed. See [50] and [71] for more details and, in particular, for applications to independence results.

Hirst and Simpson recognized the importance of combinatorial properties of structures of the form $(I, \mathrm{Cod}(M/I))$ in the analysis of fragments of second-order arithmetic and their applications to Reverse Mathematics. Much of the work of Kirby and Paris, reconstructed in this context, together with new results are in Hirst's Ph.D. thesis [55]. Full discussion of the role of König's Lemma and Ramsey's Theorem(s) in second-order arithmetic is given in [187].

Kirby & Paris asked in [83] whether the assumption that $(I, \mathrm{Cod}(M/I)) \models \mathsf{RT}^2_\infty$ implies that I is strong. The problem is related to an earlier question of Jockusch [62] concerning effective versions of Ramsey's Theorem : is there a recursive partition of $[\mathbb{N}]^2$ into two pieces such that $0'$ is recursive in any infinite homogeneous set? The negative solution was given by Seetapun [181]. Seetapun's result implies that RT^2_∞ does not prove ACA_0 over RCA_0. Some improvements and a comprehensive account of the problem are given in [21].

Proposition 7.2.5 ($\mathsf{RT}^2_2 \implies \mathsf{RT}^1_\infty$) and Theorem 7.2.2 (there are semiregular cuts which are not regular) imply that WKL_0 does not prove RT^2_2. This and other related results are independently due to Hirst [55], see [21].

Clote [22] contains various results concerning cuts, partition properties, and collection schemas for fragments of arithmetic.

Theorem 7.2.2 is from [105]. A generalization to higher levels of the arithmetic hierarchy is due to Kanovei [65]. Kanovei's generalization requires the model to be recursively saturated. It is open whether this assumption is necessary. A shorter proof of Kanovei's result using partial inductive satisfaction classes is published in [99].

The example in Exercise 7.5.5 is from [80].

Theorem 7.4.2 is due to Kaye [70]. Part of Lemma 7.4.5, stating that every countable model of $B\Sigma_n + \mathsf{exp} + \neg I\Sigma_n$, $n > 0$ has continuum many automorphisms, was proved earlier in [94] (with a correction in [95]). For interesting complementary results on cofinal extensions of models fragments of arithmetic see [72].

The elegant Theorem 7.4.6 on schemes axiomatizing arithmetic is due to Wilkie [212]. There are some fine points of this result that need to be mentioned. First, the finite fragment T_1 in Theorem 7.4.6 might not be provable in PA. In fact, the second theorem in [212] states:

> *There is a restricted formula $\Phi(X)$ such that $\forall X \Phi(X)$ is categorical for $\mathbb{N}$ (over some finite* PA*-provable $\mathcal{L}_{\mathsf{PA}}$ theory), but such that for no $n < \omega$ do we have $I\Sigma_n + \Phi(\mathrm{Def}) \vdash \mathsf{PA}$.*

Concerning the question of categorical axiomatizations which are not expressed in the restricted form, Paris gave a counterexample which is our Exercise 7.5.13 (the counterexample is given in [212]).

8

AUTOMORPHISMS OF RECURSIVELY SATURATED MODELS

Every countable recursively saturated first-order structure is strongly ω-homogeneous, which means that if $\bar{a}, \bar{b}$ are finite tuples from the structure and $\text{tp}(\bar{a}) = \text{tp}(\bar{b})$, then there is an automorphism f such that $f(\bar{a}) = \bar{b}$. Also, each infinite definable set A has infinitely many elements realizing the same type (DO IT!). This shows that countable recursively saturated structures have rich automorphism groups. In this chapter we show that some interesting properties of countable recursively saturated models of PA can be expressed in terms of their automorphisms and automorphism groups. In particular, a countable recursively saturated model is arithmetically saturated iff it has an automorphism moving all undefinable elements. Much of this chapter is devoted to the proof of this and several other characterizations of arithmetic saturation. We also study the relationship between the type of an element and its stabilizer in the automorphism group. As elsewhere in this book, the results apply to models of PA*. However, since we now have to pay attention to the way the syntax is arithmetized, *in this chapter we consider* PA* *for finite languages only.*

8.1 Moving undefinable elements

Let M be a countable recursively saturated model. Which elements of M can be moved by an automorphism? What do orbits of elements of M look like? Which elements can be moved if some set of elements is fixed? Let us start with two simple observations.

A direct argument shows that every undefinable element of a countable recursively saturated model can be moved by an automorphism. If M is recursively saturated, then so is (M, a), for each $a \in M$. Hence, we have the following basic proposition:

Proposition 8.1.1 *Let M be a countable recursively saturated model. Then for every $a, b \in M$, if $b \notin \text{Scl}(a)$, then there is an automorphism f such that $f(a) = a$ and $f(b) \neq b$.* □

Let f be a nontrivial automorphism of a model M. Since $f(0) = 0$ and for every $x \in M$, if $f(x) = x$, then $f(x+1) = x+1$. Then f is not inductive, and, consequently, f is not definable in M.

By Ehrenfeucht's Lemma 1.7.2, if $\text{tp}(a) = \text{tp}(b)$ and $a \neq b$, then $b \notin \text{Scl}(a)$. In particular, if $f(a) \neq a$, then $f(a) \notin \text{Scl}(a)$; hence, f is not 0-definable at a single (moved) point. This, as we will now prove, implies that nontrivial automorphisms of recursively saturated models are not even $\mathcal{L}_{\infty\omega}$-definable. The key to this result is the following lemma.

For a subset X of a first-order structure $\mathfrak{A}$, let

$$\mathfrak{a}_{\mathfrak{A}}(X) = |\{f(X) : f \in \text{Aut}(\mathfrak{A})\}|.$$

Lemma 8.1.2 (Kueker–Reyes Lemma) *Let $\mathfrak{A}$ be a countable first-order structure and $X \subseteq \mathfrak{A}$. If for any finite $A \subseteq \mathfrak{A}$, there is $g \in \text{Aut}((\mathfrak{A}, a)_{a\in A})$ such that $g(X) \neq X$, then $\mathfrak{a}_{\mathfrak{A}}(X) = 2^{\aleph_0}$.* □

Let f and g be automorphisms of a model M. Then f^g, the conjugate of f by g, is gfg^{-1}. Notice that the image of the graph of f under g is the graph of f^g (DO IT!).

Proposition 8.1.3 *Every nontrivial automorphism of a countable recursively saturated model has continuum many conjugates.*

Proof Let f be a nontrivial automorphism of a model M. Using the Kueker–Reyes Lemma, we will show that if $X = \{\langle x, f(x)\rangle : x \in M\}$, then $\mathfrak{a}_M(X) = 2^{\aleph_0}$. Since $M \models \mathsf{PA}^*$, it is enough to consider one-element subsets $A = \{a\} \subseteq M$. There are two cases.

Case 1: $f(a) = a$. Let b be such that $f(b) \neq b$. Then, by Ehrenfeucht's Lemma applied to the model (M, a), $f(b) \notin \text{Scl}(a, b)$. By Proposition 8.1.1, there is g such that $g(a) = a$, $g(b) = b$, and $g(f(b)) \neq f(b)$. Hence, $g(X) \neq X$.

Case 2: $f(a) \neq a$. Then $f(a) \notin \text{Scl}(a)$. In this case, let g be such that $g(a) = a$ and $g(f(a)) \neq f(a)$. So, $g(X) \neq X$. □

It follows from Proposition 8.1.3 that if f is a nontrivial automorphism of a countable recursively saturated model M, then f is not definable in M by a formula with a finite number of parameters in any extension of first-order logic. In particular, f is not definable in $\mathcal{L}_{\infty\omega}$.

There are countable recursively saturated linearly ordered structures which have nontrivial definable automorphisms. If a structure $\mathfrak{A}$ has a nontrivial 0-definable automorphism then so does each structure elementarily equivalent to $\mathfrak{A}$. Thus, to get an example, consider a recursively saturated model of $\text{Th}(\mathbb{Z}, <)$.

8.2 Moving cuts and classes

In the previous section we showed that if X is the graph of a nontrivial automorphism of a recursively saturated model M, then $\mathfrak{a}_M(X) = 2^{\aleph_0}$. What other sets have continuum many automorphic images? If D is an undefinable subset of a countable first-order structure $\mathfrak{A}$ and $(\mathfrak{A}, D)$ is recursively saturated, then

$\mathfrak{a}_{\mathfrak{A}}(D) = 2^{\aleph_0}$ (Do IT!). In general, however, there are undefinable sets D for which $\mathfrak{a}_{\mathfrak{A}}(D)$ is small. For example, in every model M of PA, $\mathfrak{a}_M(\mathbb{N}) = 1$.

The next exercise is another application of the Kueker–Reyes Lemma.

Exercise 8.2.1 Let I be a proper cut of a countable recursively saturated model M. Then, $\mathfrak{a}_M(I) < 2^{\aleph_0}$ iff there is $a \in M$ such that either $I \cap \mathrm{Scl}(a)$ is cofinal in I or $(M \setminus I) \cap \mathrm{Scl}(a)$ is coinitial in $M \setminus I$.

Corollary 8.2.2 *For any cut I in a countable recursively saturated model M, $\mathfrak{a}_M(I) \in \{1, \aleph_0, 2^{\aleph_0}\}$.* □

Theorem 8.2.3 *Let X be an undefinable class in a countable recursively saturated model M. Then $\mathfrak{a}_M(X) = 2^{\aleph_0}$.*

Proof Let $a \in M$ be given. For each $b \in M$, let $X_b = X \cap b_M$, and let $p_b(u, v)$ be the recursive type:

$$\{\varphi(u, a) \longleftrightarrow \varphi(v, a) : \varphi \in \mathsf{Form}\} \cup \{(v < b) \wedge (u \in X_b) \wedge (v \notin X_b)\}.$$

By the Kueker–Reyes Lemma, it is enough to show that there is a b for which $p_b(u, v)$ is finitely realizable. Assume, to the contrary, that there is no such b. Then for every $b \in M$, there is a formula $\theta_b(u, a)$ such that

$$M \models \forall u < b\ [\theta_b(u, a) \longleftrightarrow u \in X_b].$$

Now, suppose $\mathrm{Scl}(a) < b_0 < b_1$. Then

$$M \models \forall u < b_0\ [\theta_{b_0}(u, a) \longleftrightarrow \theta_{b_1}(u, a)],$$

hence, for $K = \sup(\mathrm{Scl}(a))$,

$$K \models \forall u\ [\theta_{b_0}(u, a) \longleftrightarrow \theta_{b_1}(u, a)],$$

and it follows that $\theta_{b_0}(u, a)$ and $\theta_{b_1}(u, a)$ define the same subset of M. Since b_1 was arbitrary, it follows that $\theta_{b_0}(u, a)$ defines X, contradicting the undefinability of X. □

We need a stronger version of the previous theorem. In its proof we will make use of the topology of $\mathrm{Aut}(M)$. See Section 8.8 for background.

Theorem 8.2.4 *Let $\{X_i : i < \omega\}$ be a collection of undefinable classes of a countable recursively saturated model M. Then there is an $f \in \mathrm{Aut}(M)$ such that for all $i, j < \omega$, $f(X_i) \neq X_j$.*

Proof For any $i, j < \omega$, the set

$$U_{i,j} = \{f \in \mathrm{Aut}(M) : f(X_i) \neq X_j\}$$

is dense and open in $\mathrm{Aut}(M)$. To see this, suppose that $g(a) = b$ for some $g \in \mathrm{Aut}(M)$. If $g(X_i) \neq X_j$, we are done. So suppose $g(X_i) = X_j$. Then let $f \in \mathrm{Aut}(M, b)$, given by Theorem 8.2.3, be such that $f(b) = b$ and $f(X_j) \neq X_j$. Then $fg(a) = b$ and $fg(X_i) \neq X_j$. This shows that $U_{i,j}$ is dense. It is easy to see that it is open.

By the Baire Category Theorem, $U = \bigcap_{i,j<\omega} U_{i,j}$ is comeager. Any $f \in U$ has the required property. □

Corollary 8.2.5 *Every countable recursively saturated model M has two countable recursively saturated elementary end extensions M_0 and M_1 such that*

$$\mathrm{Cod}(M_0/M) \cap \mathrm{Cod}(M_1/M) = \mathrm{Def}(M).$$

Proof Let M_0 be a countable recursively saturated elementary end extension of M and let $\mathfrak{X} = \mathrm{Cod}(M_0/M) \setminus \mathrm{Def}(M)$. Let $f \in \mathrm{Aut}(M)$ be such that for all $A \in \mathfrak{X}$, $f(A) \notin \mathfrak{X}$. Then $f : M \longrightarrow M$ is an elementary embedding of M onto an elementary cut of M_0. After identifying M with its image $f(M)$, the embedding gives a recursively saturated elementary end extension $M \prec_{\mathsf{end}} M_1$. Then for each undefinable $A \subseteq M$, if A is coded in M_0, it is not coded in M_1; hence $\mathrm{Cod}(M_0/M) \cap \mathrm{Cod}(M_1/M) = \mathrm{Def}(M)$. □

8.3 Moving gaps

If a and b are elements of a countable recursively saturated model M and for all $n < \omega$, $a + n < b$, then there are automorphisms of M which fix a and b and move some elements inside the interval $[a, b]$. In particular, for every nonstandard $a \in M$, there are an automorphism f and an element $c \in \mathrm{gap}(a)$ such that $f(c) \neq c$ and $f(c) \in \mathrm{gap}(a)$. Now, if $c \in \mathrm{gap}(a)$ and $f(c) \in \mathrm{gap}(a)$, then $f(\mathrm{gap}(a)) = \mathrm{gap}(a)$. We also have a more interesting result.

Proposition 8.3.1 *Let f be an automorphism of a model M and let $a \in M$ be such that $f(a) \in \mathrm{gap}(a)$. Then there is $c \in \mathrm{gap}(a)$ such that $f(c) = c$.*

Proof Suppose that $b = f(a) \in \mathrm{gap}(a)$. By the Blass–Gaifman Lemma 1.7.1, there are $c \in \mathrm{gap}(a)$ and a Skolem term $t(v)$ such that $t(a) = t(b) = c$. Then $f(c) = f(t(a)) = t(f(a)) = t(b) = c$. □

One can ask whether there is a nontrivial automorphism of M such that $f(a) \in \mathrm{gap}(a)$ for every a. The result known as The Moving Gaps Lemma implies that there is none.

Theorem 8.3.2 (Moving Gaps Lemma) *Suppose f is an automorphism of a countable recursively saturated model M. Then for every d such that $f(d) \neq d$ and for all $b, c \in M$ such that $d < \mathrm{gap}(b) < \mathrm{gap}(c)$, there is an a such that $\mathrm{gap}(b) < a < \mathrm{gap}(c)$ and $f(a) \notin \mathrm{gap}(a)$.*

Proof Let $d \in M$ be such that $f(d) \neq d$. Let $\varphi(x, d) = \exists y(\langle d, y \rangle = x)$. By Theorem 3.2.11, there is $a \in M$ such that $\mathrm{tp}(a)$ is rare and $M \models \varphi(d, a)$. By recursive saturation, we can assume that $\mathrm{gap}(b) < a < \mathrm{gap}(c)$. Then $d \in \mathrm{Scl}(a)$. Hence, for every $g \in \mathrm{Aut}(M)$, if $g(d) \neq d$, then $g(a) \neq a$. Thus, $f(a) \neq a$. Since $\mathrm{tp}(a)$ is rare, it follows that $f(a) \notin \mathrm{gap}(a)$. □

Corollary 8.3.3 *Let M be a recursively saturated model and let $f \in \mathrm{Aut}(M)$ be such that for all $a \in M$, $f(a) \in \mathrm{gap}(a)$. Then $f = \mathrm{id}$.* □

8.4 Back-and-forth

Almost every theorem about automorphisms of countable recursively saturated model is proved by a back-and-forth construction. One enumerates the model and then proceeds with the construction of a sequence $\sigma_0 \subseteq \sigma_1 \subseteq \cdots$ of finite partial automorphisms making sure that all elements in M are in the domain and in the range of $\bigcup_{n<\omega} \sigma_n$. Crucial in each such construction is a back-and-forth lemma which guarantees that the process can be continued preserving some required property. In other words, if σ is a partial finite automorphism with the required property, we need to show that for any $a \in M$, there is $b \in M$ such that the extension $\sigma \cup \{(a, b)\}$ is a partial automorphism with the same property. This is the "forth" step. Similarly, we must be able to extend σ by choosing b first and then finding the appropriate a. This constitutes the "back" step. Typically, the "back" and the "forth" steps are symmetric, so only one is explained in detail. The same approach is used for constructing isomorphisms between two countable structures. In this case, sequences of partial finite isomorphisms are built using enumerations of both structures.

In this section, we present one important back-and-forth construction. It is a prototype for many similar constructions.

Definition 8.4.1 If f is an automorphism of a model M, then

$$I_{\mathrm{fix}}(f) = \{a \in M : \forall x < a(f(x) = x)\}.$$

For every automorphism f, $I_{\mathrm{fix}}(f)$ is a cut of M. Suppose $f(x) = x$ for all $x < a$. Consider an element $b < 2^a$. Let $b = \Sigma_{i=1}^{c} 2^{x_i}$ be the binary expansion of b, where $x_1 < \cdots < x_c < a$, and $c \leq a$. Then

$$f(b) = \Sigma_{i=1}^{f(c)} 2^{f(x_i)} = \Sigma_{i=1}^{c} 2^{x_i} = b.$$

This proves that every cut of the form $I_{\mathrm{fix}}(f)$ is closed under exponentiation. This turns out to be the only restriction.

Theorem 8.4.2 *Suppose M is a countable recursively saturated model and I is a cut closed under exponentiation. Then there is $f \in \mathrm{Aut}(M)$ such that $I_{\mathrm{fix}}(f) = I$.* □

Theorem 8.4.2 is a consequence of the following two back-and-forth lemmas.

Recall that, according to our conventions on coding, if $a \in M$ and a set $x \subseteq a_M$ is coded in M, then x has a code below 2^a.

Lemma 8.4.3 *Let M be recursively saturated and suppose that $a, b, c \in M$ are such that for all $x < 2^{2^c}$, $(M, x, a) \equiv (M, x, b)$. Then for each $a' \in M$ there is $b' \in M$ such that, for all $x < c$, $(M, x, a, a') \equiv (M, x, b, b')$.*

Proof Let a' be given. Consider the recursive type $p(w)$ with parameters a, a', b, c consisting of all those formulas

$$\forall x < c\ [\varphi(x, a, a') \longleftrightarrow \varphi(x, b, w)],$$

where $\varphi(x, y, w)$ is a ternary $\mathcal{L}$-formula.

We will finish the proof when we show that $p(w)$ is finitely realizable. Let $\varphi_0(u, v, w), \ldots, \varphi_n(u, v, w)$ be given. Let

$$D_i = \{x < c : M \models \varphi_i(x, a, a')\}.$$

Then $D_i < 2^c$ and

$$M \models \exists w \bigwedge_{i=0}^{n} [\forall x < c(\varphi_i(x, a, w) \longleftrightarrow x \in D_i)], \tag{$*$}$$

since a' is a witness. Since the set of parameters in $(*)$ has a code which is smaller than 2^{2^c}, by the hypothesis of the lemma, $(*)$ is true with a replaced by b, and the result follows. □

Lemma 8.4.3 can be used to construct an automorphism fixing every element of I and sending a to b. To finish the proof of Theorem 8.4.2 we need another lemma which tells us how to make sure that the automorphism moves arbitrarily small elements above I.

Lemma 8.4.4 *Let M be recursively saturated. Then for every $a \in M$ and every nonstandard c there are $a', a'' < 2^{c^2}$ such that $a' \neq a''$ and for all $x < c$, $(M, x, a, a') \equiv (M, x, a, a'')$.*

Proof For $n < \omega$, let $\mathrm{tp}^n(x)$ be the type of x restricted to the first n formulas in some fixed recursive enumeration of all unary formulas. Consider the recursive type $q(v, w)$ with parameters a, c, d:

$$\{\forall x < c\ (\mathrm{tp}^n(x, a, v) = \mathrm{tp}^n(x, a, w)) : n < \omega\} \cup \{(v \neq w)\} \cup \left\{v, w < 2^{c^2}\right\}.$$

We will show that $q(v,w)$ is finitely realizable. If not, then there is $n<\omega$ such that

$$M \models \forall v, w < 2^{c^2} [\forall x < c\ (\mathrm{tp}^n(x,a,v) = \mathrm{tp}^n(x,a,w) \longrightarrow v = w)]. \qquad (*)$$

Now consider the equivalence relation on $[0, 2^{c^2}-1]$ defined by

$$v \sim w \Longleftrightarrow \forall x < c\ (\mathrm{tp}^n(x,a,v) = \mathrm{tp}^n(x,a,w)).$$

According to $(*)$, this relation has exactly 2^{c^2} equivalence classes. However, it is clear from the definition that there cannot be more than 2^{cn} such classes, which is a contradiction. □

Proof of Theorem 8.4.2 Let I be a cut of M closed under exponentiation. We enumerate $M \setminus I$ and construct an automorphism f, such that $f(x) = x$ for all $x \in I$, by a back-and-forth procedure involving Lemma 8.4.3. Each couple of back-and-forth steps is followed by an extra step to make sure that f moves points arbitrarily close to I. We fix a sequence $d_0 > d_1 > \cdots$ coinitial in $M \setminus I$. Suppose that $a_i \mapsto b_i$, $i < n < \omega$, is a finite partial automorphism and there is $c > I$ such that for all $x < c$, $(M, x, \bar{a}) \equiv (M, x, \bar{b})$. (For $n = 0$, any $c > I$ will do.) First, we use Lemma 8.4.4 to get $c' > I$ and $a', a'' < d_n$ such that $a' \neq a''$ and for all $x < c'$,

$$(M, x, \bar{a}, a') \equiv (M, x, \bar{a}, a''). \qquad (*)$$

Notice that there is such $c' > I$, because I is closed under exponentiation. Applying Lemma 8.4.3, we get b_n and $c'' > I$ such that for all $x < c''$, $(M, x, \bar{a}, a') \equiv (M, x, \bar{b}, b_n)$. Then we define $a_n = a'$ if $a' \neq b_n$ and $a_n = a''$ otherwise. □

With possibly one restriction, we can improve the statement of Theorem 8.4.2 a bit. We say that a cut $I \subseteq_{\mathsf{end}} M$ is *downward ω coded by a* if there is $a \in M$ such that $I = \inf\{(a)_n : n < \omega\}$. A cut is *downward ω coded* in M if it is downward ω coded by some $a \in M$.

Theorem 8.4.5 *Let M be a countable recursively saturated model. Suppose that $I \subseteq_{\mathsf{end}} M$ is a cut closed under exponentiation and not downward ω coded in M, and let $a, b \in M$. If $(M, x, a) \equiv (M, x, b)$ for all $x \in I$, then there is $f \in \mathrm{Aut}(M)$ such that $I_{\mathrm{fix}}(f) = I$ and $f(a) = b$.*

Proof Let $\langle \varphi_i(x, y) : i < \omega \rangle$ be a recursive enumeration of the formulas of $\mathcal{L}$ with the free variables as shown. Let

$$d_n = \max\Big\{d : d < a \wedge \forall x < d \bigwedge_{i \le n} [\varphi_i(x, a) \longleftrightarrow \varphi_i(x, b)]\Big\}.$$

By the assumption, $d_n > I$ for each $n < \omega$. By recursive saturation, there is $d \in M$ such that $(d)_n = d_n$ for all $n < \omega$. Hence, there is $e > I$ such that for all $n < \omega$, $d_n > e$, and the theorem follows from Lemmas 8.4.3 and 8.4.4 (Do it!). $\square$

The assumption that I is not downward ω coded cannot be eliminated from Theorem 8.4.5. Let a, b realize the same minimal type in M and assume that $\mathrm{gap}(a) < b$. Then for $I = \inf(\mathrm{gap}(a))$, $(M, x, a) \equiv (M, x, b)$ for all $x \in I$. However, if $f \in \mathrm{Aut}(M)$ is such that $I_{\mathrm{fix}}(f) = I$, then $f(\mathrm{gap}(a)) = \mathrm{gap}(a)$. Hence $f(a) \neq b$.

The following simple corollary will be used in Chapter 9.

Corollary 8.4.6 *Let M be a countable recursively saturated model. Suppose $I \prec_{\mathrm{end}} M$ is not downward ω coded. Then for all $a, b, c, d \in M$ such that $I < a < b$ and $I < \mathrm{gap}(c) < \mathrm{gap}(d)$, there is $f \in \mathrm{Aut}(M)$ such that $I = I_{\mathrm{fix}}(f)$ and $c < f(a) < f(b) < d$.*

Proof Since I is not downward ω coded, $I \neq \inf(\mathrm{gap}(e))$ for all $e \in M$. Then there are a_0, a_1, a_2, b_0, b_1 which realize the same minimal type in M such that $I < a_0 < a_1 < a < b < a_2$ and $c < b_0 < b_1 < d$. Since minimal types are strongly indiscernible, $(M, x, a_1, a_2) \equiv (M, x, b_1, b_2)$ for all $x < a_0$. Hence, the result follows from Theorem 8.4.5. $\square$

We finish this section with an application of Theorem 8.4.5.

Theorem 8.4.7 *Let I be a cut of a model M, and assume that I is closed under exponentiation. If $a \in M$ is definable in the structure (M, I), then $a \in \mathrm{Scl}(b)$ for some $b \in I$.*

Proof Suppose to the contrary that for every $b \in I$, $a \notin \mathrm{Scl}(b)$. The same is true in every structure which is elementarily equivalent to (M, I, a). Hence, without loss of generality, we can assume that (M, I) is countable and recursively saturated. In particular, I is not downward ω coded. Let $p(x)$ be the recursive type

$$\{x \neq a\} \cup \{\forall y \in I[\varphi(a, y) \longleftrightarrow \varphi(x, y)] : \varphi \in \mathsf{Form}\}.$$

We will show that $p(x)$ is finitely realizable in (M, I). To this end, let $\varphi_i(x, y)$, for $i \leq n$ be given. For $c \in M$ let D_c be the code of the set

$$\{\langle y, i\rangle : M \models y < c \wedge \varphi_i(a, y)\}.$$

Since I is closed under exponentiation, $D_c \in I$ for each $c \in I$. Hence, for all $c \in I$,

$$M \models \exists x[x \neq a \wedge \forall y < c(\bigwedge_{i \leq n}(\langle y, i\rangle \in D_c \longleftrightarrow \varphi_i(x, y))], \qquad (*)$$

because otherwise the formula $\forall y < c(\bigwedge_{i \leq n}(\langle y, i\rangle \in D_c \longleftrightarrow \varphi_i(x, y))$ would define a in M from the parameter $D_c \in I$. Now, by overspill, there is $c > I$ for which $(*)$ holds. This shows that $p(x)$ is finitely realizable. Let $b \in M$ realize $p(x)$. By Theorem 8.4.5, there is $f \in \mathrm{Aut}(M, I)$ such that $f(a) = b$. Hence a is not definable in (M, I), which is a contradiction. □

8.5 Extending automorphisms

If f is an automorphism of M and $M \prec_{\mathsf{end}} N$, is there an automorphism g of N which extends f? If f has such an extension, then, for every $X \subseteq M$, X is coded in N iff $f(X)$ is (Do it!). In other words, if f extends g, then it induces a permutation of $\mathrm{Cod}(N/M)$. Hence, we can identify f with an automorphism of the second-order structure $(M, \mathrm{Cod}(N/M))$. We prove that, with an exception, the converse holds. In the discussion below, we think of $\mathrm{Aut}((M, \mathrm{Cod}(N/M)))$ as a subgroup of $\mathrm{Aut}(M)$.

Theorem 8.5.1 *Let N be countable and recursively saturated and suppose that $M \prec_{\mathsf{end}} N$ and M is not downward ω coded in N. Then every $f \in \mathrm{Aut}((M, \mathrm{Cod}(N/M)))$ extends to an automorphism of N.*

Proof The proof can be carried out in the usual back-and-forth fashion, provided we prove that for all $\bar{a}, \bar{b}, a \in N$, if for all $x \in M$, $(N, \bar{a}, x) \equiv (N, \bar{b}, f(x))$, then there is b such that for all $x \in M$, $(N, \bar{a}, a, x) \equiv (N, \bar{b}, b, f(x))$. Let $M < e < N$. By the recursive saturation there is an N-finite set α such that for any formula $\varphi(\bar{u}, v, x)$,

$$N \models \forall x < e[\langle \varphi, x\rangle \in \alpha \longleftrightarrow \varphi(\bar{a}, a, x)].$$

Let $S = \alpha \cap M$. Since $S \in \mathrm{Cod}(N/M)$ we let $S' = f(S) \in \mathrm{Cod}(N/M)$ and let N-finite β be such that $S' = \beta \cap M$. For every nonstandard $c \in M$, a is a witness to the existential statement

$$N \models \exists u \forall x < c[\langle \varphi, x\rangle \in S \cap [0, c^2] \longleftrightarrow \varphi(\bar{a}, u, x)].$$

By the assumption on $\bar{a}$ and $\bar{b}$, we get that for every nonstandard $c \in M$

$$N \models \exists u \forall x < f(c)[\langle \varphi, x\rangle \in S' \cap [0, f(c)^2] \longleftrightarrow \varphi(\bar{b}, u, x)].$$

Since f is an automorphism, this implies that for every nonstandard $c \in M$

$$N \models \exists u \forall x < c[\langle \varphi, x\rangle \in S' \cap [0, c^2] \longleftrightarrow \varphi(\bar{b}, u, x)].$$

Then the same is true with S' replaced with β. For each $\varphi(\bar{u}, v, x)$, define d_φ to be the largest $d \in N$ such that

$$N \models \exists u \forall x < d[\langle \varphi, x\rangle \in \beta \cap [0, d^2] \longleftrightarrow \varphi(\bar{b}, u, x)].$$

By overspill, each $d_\varphi > M$ and then by the fact that M is not downward ω coded, there is $d \in N$ such that $M < d < d_\varphi$ for each φ (DO IT!). Thus,

$$N \models \exists u \forall x < d[\langle \varphi, x\rangle \in \beta \cap [0, d^2] \longleftrightarrow \varphi(\bar{b}, u, x)],$$

for every φ. By recursive saturation of N, there is $b \in N$ which simultaneously witnesses all of these sentences. Thus, for every φ and every $e \in M$,

$$N \models \langle \varphi, e\rangle \in S' \longrightarrow \varphi(\bar{b}, b, e).$$

Now we want to show that this b works. Indeed, for each $e \in M$, $N \models \varphi(\bar{a}, a, e)$ $\Longleftrightarrow N \models \langle \varphi, e\rangle \in S \Longleftrightarrow N \models \langle \varphi, f(e)\rangle \in S' \Longleftrightarrow N \models \varphi(\bar{b}, b, f(e))$.

Notice that in the beginning stages of the proof we considered nonstandard c. Hence the proof works for nonstandard M. If M is standard, replace c^2 with nc, where $n < \omega$ is chosen so that $n > \varphi$. □

Essentially the same proof also gives the following theorem.

Theorem 8.5.2 *Suppose $M_0 \prec_{\mathsf{end}} N_0$, $M_1 \prec_{\mathsf{end}} N_1$, N_0 and N_1 are countable and recursively saturated, and M_i is not downward ω coded in N_i for $i = 0, 1$. If $f : M_0 \longrightarrow M_1$ is an isomorphism such that for all $X \subseteq M_0$, X is coded in N_0 iff $f(X)$ is coded in N_1, then f can be extended to an isomorphism $g : N_0 \longrightarrow N_1$.* □

Theorem 8.5.1 gives an extendability criterion, but it has nothing to say about the question whether there are any nontrivial automorphisms of M which extend to an automorphism of N? As the next theorem shows, the picture is rather complex.

Theorem 8.5.3 *Let G be the automorphism group of a countable linearly ordered first-order structure. Then every countable recursively saturated model M has an elementary cut I such that* $\mathrm{Aut}(I, \mathrm{Cod}(M/I)) \cong G$. □

Theorem 8.5.3 is closely related to the results in Section 5.4. Here is one application. Let G be $\mathrm{Aut}(\mathbb{Z}, <)$, and let I be as in Theorem 8.5.3 for this G. Let $F : \mathrm{Aut}(I, \mathrm{Cod}(M/I)) \longrightarrow G$ be an isomorphism. There is $f \in \mathrm{Aut}(M)$ such that $F(f \restriction I)$ is the automorphism $n \mapsto n+1$ of $\mathbb{Z}$. Then there is no $h \in \mathrm{Aut}(M)$ such that $f = h^2$ (DO IT!). Hence, we have the following corollary.

Corollary 8.5.4 *For every countable recursively saturated model M, the automorphism group of M is not divisible.* □

Theorem 8.5.3 can also be used to prove the following.

Theorem 8.5.5 *If G is the automorphism group of a countable recursively saturated model, then* $\mathrm{Th}(G)$ *is essentially undecidable.* □

Complete proofs of Theorems 8.5.3 and 8.5.5 are given in [174].

Theorem 8.5.3 implies that every countable recursively saturated model M has a recursively saturated elementary end extension N such that no nontrivial automorphism of M extends to N. The following related question remains open. See Exercise 8.11.20 for an answer in a special case.

Problem 8.5.6 *Let M be countable and recursively saturated. Is there a nontrivial automorphism of M which cannot be extended to an automorphism of any countable recursively saturated elementary end extension of M?*

We finish this section with a proof that the assumption that M is not downward ω coded cannot be eliminated in Theorem 8.5.1.

Theorem 8.5.7 *Let M be a countable recursively saturated model. Then there are $K \prec_{\mathsf{end}} M$ and $f \in \mathrm{Aut}(K, \mathrm{Cod}(M/K))$ which has no extension to an automorphism of M.*

Proof By Theorem 3.2.11, there are $s, e \in M$ such that $s < \mathrm{gap}(e)$, $s \in \mathrm{Scl}(e)$, and $\mathrm{tp}(e)$ is rare. Let us fix such s and e, and let $K = \inf(\mathrm{gap}(e))$. Then every automorphism of M which moves s must move e to another gap; hence, it also moves K setwise.

Now we outline the rest of the proof. The details are provided in the three lemmas below. By Lemma 8.5.10, there are $a \in \mathrm{gap}(e)$ and a model $N \subseteq_{\mathsf{end}} M$ such that $K = \inf\{(a)_n : n < \omega\}$, $K \prec_{\mathsf{end}} N$, N is recursively saturated, and for each $n < \omega$, $\mathrm{gap}^N((a)_{n+1}) < (a)_n$. Clearly, $\mathrm{tp}^M(a) \neq \mathrm{tp}^N(a)$. Without loss of generality, we can assume that all $(a)_n$, for $n < \omega$, realize the same indiscernible type (DO IT!). Let $s' \in K$ be such that $s \neq s'$ and $\mathrm{tp}(s) = tp(s')$. Consider the recursive type $p(v, s, s')$

$$\{(v)_n = (a)_n : n < \omega\} \cup \{\varphi(v, s) \longleftrightarrow \varphi(v, s') : \varphi(v, w) \in \mathsf{Form}\}.$$

Since every indiscernible type is strongly indiscernible, $p(v, s, s')$ is finitely realizable, so, without loss of generality, we can assume that a realizes $p(v, s, s')$; hence $\mathrm{tp}^N(a, s) = \mathrm{tp}^N(a, s')$. Let $g \in \mathrm{Aut}(N, a)$ be such that $g(s) = g(s')$. Notice that $g(K) = K$. Let $f = g \restriction K$. Since M and N code the same subsets of K, $f \in \mathrm{Aut}(K, \mathrm{Cod}(M/K))$, but f cannot be extended to an automorphism of M, because, as noted earlier, each such automorphism moves K setwise. □

In the lemmas below we use the auxiliary functions F_n. Recall that Tr_{Σ_n} is the universal Σ_n truth formula (See Corollary 1.11.2). Let

$$F_n(x) = \min\{y : \forall\varphi, u < x[\exists v\, \mathsf{Tr}_{\Sigma_n}(\varphi, \langle u, v\rangle) \longrightarrow \exists v < y\, \mathsf{Tr}_{\Sigma_n}(\varphi, \langle u, v\rangle)]\}.$$

Notice that the formula $F_n(x) = y$ is Σ_{n+1} in PA. Also, if t is a Σ_n Skolem term, then, for every model M and all $a \in M$, $M \models \forall x < a[t(x) < F_n(a)]$.

We use the symbol $\equiv_n$ to denote elementary equivalence for Σ_n formulas.

The proof of the first lemma is left as an exercise.

Lemma 8.5.8 *Let M be a model. If $a, b \in M$ are such that $M \models F_n(a) < b$, then there are arbitrarily large $c \in M$ such that $(M, a, b) \equiv_n (M, a, c)$.* □

In the next lemma, contrary to the usual convention, the finite sequences $a_0, \ldots, a_n$ and $b_0, \ldots, b_n$ are decreasing.

Lemma 8.5.9 *Let M be a countable model and let $a_0, \ldots, a_n \in M$ be such that for all i, with $0 < i \leq n$, $M \models F_{2i}(a_i) < a_{i-1}$. Then there are $b_0, \ldots, b_n \in M$ such that $b_n = a_n$, $\mathrm{gap}(b_n) < \mathrm{gap}(b_{n-1}) < \cdots < \mathrm{gap}(b_0)$, and*

$$(M, a_0, \ldots, a_n) \equiv_1 (M, b_0, \ldots, b_n).$$

Proof We proceed by induction on n. For $n = 0$, let $b_0 = a_0$. Suppose that the statement in the lemma is true for some n and the sequence $a_0, \ldots, a_n, a_{n+1}$ satisfies the assumption of the lemma. In particular, $M \models F_{2n+2}(a_{n+1}) < a_n$. Applying Lemma 8.5.8 to $a = a_{n+1}$ and $b = \langle a_0, \ldots, a_n\rangle$, we get $\langle b'_0, \ldots, b'_n\rangle$ such that $\mathrm{Scl}(a_{n+1}) < b'_n$ and

$$(M, a_0, \ldots, a_n, a_{n+1}) \equiv_{2n+1} (M, b'_0, \ldots, b'_n, a_{n+1}).$$

Since the formula $F_{2i}(x) < y$ is Σ_{2i+1}, it follows that the inductive assumptions are satisfied for the sequence $b'_0, \ldots, b'_{n-1}, \langle b'_n, a_{n+1}\rangle$. Hence, there is a sequence $b_0, \ldots, b_n$, where $b_n = \langle b'_n, a_{n+1}\rangle$, which satisfies the conclusion of the lemma. Then the sequence $b_0, \ldots, b_{n-1}, b'_n, a_{n+1}$ has the required properties. □

Lemma 8.5.10 *Let $a \in M$, and let $K = \inf(\mathrm{gap}(a))$. If for all $i < \omega$, $M \models F_{2i}((a)_i) < (a)_{i-1}$ and $K = \inf\{(a)_n : n < \omega\}$, then there is a recursively saturated model N such that:*

(1) $a \in N \subseteq_{\mathsf{end}} M$;

(2) *for all* $n < \omega$, $\mathrm{Scl}^N((a)_{n+1}) < (a)_n$;

(3) $K \prec_{\mathsf{end}} N$.

Proof Let T be the recursive theory in $\mathcal{L} \cup \{I, a\}$:

$$\{a \in I \subseteq_{\mathsf{end}} M\} \cup \{I \models \mathsf{PA}^*\} \cup \{I \models F_n((a)_{k+1}) < (a)_k : k, n < \omega\}.$$

We will show that T is consistent. Let ψ be a finite fragment of T and let k be the largest such that $(a)_k$ occurs in ψ. By Lemma 8.5.9, there are $b_0, \dots, b_k$ such that

$$(M, (a)_0, \dots, (a)_k) \equiv_1 (M, b_0, \dots, b_k)$$

and $\mathrm{gap}(b_k) < \mathrm{gap}(b_{k-1}) < \cdots < \mathrm{gap}(b_0)$. Let $J = \sup(\mathrm{gap}(b_0))$. By Friedman's Embedding Theorem (Theorem 1.13.1) there is $I \subseteq_{\mathsf{end}} M$ such that $(a)_0 \in I$ and

$$(J, b_0, \dots, b_k) \cong (I, (a)_0, \dots, (a)_k).$$

Then $(M, I, a) \models \psi$.

By chronic resplendency, there is $N \subseteq_{\mathsf{end}} M$ such that (M, N, a) is a recursively saturated model of T. Since for every $b \in K$, $\mathrm{gap}^N(b) \subseteq K$, $K \prec_{\mathsf{end}} N$, and the result follows. □

8.6 Maximal automorphisms

The Moving Gaps Lemma implies that the identity is the only automorphism that fixes all gaps. Now we consider a dual problem: can there be an automorphism that moves all gaps other than least one? It turns out that the countable recursively saturated models with such automorphisms are exactly the arithmetically saturated models.

Recall that an element a of a structure $\mathfrak{A}$ is *algebraic* if there is a formula φ of the language of $\mathfrak{A}$ such that $\varphi(\mathfrak{A})$ is finite and $\mathfrak{A} \models \varphi(a)$. If M is a model of PA^*, then the algebraic elements of M are the definable elements.

Definition 8.6.1 If $\mathfrak{A}$ is a first-order structure and f is an automorphism of $\mathfrak{A}$, then we say that f is *maximal* if $f(a) \neq a$ for all nonalgebraic elements $a \in \mathfrak{A}$.

The main result of this section is that a countable recursively saturated model has a maximal automorphism iff it is arithmetically saturated. This follows from Theorem 8.6.3 and Corollary 8.6.7 below. We state these results separately because of the nature of their proofs and because of their further applications. We begin with a back-and-forth lemma.

Lemma 8.6.2 *Suppose that M is arithmetically saturated and $a, b \in M$ are such that $\mathrm{tp}(a) = \mathrm{tp}(b)$ and for each Skolem term $t(v)$ either $t(a) \in \mathrm{Scl}(0)$ or $t(a) \neq t(b)$. Then, for each $a' \in M$, there is $b' \in M$ such that $\mathrm{tp}(a, a') = \mathrm{tp}(b, b')$ and for each Skolem term $t(v, x)$, either $t(a, a') \in \mathrm{Scl}(0)$ or $t(a, a') \neq t(b, b')$.*

Proof Let M, a, a', and b be as in the lemma. To find b', we consider the type $p(x)$ with parameters a, a', b

$$\{\varphi(a,a') \longleftrightarrow \varphi(b,x) : \varphi(v,x) \in \mathsf{Form}\} \cup$$

$$\{t(a,a') \neq t(b,x) : t(v,x) \in \mathsf{Term} \text{ and } t(a,a') \notin \mathrm{Scl}(0)\}\,.$$

Since the type $p(x)$ is arithmetic in $\mathrm{tp}(a,a')$, to finish the proof it suffices to show that it is finitely realizable in M. Suppose it is not. Then there are a formula $\varphi(v,x)$ and Skolem terms $t_0(v,x)$,...,$t_k(v,x)$ such that $M \models \varphi(a,a')$ and $t_i(a,a') \notin \mathrm{Scl}(0)$ for $i \leq k$ and

$$M \models \forall x[\varphi(b,x) \longrightarrow \bigvee_{i=0}^{k}(t_i(b,x) = t_i(a,a'))].$$

Let us assume that k is the least number for which there are such a formula $\varphi(v,x)$ and such terms $t_0(v,x), \ldots, t_k(v,x)$. Since there is $b' \in M$ such that $\mathrm{tp}(a,a') = \mathrm{tp}(b,b')$, some terms $t(x,v)$ must contribute to inconsistency of $p(x)$; hence $k \geq 0$.

First notice that

$$M \models \exists y_0, \ldots, y_k \forall x[\varphi(b,x) \longrightarrow \bigvee_{i=0}^{k}(t_i(b,x) = y_i)].$$

Thus, there are Skolem terms $s_0(v), \ldots, s_k(v)$ such that

$$M \models \forall x[\varphi(b,x) \longrightarrow \bigvee_{i=0}^{k}(t_i(b,x) = s_i(b))]. \tag{$*$}$$

Since $\mathrm{tp}(a) = \mathrm{tp}(b)$, the same is true when b is replaced by a. Hence, $t_i(a,a') = s_i(a)$ for some $i \leq k$; without loss of generality, we can assume that $i = 0$.

Let $\varphi'(v,x)$ be $\varphi(v,x) \wedge t_0(v,x) = s_0(v)$. Then $M \models \varphi'(a,a')$ and we claim that

$$M \models \forall x[\varphi'(b,x) \longrightarrow \bigvee_{i=1}^{k}(t_i(b,x)) = t_i(a,a')].$$

Suppose it is not. Then it follows from $(*)$ that there must be $c \in M$ such that

$$M \models \varphi(b,c) \wedge t_0(b,c) = s_0(b) \wedge t_0(b,c) = t_0(a,a').$$

Then $s_0(a) = s_0(b)$ and, by the assumption of the lemma, $s_0(a) \in \mathrm{Scl}(0)$. Hence $t_0(a, a') \in \mathrm{Scl}(0)$, which contradicts the assumption that $t_0(a, a') \notin \mathrm{Scl}(0)$.

Thus, we have proved the claim. Since k was chosen to be minimal, we get a contradiction; hence $p(x)$ is finitely realizable. □

Theorem 8.6.3 *Every countable arithmetically saturated model has a maximal automorphism.*

Proof Using Lemma 8.6.2 we can construct an $f \in \mathrm{Aut}(M)$ in a back-and-forth process in which the following inductive condition is satisfied: if $f(\bar{a}) = \bar{b}$, then for each Skolem term $t(v)$, either $t(\bar{a}) \in \mathrm{Scl}(0)$ or $t(\bar{a}) \neq t(\bar{b})$. Clearly, this f is maximal. □

If f is an automorphism of a model M, then the *fixed point set* of f, $\mathrm{fix}(f) = \{x \in M : f(x) = x\}$, is closed under Skolem terms and hence, is a universe of an elementary submodel of M. Our goal now will be to, at least partially, classify those $K \prec M$ which are of the form $\mathrm{fix}(f)$ for some $f \in \mathrm{Aut}(M)$, for countable recursively saturated M.

Theorem 8.6.3 implies that every finitely generated submodel of a countable arithmetically saturated model is a fixed point set. Hence each such model has at least countably many nonisomorphic fixed point sets. Surprisingly, if M is countable recursively saturated but not arithmetically saturated, then there is only one isomorphic type of the fixed point sets of M, namely the type of M, that is, the fixed point set of the identity. To prove this we need a lemma which also has other applications.

Lemma 8.6.4 *Let M be a nonstandard model, and suppose that $e \in M$ has the property that for each nonstandard n there is $i < \omega$ such that $(e)_i < n$. Then for each $f \in \mathrm{Aut}(M)$, there is a nonstandard n such that for all $i < \omega$, if $(e)_i < n$, then $f((e)_i) = (e)_i$.*

Proof Let $e \in M$ be as in the lemma, and let $f \in \mathrm{Aut}(M)$. Let $e' = f(e)$. Then for each $n < \omega$,

$$M \models \forall i < n[(e)_i < n \longrightarrow (e')_i = (e)_i].$$

Hence, by overspill, the same is true for some nonstandard n. Let $i < \omega$ be such that $(e)_i < n$. Then $(e')_i = (e)_i$ and $f((e)_i) = (e')_{f(i)} = (e')_i = (e)_i$. □

If $\mathbb{N}$ is not strong in a model M, then there is $e \in M$ such that for each nonstandard n, there is $i < \omega$ such that $\mathbb{N} < (e)_i < n$. Hence, we get the following corollary:

Corollary 8.6.5 *Let M be a nonstandard model in which $\mathbb{N}$ is not strong. Then for every $f \in \mathrm{Aut}(M)$, there is arbitrarily small nonstandard d such that $f(d) = d$.* □

Theorem 8.6.6 *Let M be a countable recursively saturated model, and let $f \in \mathrm{Aut}(M)$ be such that for every nonstandard $n \in M$, there is d such that $\mathbb{N} < d < n$ and $f(d) = d$. Then $\mathrm{fix}(f) \cong M$.*

Proof Our task is to show that $\mathrm{fix}(f)$ is recursively saturated and $\mathrm{SSy}(\mathrm{fix}(f)) = \mathrm{SSy}(M)$. This is accomplished by showing that for every $a \in \mathrm{fix}(f)$, every complete type $p(v, a)$ realized in M is realized by some element $e \in \mathrm{fix}(f)$. Let $\langle \varphi_n(v, a) : n < \omega \rangle$ be an enumeration of $p(v, a)$ coded in M. Then the sequence $c_n = \min\{x \in M : M \models \bigwedge_{i \le n} \varphi_i(x, a)\}$ is well-defined (since $p(v, a)$ is realized in M) and is coded in M. Let c be a code and let $d = f(c)$. Then $c_n = (c)_n = (d)_n$ for each $n \in \mathbb{N}$. By recursive saturation and the assumption on f, there is a nonstandard $m \in M$ such that $f(m) = m$, $(c)_m = (d)_m$, and $M \models \varphi_n((c)_m, a)$ for every $n \in \mathbb{N}$. Thus, $(c)_m$ realizes $p(v, a)$, and $f((c)_m) = (d)_{f(m)} = (d)_m = (c)_m$. □

Corollary 8.6.7 *Let M be a countable recursively saturated model which is not arithmetically saturated. Then for every $f \in \mathrm{Aut}(M)$, $\mathrm{fix}(f) \cong M$.*

Proof Directly from Corollary 8.6.5 and Theorem 8.6.6. □

From Theorem 8.6.3 and Corollary 8.6.7, we obtain a list of conditions characterizing arithmetic saturation.

Corollary 8.6.8 *Let M be countable and recursively saturated. Then the following conditions are equivalent:*

(1) *M is arithmetically saturated;*
(2) *There is $f \in \mathrm{Aut}(M)$ such that $\mathrm{fix}(f) = \mathrm{Scl}(0)$;*
(3) *For every finitely generated $K \prec M$, there is $f \in \mathrm{Aut}(M)$ such that $\mathrm{fix}(f) = K$.*
(4) *There is $f \in \mathrm{Aut}(M)$ such that $\mathrm{fix}(f) \ncong M$.* □

8.7 Fixing strong cuts

By the results of the previous section, if M is a countable recursively saturated model of TA, then there is an $f \in \mathrm{Aut}(M)$ such that $\mathrm{fix}(f) = \mathbb{N}$ iff $\mathbb{N}$ is strong in M. In this section, we generalize this to arbitrary elementary cuts in an arbitrary countable recursively saturated model. A proof similar to the proof of Lemma 8.6.4 can be given to show that if $I \subseteq_{\mathsf{end}} M$ is not strong and $f \in \mathrm{Aut}(M)$ is such that $I \subseteq \mathrm{fix}(f)$, then $I \neq \mathrm{fix}(f)$ (Do it!). Hence we have the following:

Proposition 8.7.1 *Let I be a cut of a model M and suppose that $f \in \mathrm{Aut}(M)$ is such that $\mathrm{fix}(f) = I$. Then I is strong in M .* □

The main result of this section is the converse of Proposition 8.7.1 for countable recursively saturated models. We will prove an even stronger result.

Theorem 8.7.2 *Let I be a strong elementary cut of a countable recursively saturated model M. Then there is $f \in \mathrm{Aut}(M)$ such that $\mathrm{fix}(f) = I$ and for all $x > I$, $f(x) > x$.*

Proof Since I is strong, $\mathrm{Cod}(M/I)$ is closed under arithmetic definability. Therefore, $\mathcal{I} = (I, X_0, X_1, \dots) \models \mathsf{PA}^*$, where $X_0, X_1, \dots$ is an enumeration of all sets in $\mathrm{Cod}(M/I)$. Let $\mathcal{N} = (N, Y_0, Y_1, \dots)$ be a canonical $\mathbb{Z}$-extension of $\mathcal{I}$. Since the extension is conservative, $\mathrm{Cod}(M/I) = \mathrm{Cod}(N/I)$. We claim that N is recursively saturated. To this end, let S be a partial inductive satisfaction class for M (see Definition 1.9.1). Then there is i such that $S \cap M = X_i$, hence Y_i is a partial inductive satisfaction class for N. The claim now follows from Proposition 1.9.4. By Theorem 8.5.2, the identity function on I can be extended to an isomorphism $f : M \cong N$.

Since $\mathcal{N}$ is a $\mathbb{Z}$-extension, it has a set of generators $\langle a_i : i \in \mathbb{Z}\rangle$ over I, with all a_i realizing the same minimal type. Moreover, $a_i < a_j$ iff $i < j$, and the map $a_i \mapsto a_{i+1}$ extends to an automorphism f of $\mathcal{N}$ such that $f \restriction I = \mathrm{id}$. Then $\mathrm{fix}(f) = I$ and for all $x > I$, $f(x) > x$. Since $(M, I) \cong (N, I)$, the result follows. $\square$

If M is a model having nonstandard definable elements, then there is no $f \in \mathrm{Aut}(M)$ such that $f(x) > x$ for all undefinable x. Indeed, if $a < b$, where $a \notin \mathrm{Scl}(0)$ and $b \in \mathrm{Scl}(0)$, then, for every automorphism f, $f(a) > a$ iff $f(b-a) < b - a$. One can prove more:

Exercise 8.7.3 Let M be recursively saturated, and suppose that a be an element of $\mathrm{gap}(0) \setminus \mathrm{Scl}(0)$. Let Ω be a maximal convex subset of $\mathrm{gap}(0)$ such that $a \in \Omega$ and $\Omega \cap \mathrm{Scl}(0) = \emptyset$. If $f \in \mathrm{Aut}(M)$ is such that $f(a) > a$, then there is $b \in \Omega$ such that $f(b) < b$. (Hint 1: take a coded decreasing sequence $\langle a_n : n < \omega\rangle$ such that $\inf\{a_n : n < \omega\} = \sup(\Omega)$, and consider the sequence defined by $b_n = a_n - a$. Hint 2: we can assume that for arbitrarily small nonstandard $x \in M$, $f(x) \leq x$.)

Let us finish this section with a generalization of Theorem 8.7.2.

Theorem 8.7.4 *If I is a strong elementary cut of a recursively saturated model M and $f \in \mathrm{Aut}(I, \mathrm{Cod}(M/I))$, then there is $g \in \mathrm{Aut}(M)$ such that $f \subseteq g$ and $\mathrm{fix}(f) = \mathrm{fix}(g)$.*

Proof By Theorem 8.5.1, there is $h \in \mathrm{Aut}(M)$ such that $f \subseteq h$. Let us fix such an h.

Let $\mathcal{N}$ be as the proof of Theorem 8.7.2 except that instead of requiring $\mathcal{N}$ to be a canonical $\mathbb{Z}$-extension of $\mathcal{I}$, we now assume that $\mathcal{I}$ is a canonical $\mathbb{Q}$-extension. So suppose that $\mathcal{N} = \mathrm{Scl}^{\mathcal{N}}(I \cup \{a_q : q \in \mathbb{Q}\})$, a_q realizing the same minimal type, where $a_q < a_r$ iff $q < r$. Then $N \setminus M$ is the union of $\{\mathrm{gap}^*(a_q) : q \in \mathbb{Q}\}$, where $*$ indicates that the gaps are in the sense of $\mathcal{N}$.

Now we define an order preserving one-to-one map $\sigma : \mathbb{Q} \longrightarrow \mathbb{Q}$ by a back-and-forth construction. Suppose the partial mapping $(q_0, q_1, \dots, q_{n-1}) \mapsto (r_0, r_1, \dots, r_{n-1})$ satisfies the inductive assumption: for all $i, j < n$, $q_i < q_j$ iff $r_i < r_j$, and $f(a_{r_i}) \notin \text{gap}^*(a_{q_i})$. In the "forth" step, pick the first element q_n (in a fixed enumeration of $\mathbb{Q}$) not among $q_0, \dots, q_{n-1}$ and consider $f(a_{q_n})$. There are infinitely many r which we can chose for the extension $(q_0, q_1, \dots, q_{n-1}, q_n) \mapsto (r_0, r_1, \dots, r_{n-1}, r)$ to preserve ordering. We have $f(a_r) \in \text{gap}^*(a_{q_n})$ for at most one such r. Hence, we can define r_n so that $f(a_{r_n}) \notin \text{gap}^*(a_{q_n})$. The "back" step is symmetric.

Now, let g' be the unique extension of σ to an automorphism of $\mathcal{N}$ such that $g' \restriction I = \text{id}$, and let $g = fg'$. Then $f \circ (g \restriction I) = f$ and for each $i < \omega$,

$$g(a_{q_i}) = f(a_{r_i}) \notin \text{gap}^*(a_{q_i}).$$

Hence, $\text{fix}(g) = \text{fix}(f)$. □

8.8 Topology on the automorphism group

Throughout this section let M be a fixed countable recursively saturated model, and let $G = \text{Aut}(M)$.

Recall that for $a \in M$, the stabilizer of a is

$$G_a = \{f \in G : f(a) = a\}.$$

In this section we describe some relationships between types of elements and group-theoretic properties of their stabilizers. Let us also recall the following notation already used in the discussion on the automorphism groups of canonical I-extensions in Chapter 3. The *setwise stabilizer* of $X \subseteq M$ is

$$G_{\{X\}} = \{f \in G : f(X) = X\} = \text{Aut}(M, X).$$

The *pointwise stabilizer* of $X \subseteq M$ is

$$G_{(X)} = \{f \in G : \forall x \in X \ f(x) = x\} = \text{Aut}((M, a)_{a \in X}).$$

There is a natural topology on G. The *basic open subgroups* are pointwise stabilizers of finite sets. The *basic open subsets* are cosets of basic open subgroups. For models of PA^*, the topology is determined by cosets of stabilizers of single elements. For $f \in G$ and $H \subseteq G$, f is in the topological closure of H iff for any finite $A \subseteq M$, there is $g \in H$ such that $f \restriction A = g \restriction A$ (Do it!).

Let $\{a_0, a_1, \dots\}$ be an enumeration of M. We define a metric on G, by letting for $f \neq g$, $d(f, g) = 2^{-k}$, where k is the smallest such that $f(a_k) \neq g(a_k)$. One can verify that the topology on G defined by this metric agrees with the one defined via stabilizers of finite sets (Do it!). It is also easy to verify that G with this topology is complete and separable; hence G is a Polish space (DO IT!). Multiplication and inverse are continuous operations (DO IT!), so G is a Polish group.

The proposition below lists some useful properties of open subgroups of G. These facts apply to arbitrary automorphism groups, and they follow directly from the definitions.

Proposition 8.8.1 *Let H be a subgroup of G. Then:*

(1) *H is open iff H contains a basic open subgroup;*
(2) *If H is open, then it is closed;*
(3) *If $G_a \leq H$, $f \in H$, and $f(a) = b$, then $g \in H$ for every $g \in G$ such that $g(a) = b$;*
(4) *If $a, b \in M$, $\mathrm{tp}(a) = \mathrm{tp}(b)$, and $g \in H$ for every $g \in G$ such that $g(a) = b$, then $G_a, G_b \leq H$.* □

Since the automorphism group of any expansion of M is a closed subgroup of $\mathrm{Aut}(M)$, it easily follows that there are closed subgroups of $\mathrm{Aut}(M)$ which are not open (DO IT!).

To illustrate the use of topology, we prove a result concerning pointwise stabilizers of cuts in countable recursively saturated models.

If $I \subseteq_{\mathsf{end}} M$, then we define

$$G_{(>I)} = \bigcup \{G_{(J)} : I < J \subseteq_{\mathsf{end}} M\}.$$

The following proposition is a direct corollary of Theorem 8.4.2.

Proposition 8.8.2 *If I and J are cuts in M, $I < J$, and I is closed under exponentiation, then $G_{(J)} < G_{(I)}$.* □

Recall that $\exp^0(x) = x$ and for $n < \omega$,

$$\exp^{n+1}(x) = \exp^n(2^x).$$

For $a \in M$, let

$$I_{\log}(a) = \{x \in M : \forall n < \omega \ \exp^n(x) < a\},$$
$$I_{\exp}(a) = \{x \in M : \exists n < \omega \ x < \exp^n(a)\}.$$

Thus, in the gap terminology, $I_{\exp}(a) \setminus I_{\log}(a)$ is the exponential gap of a. Both $I_{\log}(a)$ and $I_{\exp}(a)$ are closed under exponentiation.

Let $I \subseteq_{\mathsf{end}} M$ be a cut of M such that $I_{\log}(a) \subseteq_{\mathsf{end}} I \subseteq_{\mathsf{end}} I_{\exp}(a)$ for some $a \in M$. Then $I_{\exp}(a)$ is the smallest cut which is exponentially closed and contains

I and $I_{\log}(a)$ is the largest cut which is exponentially closed and does not contain I. Moreover, $G_{(>I)} = G_{(I_{\exp}(a))}$ (DO IT!). If I is closed under exponentiation and $I \neq I_{\log}(a)$ for every $a \in M$, then for each $c > I$, there is a cut J which is closed under exponentiation and such that $J < c$ and $I < J$ (DO IT!).

Proposition 8.8.3 *Let $I \subseteq_{\mathsf{end}} M$ be closed under exponentiation, and assume that $I \neq I_{\log}(a)$ for all $a \in M$. Then the topological closure of $G_{(>I)}$ is $G_{(I)}$.*

Proof We will show that for all $a, b \in M$, if there exists $f \in G_{(I)}$ such that $f(a) = b$, then there are a cut J and $g \in G_{(J)}$ such that $I < J$ and $g(a) = b$. By Lemma 8.4.4 and the remark preceding the proposition, it suffices to prove that there is $c > I$ such that for all $x < c$, $(M, x, a) \equiv (M, x, b)$.

For each $n < \omega$ define $d_n = \min\{x : \mathrm{tp}^n(x, a) \neq \mathrm{tp}^n(x, b)\}$. For each $n < \omega$, $d_n > I$ and $d_{n+1} \leq d_n$. We will finish the proof by showing that $\inf\{d_n : n < \omega\} > I$. By recursive saturation, there is $d \in M$ such that $(d)_n = d_n$ for all $n < \omega$. By overspill, it follows that $\inf\{d_n : n < \omega\} > \mathbb{N}$. Let us assume then that $\mathbb{N} < I$. Let $d' = f(d)$. Suppose $\inf\{d_n : n < \omega\} = I$. Then there is a nonstandard $e \in I$ such that for all nonstandard $i < e$, $(d)_i \in I$. It follows that for each $i < e$, $(d')_i = (d)_i$. Hence, for some standard n, $(d')_n = (d)_n$. Then $f(d_n) = d_n$, which is a contradiction, because $f(a) = b$ and $\mathrm{tp}(d_n, a) \neq \mathrm{tp}(d_n, b)$.
□

8.9 Maximal point stabilizers

Throughout this section let M be a fixed countable recursively saturated model, and let $G = \mathrm{Aut}(M)$.

In this section we give a group-theoretic characterizations of stabilizers of elements realizing unbounded selective types and minimal types.

For a group G and $A \subseteq G$, $\langle A \rangle$ denotes the subgroup of G generated by A.

Theorem 8.9.1 *Let $a \in M$ be such that $a > \mathrm{Scl}(0)$. Then G_a is a maximal subgroup of G iff $\mathrm{tp}(a)$ is selective. Moreover, if $\mathrm{tp}(a)$ is selective, then for every $f \in G \setminus G_a$, either*

$$G = \{g_0 f g_1 f^{-1} g_2 : g_0, g_1, g_2 \in G_a\}$$

or

$$G = \{g_0 f^{-1} g_1 f g_2 : g_0, g_1, g_2 \in G_a\}.$$

Proof If $\mathrm{tp}(a)$ is not selective, then $\mathrm{Scl}(a)$ is not a minimal extension of $\mathrm{Scl}(0)$. Hence, there is $c \in \mathrm{Scl}(a) \setminus \mathrm{Scl}(0)$ such that $a \notin \mathrm{Scl}(c)$. Then $G_a < G_c$. Hence, G_a is not maximal.

Assume now that $\mathrm{tp}(a)$ is selective. Pick an $f \in G \setminus G_a$ and suppose that $f(a) = b < a$. Under this assumption the first of the equalities of the theorem will be proved. In the other case the same argument works with f^{-1} instead of f; proving the second equality. The type of a is unbounded and selective, and hence, by Corollary 3.2.4, it is rare. Then, since $f(a) = b < a$, we have $b < \mathrm{gap}(a)$.

Now consider arbitrary $g \in G$ such that $g(a) = c$. Let $p(v)$ be the recursive type expressing that $\mathrm{tp}(v, a) = \mathrm{tp}(b, a)$ and $\mathrm{tp}(v, a) = \mathrm{tp}(v, c)$. We will show that $p(v)$ is finitely realizable.

Suppose $M \models \varphi(b, a)$. Let $m = \min\{x : \varphi(x, a)\}$. Since $m \leq b < \mathrm{gap}(a)$ and $\mathrm{tp}(a)$ is selective, $m \in \mathrm{Scl}(0)$. It follows that $\mathrm{tp}(m, a) = \mathrm{tp}(m, c)$, as $g(a) = c$. Thus, $p(v)$ is finitely realizable.

Let d be an element realizing $p(v)$ in M. Then

$$\mathrm{tp}(b, a) = \mathrm{tp}(d, a) = \mathrm{tp}(d, g(a)) = \mathrm{tp}(g^{-1}(d), a).$$

It follows that there are $g_0, g_2 \in G_a$ such that $g_0(b) = d$ and $g_2(g^{-1}(d)) = b$. Then $f^{-1}g_0^{-1}gg_2^{-1}f(a) = a$, and we let $g_1 = f^{-1}g_0^{-1}gg_2^{-1}f$. □

Theorem 8.9.1 implies that the stabilizer of any element realizing a minimal type is a maximal subgroup of G. The next result improves this for 2-indiscernible types.

Definition 8.9.2 We say that a subgroup $H < G$ is *strongly maximal* if for every $f \in G \setminus H$

$$G = H \cup \{g_0 f g_1 : g_0, g_1 \in H\} \cup \{g_0 f^{-1} g_1 : g_0, g_1 \in H\}.$$

By Theorem 3.2.15, there are selective types which are not 2-indiscernible; hence the next proposition shows that the difference between the form of representation of G in the above definition and in Theorem 8.9.1 is essential.

Theorem 8.9.3 *For $a \in M$, G_a is strongly maximal iff* $\mathrm{tp}(a)$ *is 2-indiscernible.*

Proof First assume that $\mathrm{tp}(a)$ is 2-indiscernible. Let $f \in G$ be such that $f(a) = b \neq a$. Let g be another automorphism such that $g(a) = c \neq a$. By considering f^{-1} instead of f, if necessary, we can assume that $a < b \longleftrightarrow a < c$. Then $\mathrm{tp}(a, b) = \mathrm{tp}(a, c)$. So there is $g_0 \in G_a$ such that $g_0(b) = c$. Then $fg_0^{-1}g = g_1 \in G_a$. Hence $g = g_0 f^{-1} g_1$.

Now suppose that G_a is a strongly maximal subgroup of G. Let b, c be such that $a, b < c$ and $\mathrm{tp}(a) = \mathrm{tp}(b) = \mathrm{tp}(c)$. Let $f, g \in G$ be such that $f(a) = b$ and $g(a) = c$. By strong maximality of G_a, there are g_0, g_1 such that $g = g_0 f g_1$. Notice that the other case is ruled out since it would imply that $g(a) < a$. Then, $g_0(b) = c$; hence $\mathrm{tp}(a, b) = \mathrm{tp}(a, c)$.

Now, let b, c, d realizing $\text{tp}(a)$ be such that $a < b$ and $c < d$. Let $h \in G$ be such that $h(c) = a$. Then $\text{tp}(c, d) = \text{tp}(a, h(d))$ and, by the previous argument $\text{tp}(a, h(d)) = \text{tp}(a, b)$, which proves that $\text{tp}(a)$ is 2-indiscernible. □

We conclude this section by showing that the correspondence between unbounded selective types and maximal stabilizers in Theorem 8.9.1 does not hold for bounded types.

Theorem 8.9.4 *Suppose* $\text{Th}(M)$ *does not have a standard model. Then there is* $a \in M$ *such that* $\text{tp}(a)$ *is selective and* G_a *is not a maximal subgroup of* G.

Proof Let $p(x)$ be the bounded selective type constructed in the proof of Theorem 3.2.15. The type $p(x)$ is determined by a descending sequence of definable sets of the form $\Pi_{i \in I} A_i$, where $k = \text{card}^M(I)$ is nonstandard, and for a nonstandard m and all $i \in I$, $m \leq \text{card}^M(A_i)$. We call such sets k-dimensional m-boxes. The construction of $p(x)$ can be made effective in $\text{Th}(M)$; hence $p(x)$ is realized in M. The argument below shows that if a realizes $p(x)$, then G_a is not a maximal subgroup of G. Consider the following type $q(x, y, z)$

$$p(x) \cup p(y) \cup p(z) \cup \{\exists! i\ [(x)_i \neq (y)_i]\} \cup \{\text{card}\,\{i : (x)_i \neq (z)_i\} > n : n < \omega\}.$$

For every formula $\varphi(x) \in p(x)$, $\varphi(M)$ contains a k-dimensional m-box for some nonstandard k and m. Hence, $q(x, y, z)$ is finitely realizable, and, because it is recursive in $p(x)$, it is realized in M. Suppose the triple (a, b, c) realizes $q(x, y, z)$ in M. Let $g, h \in G$ be such that $g(a) = b$ and $h(a) = c$. For every $x, y \in M$, if the set $\{i : (x)_i \neq (y)_i\}$ is finite, then so is the set $\{i : (f(x))_i \neq (f(y))_i\}$ for any $f \in \text{Aut}(M)$. Then it follows that for every $f \in \langle G_a \cup \{g\}\rangle$, the set $\{i : (a)_i \neq (f(a))_i\}$ is finite (DO IT!). Hence, for every such f, $f(a) \neq c$. Thus, $h \notin \langle G_a \cup \{g\}\rangle$, proving that G_a is not a maximal subgroup of G. □

8.10 Arithmetic saturation and open subgroups

In this section, we expand the list of properties characterizing arithmetic saturation for countable models. Most arguments are variations of the proof of Theorem 8.9.1. We continue with the assumption of the previous section that M is a countable recursively saturated model and $G = \text{Aut}(M)$.

For $a, b \in M$, we say that $\text{tp}(a, b)$ is an *heir* of $\text{tp}(a)$, if for every formula $\varphi(x, y) \in \text{tp}(a, b)$, there is $k \in \text{Scl}(0)$ such that $\varphi(x, k) \in \text{tp}(a)$.

Lemma 8.10.1 *Suppose* M *is arithmetically saturated. Then for all* $a, b \in M$, *there is* $b' \in M$ *such that* $\text{tp}(b') = \text{tp}(b)$ *and* $\text{tp}(a, b')$ *is an heir of* $\text{tp}(a)$.

Proof For given $a, b \in M$, consider the type $p(v)$:

$$\text{tp}(b) \cup \{\varphi(a, v) : \forall k \in \text{Scl}(0)\ M \models \varphi(a, k)\}\,.$$

Clearly, $p(v)$ is arithmetic in $\mathrm{tp}(a,b)$ and it is finitely realizable in M. If b' realizes $p(v)$, then $\mathrm{tp}(b') = \mathrm{tp}(b)$, and $\mathrm{tp}(a,b')$ is an heir of $\mathrm{tp}(a)$. □

Lemma 8.10.2 *If $a, b \in M$ and $\mathrm{tp}(a,b)$ is an heir of $\mathrm{tp}(a)$, then $\langle G_a \cup G_b\rangle = G$.*

Proof Consider some $f \in G$ and suppose that $f(a) = c$. Let $p(v)$ be the type expressing that $\mathrm{tp}(a,b) = \mathrm{tp}(a,v) = \mathrm{tp}(c,v)$. To see that $p(v)$ is finitely realizable, consider $\varphi(x,y) \in \mathrm{tp}(a,b)$. Since $\mathrm{tp}(a,b)$ is an heir of $\mathrm{tp}(a)$, there is $k \in \mathrm{Scl}(0)$ such that $M \models \varphi(a,k)$. Since $\mathrm{tp}(a) = \mathrm{tp}(c)$, we also have $M \models \varphi(c,k)$.

Let $d \in M$ realize $p(v)$. There are $g \in G_a$ and $h \in G_d$ such that $g(b) = d$ and $h(a) = c$. Then $k = g^{-1}hg$ is in G_b. Since $l = h^{-1}f$ is in G_a, we have $f = hl = gkg^{-1}l$, hence $f \in \langle G_a \cup G_b\rangle$. □

Now we are ready for the main result of this section.

Theorem 8.10.3 *M is arithmetically saturated iff whenever $H < G$ is open, then there is $f \in G$ such that $\langle H \cup \{f\}\rangle = G$.*

Proof Assume that M is arithmetically saturated and let $H < G$ be open. Without loss of generality, we can assume that $H = G_a$ for some $a \in M$. By Lemma 8.10.1, there is $b \in M$ be such that $\mathrm{tp}(a) = \mathrm{tp}(b)$ and $\mathrm{tp}(a,b)$ is an heir of $\mathrm{tp}(a)$. Let $f \in G$ be such that $f(b) = a$. Then, since $f^{-1}Hf = G_b$, $\langle G_a \cup \{f\}\rangle \supseteq \langle G_a \cup G_b\rangle$, and the conclusion follows by Lemma 8.10.2.

Next, suppose that M is not arithmetically saturated. Since $\mathbb{N}$ is not strong in M, there is $a \in M$ such that for every nonstandard n, there is $i \in \mathbb{N}$ such that $\mathbb{N} < (a)_i < n$. By Proposition 1.8.4, we can assume that for all $i \in \mathbb{N}$, if $(a)_i > \mathbb{N}$, then $(a)_i \notin \mathrm{Scl}(0)$.

Let $H = G_a$ and let $f \in G$ be given. We will show that $\langle G_a \cup \{f\}\rangle \neq G$. Let $a' = f(a)$. For every $n \in \mathbb{N}$,

$$M \models \forall i < n[(a)_i < n \longrightarrow (a')_i = (a)_i].$$

By overspill, there is a nonstandard such n. In particular, for all $i \in \mathbb{N}$, if $(a)_i < n$, then $f((a)_i) = (a)_i$.

Now, let $i \in \mathbb{N}$ be such that $\mathbb{N} < (a)_i < n$. Let $c = (a)_i$. Then c is undefinable in M; hence $G_c < G$. But $c \in \mathrm{Scl}(a)$ and $f(c) = c$, so we have $G_a \cup \{f\} \subseteq G_c < G$, and the result follows. □

Corollary 8.10.4 *M is arithmetically saturated iff G is finitely generated over each of its open subgroups.*

Proof One direction is just Theorem 8.10.3. For the converse, notice that the proof of Theorem 8.10.3 works also when f is replaced with a finite set $f_0, \ldots, f_n$ of automorphisms in G (Do it!). □

Corollary 8.10.5 *M is arithmetically saturated iff for all $a, b \in M$, there is $b' \in M$ such that* $\mathrm{tp}(b') = \mathrm{tp}(b)$ *and* $\mathrm{tp}(a, b')$ *is an heir of* $\mathrm{tp}(a)$.

Proof In one direction, this is just Lemma 8.10.1. For the other direction, notice that the only property of arithmetically saturated models used in the first part of the proof of Theorem 8.10.3 is the property formulated in this corollary. □

Corollary 8.10.6 *Suppose that M is arithmetically saturated, and let H be an open proper subgroup of G. Then H is contained in maximal subgroup of G.*

Proof By Theorem 8.10.3, let $f \in G$ be such that $\langle H \cup \{f\} \rangle = G$. Let H^* be maximal with the property that $H < H^*$ and $f \notin H^*$, given by Zorn's lemma. Clearly, H^* is a proper maximal subgroup of G containing H. □

We are left with an open problem:

Problem 8.10.7 *Suppose every open proper subgroup of G is contained in a maximal subgroup of G. Is M arithmetically saturated?*

8.11 Exercises

♣8.11.1 If M is a recursively saturated model and $a, b \in M$ are such that $\mathrm{gap}(a) < b$, then there are unboundedly many $c \in M$ such that $(M, a, b) \equiv (M, a, c)$. More generally, if $K \prec_{\mathsf{end}} M$ and $b \in M \backslash K$, then there are unboundedly many c such that $(M, a, b) \equiv (M, a, c)$ for all $a \in K$.

The next four exercises should be done together.

♣8.11.2 If D is a 0-definable subset of a model M and $|D| > 1$, then there are $a, b \in D$ such that $\mathrm{tp}(a) \neq \mathrm{tp}(b)$.

♦8.11.3 If D is an unbounded definable subset of a recursively saturated model M, then there are $a, b \in D$ such that $\mathrm{tp}(a) \neq \mathrm{tp}(b)$. (HINT: see Corollary 2.1.11.)

♣8.11.4 If X is an unbounded inductive subset of a model M, then $\mathrm{Aut}(M)_{(X)}$ is trivial. (HINT: Apply Corollary 2.1.11 to (M, X).)

♣8.11.5 If $p(v)$ is an unbounded type realized in a countable recursively saturated model M, then there is an unbounded inductive set $E \subseteq p^M$.

♣8.11.6 By Theorem 8.5.3 every countable recursively saturated model has an elementary cut K whose setwise stabilizer is equal to its pointwise stabilizer. Use this result to prove the Moving Gaps Lemma.

♣8.11.7 If M is countable and arithmetically saturated, then, there is $f \in \mathrm{Aut}(M)$ such that $\mathrm{fix}(f) \not\prec M$ and $\mathrm{fix}(f)$ is not a finitely generated submodel of M.

If $I \subseteq_{\mathsf{end}} M$ and $K \prec M$, then we say that K is I- *small*, if there is $c \in M$ such that $K = \{(c)_i : i \in I\}$. In particular, if K is $\mathbb{N}$-small, we call it *small* .

♠**8.11.8** Let I be a cut of a countable recursively saturated model M, and let $K \prec M$ be I-small. There is $f \in \mathrm{Aut}(M)$ such that $\mathrm{fix}(f) = K$ iff I is strong in M.

♠**8.11.9** Körner's theorem [90] says that every countable arithmetically saturated first-order structure has a maximal automorphism. Show that Körner's theorem follows from Theorem 8.6.3.

For a model M, let

$$\mathrm{RSA}(M) = \{f \in \mathrm{Aut}(M) : (M, f) \text{ is recursively saturated}\}\,.$$

♦**8.11.10** The subgroup generated by $\mathrm{RSA}(M)$ is proper normal subgroup of $\mathrm{Aut}(M)$.

♥**8.11.11** If M is countable and recursively saturated and $I \subseteq_{\mathsf{end}} M$ is closed under exponentiation, then there are $f, g \in \mathrm{RSA}(M)$ such that $I_{\mathsf{fix}}(fg) = I$.

♣**8.11.12** Use Theorem 8.6.3 to prove the following proposition: if $a \in M$ realizes a rare type and $b \in \mathrm{gap}(a)$, then $a \in \mathrm{Scl}(b)$.

♦**8.11.13** If R is the set of all rare types realized in M, then $M = \mathrm{Scl}(\bigcup_{p \in R} p^M)$. The same is true if R is replaced by its complement in the family of complete types realized in M (♣), but it is false if R is replaced by the set of all minimal types realized in M (♥).

♦**8.11.14** Every model M which is not prime has an elementary submodel which is not a fixed point set. If M has more than one gap, then M has an elementary cut which is not a fixed point set.

♥**8.11.15** Every countable arithmetically saturated model M has continuum many nonisomorphic elementary submodels of the form $\mathrm{fix}(f)$ for some $f \in \mathrm{Aut}(M)$.

♣**8.11.16** If K is a countable model, then there is an arithmetically saturated model M and an $f \in \mathrm{Aut}(M)$ such that $\mathrm{fix}(f) \cong K$.

♦**8.11.17** If M is recursively saturated and $K \prec M$ is such that, for each $a \in M$, $K \cap \mathrm{gap}(a) \neq \emptyset$, then $K = M$.

♦**8.11.18** If M is countable and recursively saturated and $M \prec_{\mathsf{cof}} N$, then there are $c, d \in N$ such that $\mathrm{gap}(c) < d$ and $[c, d] \cap M = \emptyset$.

♦**8.11.19** If M is a countable model and $I \subseteq_{\mathsf{end}} M$ is not semiregular in M, then I has continuum many automorphisms. (HINT 1: without loss of generality, we can assume that M is recursively saturated. HINT 2: if $J = \mathrm{cf}(I)$ and $c \in M$ is such that $I = \sup\{(c)_i : i \in J\}$, then M has automorphisms which fix $J \cup \{c\}$ pointwise and move some elements of I.)

♣**8.11.20** If M is recursively saturated and $f \in \mathrm{Aut}(M)$ is such that $f \in \mathrm{Aut}(M, S)$ for some partial inductive satisfaction class S, then there are a recursively saturated model N and $g \in \mathrm{Aut}(N)$ such that $M \prec_{\mathrm{end}} N$ and $f \subseteq g$.

♦**8.11.21** If M is countable and recursively saturated $G = \mathrm{Aut}(M)$ and $f \in G$ is such that $f(a) < \mathrm{Scl}(a) \setminus \mathrm{Scl}(0)$, then $G = \langle G_a \cup \{f\} \rangle$.

8.12 Remarks & References

Automorphisms of models of PA have an extensive literature. A good survey of results until 1995 is Kotlarski [113].

Corollary 8.1.2 ($\mathcal{L}_{\infty\omega}$-undefinability of automorphisms) was proved in [75] using the Moving Gaps Lemma. The proof presented here is from [74], where the proof of Proposition 8.1.3 is attributed to Athanassios Tzouvaras.

The Moving Gaps Lemma (Theorem 8.3.2) is due to Kotlarski. The proof given in this chapter, is from [108] by Kossak & Schmerl. This paper also includes another short proof using minimal types. In this chapter, we stated and proved the lemma for gaps of elements above the least gap. The lemma has a variant concerning the action of the automorphism group on the undefinable elements of the least gap. Let M be a countable recursively saturated model. Suppose M has nonstandard definable elements. Then every element of $a \in \mathrm{gap}(0) \setminus \mathrm{Scl}(0)$ belongs to a maximal convex set consisting of undefinable elements of M, which is called the *interstice* of a. In the case of an arithmetically saturated M, each interstice is partitioned into countably many convex sets. These sets are called *interstitial gaps*, and each of them is of the form $\mathrm{gap}^{\mathcal{F}}(e)$ for some set $\mathcal{F}$ of Skolem terms. The study of the action of $\mathrm{Aut}(M)$ on the interstitial gaps results in an interesting structure theory of gap(0). The first erroneous proof of the Moving Interstitial Gaps Lemma was given in [75]. The error was corrected by Bamber & Kotlarski [4], who also introduced the interstitial terminology. Their results were later improved and generalized by Bigorajska, Kotlarski, and Schmerl in [9] and [175].

In the introduction to [193] Smoryński wrote: "Back-and-forth, arguments are so dull and familiar that nothing beyond length is gained by their inclusion." While this might be true about some arguments, a good counterexample is Theorem 8.4.2, proved by Smoryński in [194]. Lemma 8.4.3 is due independently to Kotlarski [111] and Alena Vencovská, who proved a similar result for Alternative Set Theory. Smoryński's work, in particular [193] and [194], inspired many further developments discussed in this chapter. See [193] for an interesting account of the early history of the subject.

Theorem 8.4.7 is a generalization of a result of Kanovei [66] which says that if an element of a model M is definable in $(M, \mathbb{N})$, then it is already definable in M. Theorem 8.4.7 was proved in [100] by Kossak & Bamber. For more insight concerning back-and-forth we recommend John Surman [202].

Theorem 8.5.1 is from [101] by Kossak & Kotlarski. Theorem 8.5.7 was proved in [102]. These results are about extending automorphisms to elementary end extensions. The analogous question in the territory of cofinal extensions is virtually unexplored. If $M \prec_{\mathsf{cof}} N$, then a subset X of M is coded in K if $X = a \cap M$ for some $a \in N$. It is shown in [102] that every countable recursively saturated model M has a countable cofinal extension such that every automorphism of M which sends coded sets to coded sets extends to an automorphism of N. This is the only published result on extending automorphisms to cofinal extensions. Exercise 8.11.17 is an unpublished result of Kotlarski.

Theorem 8.5.3 was proved by Schmerl in [174]. The proof is closely related to the proof of Theorem 5.4.4.

Theorems 8.6.3 and 8.7.2 are due to Kaye, Kossak, and Kotlarski [75]. Our proof of Theorem 8.7.2 differs much from the one given in [75]. It is based on a proof of a special case from [107]. In fact, the proof presented in [75] has some important details omitted. A complete back-and-forth proof of Theorem 8.7.2 is given in [96]. Recently, Enayat has found a very elegant way to prove results like Smoryński's Theorem 8.4.2 and Theorem 8.7.2. Enayat's technique, which avoids back-and-forth, is a refinement of the arguments used in the proof of Theorem 8.7.2. The result in Exercise 8.7.3 is Corollary 2.13 in [175]. Among the many other results in [175] is a general version of the Moving Gaps Lemma for intersticial gaps.

Frederike Körner [90] has generalized Theorem 8.6.3 by proving that every countable arithmetically saturated first-order structure has a maximal automorphism. Motivated by the converse for models of PA she has also asked the following question:

Problem 8.12.1 *Let T be a complete first-order theory. Is it true that either every countable recursively saturated model of T has a maximal automorphism, or for every countable recursively saturated model $\mathfrak{A}$ of T, $\mathfrak{A}$ has a maximal automorphism iff $\mathfrak{A}$ is arithmetically saturated?*

Grégory Duby [27] imporoved Körner's result by showing that every countable arithmetically saturated first-order structure has an automorphism of whose all powers are maximal.

As observed by Fredrik Engström in his Ph.D. thesis [36], the theorem on the existence of maximal automorphisms of countable arithmetically saturated models can be viewed as an omitting types result. Consider the type $p(x) = \{g(x) = x \wedge x \neq t(0) : t \in \mathsf{Term}\}$ in $\mathcal{L}_{\mathsf{PA}}$ with a function symbol g. Then for each model M and automorphism f, (M, f) omits $p(x)$ iff f is maximal. Engström studies various expandability notions and introduces a strong form of resplendence which guarantees existence of expansions omitting certain types.

Results concerning the automorphism groups of recursively saturated models from this and the next chapter can be considered as part of the general theory of automorphisms of first-order structures. The book [77] edited by Kaye & Macpherson is a good introduction to the subject and covers many topics directly related to the material discussed in this book.

The study of connections between types of elements and their stabilizers began in [75] and was continued in [103] by Kossak et al. More results concerning types realized in intersticial gaps can be found in [9], [8] by Bigorajska and other papers referenced there. Theorem 8.9.4 is from [9].

Our illustration of the use of topology on $\mathrm{Aut}(M)$, Proposition 8.8.2, is due to Kaye [74]. Proposition 8.8.2 is one of the preliminary lemmas in the proof the theorem of Kaye [74] characterizing closed normal subgroups of $\mathrm{Aut}(M)$. The result says that H is a closed normal subgroup of $\mathrm{Aut}(M)$ iff H is a pointwise stabilizer of an invariant initial segment of M. Kaye leaves open the problem of classifying all normal subgroups, conjecturing that they are all $G_{(I)}$ or $G_{(>I)}$ for invariant I. A different proof of Kaye's theorem which has some additional consequences is given in [173].

Theorems 8.9.1 and 8.9.3 are from [103]. This paper, which continues work started in [75], contains more results on open maximal subgroups. For more results in this direction see also [114]. Corollaries 8.10.4, 8.10.5, and 8.10.6 are from [107]. Kotlarski & Piekart in [116] discuss the question: When does an open subgroup of the automorphism group of a countable recursively saturated model of TA extend to a unique maximal subgroup¿

Exercise 8.11.11 is due to Jiří Sgall and Anton Sochor (unpublished). Kotlarski has found a very elegant short proof.

Exercise 8.11.19 is based on [94]. Earlier similar results were obtained by Hamid Lessan in his Ph.D. thesis [121].

In [110] and [111] Kotlarski began a systematic study of combinatorial properties of families of elementary cuts of countable recursively saturated models. In particular, he defined the notion of a closed subset of a recursively saturated model and proved several results concerning closed cuts. A subset X of a model M is *closed* if for every $x \notin X$, there is an $f \in \mathrm{Aut}(M)$ such that $f \restriction X = \mathrm{id}$ and $f(x) \neq x$. Clearly, if the type of $a \in M$ is rare, then $\inf(\mathrm{gap}(a))$ is not closed. Kotlarski showed that if $I \prec_{\mathsf{end}} M$ is not closed, then it must be of the form $\inf(\mathrm{gap}(a))$ for some a. Piekart [151] gives a complete summary of what is known about closed cuts.

9

AUTOMORPHISM GROUPS OF RECURSIVELY SATURATED MODELS

In this chapter we use the topology of the automorphism group in a more substantial way than just to formulate the results in the topological language. We present several results concerning the structure of the automorphism group of a countable recursively saturated model. One of the central notions is that of a Lascar generic automorphism. We prove that all countable arithmetically saturated models have Lascar generic automorphisms; and we use Lascar generics to prove that all countable arithmetically saturated models have the small index property. In the proofs, we take full advantage of the fact that automorphism groups of countable structures are Polish groups, and that, in particular, the Baire Category Theorem applies. In another major result in this section, we show that every countable arithmetically saturated model is determined up to isomorphism by its complete theory and its automorphism group. We also characterize countable arithmetically saturated models as those whose automorphism groups have uncountable cofinality.

9.1 Generic automorphisms

Recall that a subset of a topological space is *meager* if it is contained in a countable union of closed nowhere dense sets. A *comeager* set is the complement of a meager set. If $\mathfrak{A}$ is a countable first-order structure, then $\mathrm{Aut}(\mathfrak{A})$ with the topology whose basic open subgroups pointwise stabilizers of finite sets is a Polish (completely metrizable and separable) group, and, in particular, the Baire Category Theorem holds: every comeager subset of $\mathrm{Aut}(\mathfrak{A})$ is dense (hence, nonempty).

An automorphism f of a countable first-order structure $\mathfrak{A}$ is *generic* if its conjugacy class $[f]_G = \{f^g : g \in G\}$ is comeager in $G = \mathrm{Aut}(\mathfrak{A})$.

Generic automorphisms are an important tool in the study of automorphism groups of $\aleph_0$-categorical structures. They play a similar role in the study of arithmetically saturated models of arithmetic; however the definition of genericity needs to be altered. In Definition 9.1.4, we define the notion of *Lascar generic* automorphisms. It is an open problem, and the specifics of it are discussed later in this chapter, whether Lascar generics are generic.

Recall that a submodel $K \prec M$ is *small*, written $K \prec_{\mathsf{sm}} M$, if for some $a \in M$, $K = \{(a)_i : i \in \omega\}$. Every finitely generated elementary submodel of a recursively saturated model M is small in M (Do IT!). By compactness, every countable model K is small in some recursively saturated model M. If K is a small elementary submodel of M, then $\mathrm{Aut}^M(K)$ denotes the group of those automorphisms of K which can be extended to automorphisms of M. Notice that, using the notation introduced in the previous chapter, there is a natural isomorphism between $\mathrm{Aut}^M(K)$ and $G_{\{K\}}/G_{(K)}$.

Notice that if $K \prec_{\mathsf{sm}} M$, then $G_{(K)}$ is an open subgroup of G. Therefore $G_{\{K\}}$ is open as well. The next proposition generalizes this slightly.

Proposition 9.1.1 *If $K \prec_{\mathsf{sm}} M$, $g \in \mathrm{Aut}^M(K)$, and $G = \mathrm{Aut}(M)$, then the set of $f \in G$ such that $g \subseteq f$ is open.*

Proof Let $K = \{(a)_n : n < \omega\}$ for some $a \in M$, and suppose that $f \in G$ extends g. Then $H = \{f' \in G : f'(a) = f(a)\}$ is an open neighborhood of f, and for all $f' \in H$, $g \subseteq f'$ (Do IT!). □

Definition 9.1.2 Let $K \prec_{\mathsf{sm}} M$. We say that $f \in \mathrm{Aut}^M(K)$ is *existentially closed* if for every formula $\varphi(x, y)$ with parameters in K and for all $h \in \mathrm{Aut}(M)$, if $f \subseteq h$ and $M \models \exists x \varphi(x, h(x))$, then $K \models \exists x \varphi(x, f(x))$.

Lemma 9.1.3 *Let M be countable and arithmetically saturated. Suppose that $a, a' \in M$ are such that $\mathrm{tp}(a) = \mathrm{tp}(a')$. Then there are $K \prec_{\mathsf{sm}} M$ and an existentially closed $f \in \mathrm{Aut}^M(K)$ such that $a \in K$ and $f(a) = a'$.*

Proof By standard closure arguments one can construct a model $K \prec M$ and an automorphism f of K such that $a, a' \in K$, $f(a) = a'$, and for all $a_0, \dots, a_{i-1}$ in K and all $\varphi(x, y)$, if there are $c, d \in M$ and such that $\mathrm{tp}(a_0, \dots, a_{i-1}, c) = \mathrm{tp}(f(a_0), \dots, f(a_{i-1}), d)$, and $M \models \varphi(c, d)$, then there are such c and d in K. Now we define a type, which is arithmetic in $\mathrm{tp}(a, a')$, the construction of which emulates the construction of such K and f.

For each $i < \omega$ we define an operation $\mathrm{E}_i(X)$ such that for each $A \subseteq M$, if $A_0 = A$ and for $i < \omega$, $A_{i+1} = \mathrm{E}_i(A_i)$, then $\bigcup_{i<\omega} A_i = \mathrm{Scl}(A)$. The particular choice of E_i is not important, but we need to be more specific because we also need an effective bound for the size of $E_i(A)$ for finite A. Let $\langle t_i(x) : i < \omega \rangle$ be a recursive enumeration of all unary Skolem terms, and for $i < \omega$ and $A \subseteq M$, let

$$\mathrm{E}_i(A) = A \cup \{\langle a, b \rangle : a, b \in A\} \cup \{t_j(a) : j < i, a \in A\}.$$

We define a type $p(v, w)$ such that for c, d realizing the type, we have $K = \{(c)_i : i < \omega\} = \{(d)_i : i < \omega\} \prec_{\mathsf{sm}} M$ and $(c)_i \mapsto (d)_i$ an existentially closed automorphism of K. We define $p(v, w)$ as $\bigcup_{i<\omega} p_i(v, w)$ by induction on i. The definition involves an increasing coded ω-sequence $0 = \lambda_0 < \lambda_1 < \cdots < \omega$, with the property that for every set A such that $|A| = 2\lambda_i + 2$, $|E_i(A)| < \lambda_{i+1}$. It is easy to define such a sequence λ_i by induction. The specifics are not important, as any coded sequence satisfying the condition above will do. Think of v and w as finite approximations to an existentially closed automorphism which is built in a back-and-forth process, where at step i we have the sequences of length $\lambda_i + 1$ and we extend them to sequences of length $\lambda_{i+1} + 1$ in the next step.

Let $p_0(v,w) = \{(v)_0 = a \wedge (w)_0 = a'\}$. Suppose now that $p_i(v,w)$ has been defined. If i is odd, then we adjoin to $p_i(v,w)$ the formula

$$\mathrm{E}_i\left(\{(v)_0,\ldots,(v)_{\lambda_i},(w)_0,\ldots,(w)_{\lambda_i}\}\right) \subseteq \{(v)_0,\ldots,(v)_{\lambda_{i+1}}\},$$

and if i is even, we adjoin the formula

$$\mathrm{E}_i\left(\{(v)_0,\ldots,(v)_{\lambda_i},(w)_0,\ldots,(w)_{\lambda_i}\}\right) \subseteq \{(w)_0,\ldots,(w)_{\lambda_{i+1}}\}.$$

Then we define $q_i(v,w)$ to be $p_i(v,w)$ together with the formula we just added plus the recursive set of formulas expressing that

$$\mathrm{tp}((v)_0,\ldots,(v)_{\lambda_{i+1}}) = \mathrm{tp}((w)_0,\ldots,(w)_{\lambda_{i+1}}).$$

Although we have not finished defining $p_{i+1}(v,w)$ yet, notice that if we declared that $p_{i+1}(v,w) = q_i(v,w)$, any elements c,d realizing $p(v,w)$ would give us a small model and an automorphism $(c)_i \mapsto (d)_i$. We still need to insert an extra condition to make sure that this automorphism is existentially closed.

Let $\langle \varphi_i(x_{k_0},\ldots,x_{k_i},x,y) : i < \omega\rangle$ be a recursive enumeration of all formulas in the variables shown, where $\{k_0,\ldots,k_i\} \subseteq [0,\lambda_i]$. If for all finite conjunctions $\psi(v,w)$ of formulas in $q_i(v,w)$ and all $\theta(u,x)$,

$$M \models \exists v,w,x,y\ \{\psi(v,w) \wedge [\theta(v,x) \longleftrightarrow \theta(w,y)] \wedge \varphi_i((v)_{k_0},\ldots,(v)_{k_i},x,y)\}, \quad (*)$$

then let

$$p_{i+1}(v,w) = q_i(v,w) \cup \{\varphi_i((v)_{k_0},\ldots,(v)_{k_i},(v)_{\lambda_i+1},(w)_{\lambda_i+1})\}.$$

Otherwise let $p_{i+1}(v,w) = q_i(v,w)$. Notice that the condition $(*)$ guarantees, that if $q_i(v,w)$ is finitely realizable, then so is $p_i(v,w)$.

To finish the proof we need two observations. The first is that for each $i<\omega$, $p_i(v,w)$ is finitely realizable (DO IT!). The second is that $\bigcup_{i<\omega} p_i$ is arithmetic in $\mathrm{tp}(a,a')$; hence, by arithmetic saturation, $p(v,w)$ is realized in M, and the result follows. □

Definition 9.1.4 Let M be a model and $G = \mathrm{Aut}(M)$. An automorphism $g \in G$ is *Lascar generic* if the following conditions are satisfied:

(1) For each finite $A \subseteq M$, there is $K \prec_{\mathsf{sm}} M$ such that $A \subseteq K$, $g(K) = K$, and $g \restriction K$ is existentially closed.
(2) If $K \prec_{\mathsf{sm}} M$, and $g(K) = K$ and $g \restriction K$ is existentially closed, $K \prec L \prec_{\mathsf{sm}} M$, $f \in \mathrm{Aut}^M(L)$, and $g \restriction K \subseteq f$, then there is $h \in G_{(K)}$ such that $f \subseteq g^h$.

It follows directly from the definition that the set of Lascar generic automorphisms is closed under conjugation. Here is the fundamental property of Lascar generics.

Proposition 9.1.5 *Let f and g be Lascar generic automorphisms of a countable model M, and let $G = \mathrm{Aut}(M)$. Suppose that $M_0 \prec_{\mathsf{sm}} M$, $f \restriction M_0 = g \restriction M_0 \in \mathrm{Aut}(M_0)$, and $f \restriction M_0$ is existentially closed. Then f and g are conjugate in G.*

Proof We will define $\alpha, \beta \in G$ such that $g^\alpha = f^\beta$. The automorphisms α and β will be obtained as the limits of two sequences of automorphisms defined inductively as we "glue" f and g together in a back-and-forth process, alternating applications of (1) and (2) in Definition 9.1.4. The idea is to build an elementary chain $\langle M_i : i < \omega \rangle$ of small elementary submodels of M such that $M = \bigcup_{i<\omega} M_i$ aid for each i, the restriction of a conjugate either of f or g to M_i is an existentially closed automorphism of M_i, and then to use (2) from the definition above to glue a conjugate of f to a conjugate of g on M_i using an automorphism in $G_{(M_{i-1})}$. To make sure that $\bigcup M_{i<\omega} = M$, we let $\langle a_i : i < \omega \rangle$ be an enumeration of M. At step i we use (1) to make sure that $a_i \in M_{i+1}$. Let us describe the first three steps of this process.

In step 1, since g is Lascar generic, by (1), there is $M_1 \prec_{\mathsf{sm}} M$ such that $a_0 \in M_1$, $M_0 \prec M_1$, $f(M_1) = M_1$, and $g \restriction M_1$ is existentially closed. Fix such a model M_1. Then, since f is Lascar generic and $f \restriction M_0$ is existentially closed, then, by (2), there is $h_0 \in G_{(M_0)}$ such that $g \restriction M_1 \subseteq f^{h_0}$.

So now, in step 2, g and f^{h_0} are Lascar generic, they agree on M_1 and $g \restriction M_1$ is existentially closed. So, let $M_2 \prec_{\mathsf{sm}} M$ be such that $M_1 \prec M_2$, $a_1 \in M_2$, and f^{h_0} is existentially closed on M_2. So there is $h_1 \in G_{(M_1)}$ such that $f^{h_0} \restriction M_2 \subseteq g^{h_1}$.

In step 3, we get M_3 such that $a_2 \in M_3 \prec_{\mathsf{sm}} M$, and then get $h_2 \in G_{(M_2)}$ such that $g^{h_1} \restriction M_3 \subseteq (f^{h_0})^{h_2}$.

Continuing in the same fashion for every $i < \omega$, we get $h_i \in G_{(M_i)}$ such that for all even n,

$$g^{h_1 h_3 \cdots h_{n-1}} \restriction M_{n-1} = f^{h_0 h_2 \cdots h_n} \restriction M_{n-1}.$$

Since $\bigcup_{i<\omega} M_i = M$, the following functions are well-defined. In fact, they are automorphisms of M:

$$\alpha = \lim_{i \to \infty} h_1 h_3 \cdots h_{2i+1},$$
$$\beta = \lim_{i \to \infty} h_0 h_2 \cdots h_{2i}.$$

Then $g^\alpha = f^\beta$ (Do it!), and the result follows. □

In Definition 9.1.4, we did not assume that M is recursively saturated, but, to avoid trivial examples, this assumption is necessary. We will actually need more to establish that Lascar generics exist.

Theorem 9.1.6 *The set of Lascar generic automorphisms of a countable arithmetically saturated model M is comeager in $G = \mathrm{Aut}(M)$.*

Proof Let A be a finite subset of M. By Proposition 9.1.1, the set of all $g \in G$, such that $\exists K[A \subseteq K \prec_{\mathsf{sm}} M]$, $g(K) = K$, and $g \restriction K$ is existentially closed, is open. Proposition 9.1.3 shows that this set is dense. It follows that the set of automorphisms which satisfy the first part of the definition of Lascar generics is comeager in G.

Now suppose that K, L, and g satisfy the conditions:

(1) $K \prec_{\mathsf{sm}} M$ and $K \prec L \prec_{\mathsf{sm}} M$;
(2) $g \in \mathrm{Aut}^M(L)$ and $g(K) = K$;
(3) $g \restriction K$ is existentially closed.

Let $O(K, L, g)$ be the set

$$\{f \in G : \exists h \in G_{(K)}\ [g \subseteq f^h]\} \cup \{f \in G : \exists a \in K[g(a) \neq f(a)]\}.$$

Since there are countably many K, L, and g as above, and each set $O(K, L, g)$ is open, to prove that their intersection is comeager, we will show that each of them is dense. This follows from the claim:

Claim: Suppose that $c,\ c' \in M$ are such that for all $a \in K$, $\mathrm{tp}(a, c) = \mathrm{tp}(g(a), c')$. Then there exist $h \in G_{(K)}$ and $f \in G$ such that $g \subseteq f^h$, and $f(c) = c'$.

Proof of the claim: Let $\alpha, \beta \in M$ be such that $K = \{(\alpha)_n : n < \omega\}$ and $L = \{(\beta)_n : n \in \omega\}$. The proof of the claim is finished if we find d, d', and $f \in G$ such that $g \subseteq f$, and $f(d) = d'$, and for all $a \in K$, $\mathrm{tp}(a, c, c') = \mathrm{tp}(a, d, d')$ (DO IT!).

Then g can be extended to an automorphism of M. Let β' be the image of β under an extension of g. We apply resplendency to the set of formulas $\Gamma(v, v', f)$ with parameters $\alpha, \beta, \beta', c, c'$ in $\mathcal{L} \cup \{f\}$ expressing that f is an automorphism, $\mathrm{tp}((\alpha)_k, c, c') = \mathrm{tp}((\alpha)_k, v, v')$, for all $k < \omega$ and containing the formulas $f((\beta)_k) = (\beta')_k \wedge f(v) = v'$ for all $k < \omega$.

Since $g \restriction K$ is existentially closed, every finite fragment of Γ is satisfied by some $d, d' \in K$ and any $f \in G$ which extends g. Hence, by resplendency, there are $d, d' \in M$ and $f \in G$, satisfying Γ, which finishes the proof of the claim.

The claim shows that the intersection of all $O(K, L, g)$ for K, L, g satisfying (1), (2), and (3) above is comeager. All automorphisms in this intersection are Lascar generics. □

Theorem 9.1.6 has an immediate corollary:

Corollary 9.1.7 *In a countable arithmetically saturated models, every generic automorphism is Lascar generic.* □

9.2 Dense conjugacy classes

The set of Lascar generics is comeager and closed under conjugation, but it is not clear whether Lascar generics are generic. They are, nevertheless, locally generic. An automorphism is *locally generic* if its conjugacy class is comeager in some open subset of the automorphism group. We prove that for arithmetically saturated models of TA, all Lascar generics are generic. It follows that, in an arithmetically saturated model of TA, all Lascar generics are conjugate. First we give a condition which implies the existence of generic automorphisms of arithmetically saturated models.

Proposition 9.2.1 *If a countable arithmetically saturated model has an automorphism whose conjugacy class is dense, then every Lascar generic of the model is generic.*

Proof Suppose that $[h]$ is dense in G. Let $g_1, g_2 \in G$ be Lascar generics. Then there are $M_1, M_2 \prec_{\mathsf{sm}} M$ such that $g_1 \restriction M_1$ and $g_2 \restriction M_2$ are existentially closed. Let $a_1, a_2 \in M$ be such that $M_1 = \{(a_1)_i : i < \omega\}$ and $M_2 = \{(a_2)_i : i < \omega\}$. Let h_1, h_2 be conjugates of h such that $h_1(a_1) = g_1(a_1)$ and $h_2(a_2) = g_2(a_2)$, and let f be such that $h_2 = f^{-1}h_1f$. Then $h_1(f(a_2)) = f(g_2(a_2))$. The set of Lascar generics is dense; hence, there is a Lascar generic g agreeing with g_1 on $\{a_1, f(a_2)\}$, so that $g(a_1) = g_1(a_1)$ and $g(f(a_2)) = f(g_2(a_2))$. Thus, $f^{-1}gf(a_2) = g_2(a_2)$. Now, both g and $f^{-1}gf$ are Lascar generics, and they agree with g_1 and g_2 on M_1 and M_2, respectively. Therefore, g and g_1 are conjugates and also $f^{-1}gf$ and g_2 are conjugates. Thus, g_1 and g_2 are conjugates. □

By Theorem 9.1.6 we have the following corollary.

Corollary 9.2.2 *If a countable arithmetically saturated model M has an automorphism whose conjugacy class is dense, then for every $f \in \mathrm{Aut}(M)$, f is generic iff it is Lascar generic.* □

Notice that in Proposition 9.2.1, we could assume that M is just recursively saturated. However, at this point it could be a debatable improvement since we do not know whether a model which is recursively saturated but not arithmetically saturated has Lascar generics.

The next natural question is: which countable recursively saturated models have an automorphism whose conjugacy class is dense? The answer is not straightforward. In the proof we will use the following theorem of Hajnal [51]. To state the theorem, we need a few definitions. A *digraph* is a structure (V, E), where $E \subseteq V^2$ is such that for all $x \in V$, $(x, x) \notin E$. If $D = (V, E)$ is a digraph, then the *chromatic number* of D, denoted by $\chi(D)$, is the smallest cardinal κ

for which there is a function (coloring) $f : V \longrightarrow \kappa$ such that for all $(x, y) \in E$, $f(x) \neq f(y)$. If $D_1 = (V_1, E_1)$ and $D_2 = (V_2, E_2)$ are digraphs, then their product $D_1 \times D_2$ is the digraph $(V_1 \times V_2, E)$, where $((x_1, x_2), (y_1, y_2)) \in E$ iff $(x_1, y_1) \in E_1$ and $(x_2, y_2) \in E_2$.

> HAJNAL'S THEOREM: If D_1 and D_2 are digraphs, $\chi(D_1) \geq \aleph_0$ and $\chi(D_2) \geq \aleph_0$, then $\chi(D_1 \times D_2) \geq \aleph_0$.

Theorem 9.2.3 *Every countable recursively saturated model of* TA *has an automorphism whose conjugacy class is dense.*

Proof Let M be a countable recursively saturated model of TA. We construct an automorphism h whose conjugacy class is dense by back-and-forth. It is the usual construction, however, infinitely often we stop, and consider the partial automorphism $a_1 \mapsto a_2$ constructed so far. Think of a_1 and a_2 as codes of the finite domain and range of the partial automorphism. Consider $b_1, b_2 \in M$ such that $\mathrm{tp}(b_1) = \mathrm{tp}(b_2)$. We will find b_1', b_2' such that $\mathrm{tp}(b_1, b_2) = \mathrm{tp}(b_1', b_2')$ and $\mathrm{tp}(a_1, b_1') = \mathrm{tp}(a_2, b_2')$. Then we will declare that the automorphism we are constructing sends b_1' to b_2'. Notice that if h is built in this way, then, since there is $g \in \mathrm{Aut}(M)$ such that $g(b_1, b_2) = (b_1', b_2')$, we have $g^{-1}hg(b_1) = b_2$; hence, $[h]$ has a nonempty intersection with the basic open set $\{f \in G : f(b_1) = b_2\}$.

To find b_1' and b_2', consider the type $\Gamma(x_1, x_2, y_1, y_2)$ with parameters a_1, a_2, b_1, b_2:

$$\{\varphi(x_1, x_2) : M \models \varphi(a_1, a_2)\} \cup \{\psi(y_1, y_2) : M \models \psi(b_1, b_2)\} \\ \cup \{\mathrm{tp}^n(x_1, y_1) = \mathrm{tp}^n(x_2, y_2) : n < \omega\},$$

where $\mathrm{tp}^n(a, b)$ stands for the type of (a, b) restricted to the first n formulas in some fixed recursive enumeration of all formulas in two variables. Notice that $\Gamma(x_1, x_2, y_1, y_2)$ is recursive in $\mathrm{tp}(a_1, a_2, b_1, b_2)$.

It is enough to show that $\Gamma(x_1, x_2, y_1, y_2)$ is finitely realizable. Suppose it is not. Then there are $\varphi(x_1, x_2), \psi(y_1, y_2) \in \Gamma$ and $n < \omega$ such that

$$M \models \forall x_1, x_2, y_1, y_2[\varphi(x_1, x_2) \wedge \psi(y_1, y_2) \longrightarrow \mathrm{tp}^n(x_1, y_1) \neq \mathrm{tp}^n(x_2, y_2)]. \qquad (*)$$

Let $D_1 = (M, E_1)$, $D_2 = (M, E_2)$, where $(v_1, v_2) \in E_1$ iff $M \models \varphi(v_1, v_2)$ and $(w_1, w_2) \in E_2$ iff $M \models \psi(w_1, w_2)$. First we check that D_1, D_2 are digraphs. Suppose $M \models \exists x \varphi(x, x)$. Then let $c = \min\{x : M \models \varphi(x, x)\}$. Then from $(*)$ we get $\mathrm{tp}^n(c, b_1) \neq \mathrm{tp}^n(c, b_2)$, and this is a contradiction because $c \in \mathrm{Scl}(0)$ and $\mathrm{tp}(b_1) = \mathrm{tp}(b_2)$. The argument for D_2 is analogous.

We claim that $\chi(D_1) \geq \aleph_0$ and $\chi(D_2) \geq \aleph_0$. To prove the claim, suppose that D_1 is k-colorable for some $k < \omega$. Then every subdigraph of D_1 is k-colorable. In particular every finite subdigraph of $D_1^{\mathbb{N}} = (\mathbb{N}, E_1 \cap \mathbb{N}^2)$ has a k-coloring. Since $M \models$ TA, $D_1^{\mathbb{N}}$ is 0-definable in $\mathbb{N}$. The theorem: "If every finite subdigraph of a digraph D is k-colorable, then D is k-colorable" is provable in WKL_0. Thus, $D_1^{\mathbb{N}}$ has a 0-definable k-coloring α' (DO IT!). The same definition gives a k-coloring α of D_1. Since (a_1, a_2) is an edge of D_1, we have $\alpha(a_1) = k_1$, $\alpha(a_2) = k_2$ for some distinct $k_1, k_2 < k$. But this gives a contradiction because $\mathrm{tp}(a_1) = \mathrm{tp}(a_2)$. Thus, the claim is proved for D_1. Similar for D_2.

By Hajnal's theorem, $\chi(D_1 \times D_2) \geq \aleph_0$. However, by $(*)$, the mapping $(x, y) \mapsto \mathrm{tp}^n(x, y)$ is a 2^n-coloring of $D_1 \times D_2$. This contradiction shows that Γ is finitely realizable and finishes the proof. □

For natural numbers k and n we now consider the following statement $H(k, n)$ concerning finite digraphs:

$$\forall D_1, D_2[\chi(D_1) \geq k \wedge \chi(D_2) \geq k \longrightarrow \chi(D_1 \times D_2) \geq n].$$

It is an open problem whether for each n there is a k such that $H(k, n)$ holds. The statement $\forall n \exists k H(k, n)$ is known to be equivalent to $\exists k H(k, 4)$. Moreover, the proof when formalized in PA gives the following proposition:

Proposition 9.2.4 *For any model M the following are equivalent:*

(1) $\forall n < \omega \exists k < \omega\ M \models H(k, n)$.
(2) $\exists k < \omega\ M \models H(k, 4)$. □

Theorem 9.2.5 *Let M be a countable recursively saturated model having a nonstandard prime submodel. Then the following are equivalent:*

(1) *M has an automorphism whose conjugacy class is dense.*
(2) *For some $k < \omega$, $M \models H(k, 4)$.*

Proof To prove that (2) implies (1), we proceed exactly as in the proof of Theorem 9.2.3. Our objective is to prove that the type $\Gamma(x_1, x_2, y_1, y_2)$ is finitely realizable. The graphs D_1 and D_2 are defined the same way, but now we can not claim that $\chi(D_i) \geq \aleph_0$, $i = 1, 2$, by arguing that otherwise there would be a 0-definable finite coloring α of $D_i \cap \mathbb{N}$, since these digraphs might not be definable in $\mathbb{N}$ (and even if they were that would not be enough). But we still get a contradiction if we assume that Γ is inconsistent and *there is* a 0-definable finite coloring of either D_1 and D_2. Hence, suppose that Γ is inconsistent and neither D_1 nor D_2 has a 0-definable finite coloring. Since the theorem, stating that if every finite subdigraph of a digraph D is k-colorable, then so is D, is provable in PA, it follows that, for each $k < \omega$, we must have

$$M \models \exists D_1' \subseteq D_1 \exists D_2' \subseteq D_2[\chi(D_1') > k \wedge \chi(D_2') > k].$$

By Proposition 9.2.4, if k is large enough, then for D'_1, D'_2 whose chromatic numbers in M are bigger than k, we have $M \models \chi(D'_1 \times D'_2) > 2^n$. But, as in the previous proof, the mapping $(x, y) \mapsto \mathrm{tp}^n(x, y)$ is a 0-definable 2^n-coloring of $D'_1 \times D'_2$, contradiction.

Next we show that (1) implies (2). Assume that M has an automorphism whose conjugacy class is dense, and suppose that for all $k < \omega$, $M \models \neg H(k, 4)$.

Let $\theta(x)$ be the formula

$$\exists D_1, D_2\ [\chi(D_1) \geq x \wedge \chi(D_2) \geq x \wedge \chi(D_1 \times D_2) < 4].$$

For every standard k, $M \models \theta(k)$. By overspill, $\theta(k)$ holds in M for some nonstandard k. Pick a nonstandard definable $a \in M$, and let t be the largest $t < a$ for which $M \models \theta(2^t)$. Clearly, t is nonstandard and definable. There are 0-definable D_1, D_2 such that $\chi(D_1) > 2^t \wedge \chi(D_2) > 2^t \wedge \chi(D_1 \times D_2) < 4$. Let $D_1 = (V_1, E_1)$, $D_2 = (V_2, E_2)$, and let $\varphi_0(x), \psi_0(x), \rho_1(x_1, x_2)$, and $\rho_2(y_1, y_2)$ be the definitions of V_1, V_2, E_1, and E_2, respectively. Also, let β be a definable 3-coloring of $D_1 \times D_2$. (We could instead let β be any k-coloring, where $k < \omega_i$.)

Let $\theta_0(x), \theta_1(x), \ldots$ be a recursive list of all unary formulas. We define two complete types: $p(x) = \{\varphi_i(x) : i < \omega\}$ and $q(x) = \{\psi_i(x) : i < \omega\}$, where $\varphi_0(x)$ and $\psi_0(x)$ are as defined above. Proceeding inductively, we define $\varphi_{i+1}(x)$ to be $\theta_i(x)$ or $\neg\theta_i(x)$, preserving the largeness condition: the induced subdigraph $D_1 \cap \varphi_i(M)$ is not 2^{t-i}-colorable. Notice that this is always possible because, if $D = D_1 \cap \theta_i(M)$ is not 2^{t-i}-colorable in M and $D = A_1 \cup A_2$ is a 0-definable partition, then it cannot be that both A_1 and A_2 are 2^{t-i-1}-colorable. We follow the same procedure defining $\psi_{i+1}(x)$, considering induced subdigraphs of $D_2 \cap \psi_i(M)$, which are not 2^{t-i}-colorable.

Now consider the types:

$$P(x_1, x_2) = p(x_1) \cup p(x_2) \cup \{\rho_1(x_1, x_2)\},$$
$$Q(y_1, y_2) = q(y_1) \cup q(y_2) \cup \{\rho_2(y_1, y_2)\}.$$

Both $P(x_1, x_2)$ and $Q(y_1, y_2)$ are recursive in $\mathrm{Th}(M)$, and they are both finitely realizable in M because none of the induced subdigraphs in the construction of $p(x)$ and $q(x)$ is 1-colorable. Let (a_1, a_2) realize P and (b_1, b_2) realize Q. We have $\mathrm{tp}(a_1) = \mathrm{tp}(a_2)$ and $\mathrm{tp}(b_1) = \mathrm{tp}(b_2)$. Let h be an automorphism whose conjugacy class is dense. Then there are a'_1, a'_2, b'_1, b'_2 such that $h(a'_1) = a'_2$, $h(b'_1) = b'_2$, $\mathrm{tp}(a_1, a_2) = \mathrm{tp}(a'_1, a'_2)$, and $\mathrm{tp}(b_1, b_2) = \mathrm{tp}(b'_1, b'_2)$. Since

$$M \models \rho_1(a_1, a_2) \wedge \rho_2(b_1, b_2),$$

it follows that

$$M \models \rho_1(a'_1, a'_2) \wedge \rho_2(b'_1, b'_2).$$

Hence, $((a_1', b_1'), (a_2', b_2'))$ is an edge in $D_1 \times D_2$. Since β is 0-definable, $h(\beta(a_1', b_1')) = \beta(a_2', b_2')$. But $\beta(a_1', b_1')$ is standard, so $\beta(a_1', b_1') = \beta(a_2', b_2')$, which is a contradiction. □

9.3 Small index property

This section is devoted to a rather special topic. The question we ask is: can the topology on an automorphism group be determined completely by the group itself (as an abstract group)? In some specific cases the answer is positive, and generic automorphisms play an important role. We begin with some general facts concerning topology on automorphisms.

If $\mathfrak{A}$ is a first-order structure and $G = \mathrm{Aut}(\mathfrak{A})$, then it is fairly easy to see that a subgroup $H < G$ is open in G iff there is a finite $X \subseteq \mathfrak{A}$ such that $G_{(X)} < H$. Consequently, if the universe of $\mathfrak{A}$ is countable, then $[G : H] \leq \aleph_0$ for each open $H < G$, where $[G : H]$ is the index if H in G. We also have the following general theorem.

Theorem 9.3.1 *Let G be a Polish group. If $H < G$ is meager, then $[G : H] = 2^{\aleph_0}$.* □

A first-order structure $\mathfrak{A}$ has the *small index property* if every subgroup $H < \mathrm{Aut}(\mathfrak{A})$ such that $[\mathrm{Aut}(\mathfrak{A}) : H] \leq \aleph_0$ is open in $\mathrm{Aut}(\mathfrak{A})$. In the definition of the small index property we could replace $\leq \aleph_0$ with $< 2^{\aleph_0}$. In practice it makes little difference as it can be shown that every subgroup H of a Polish group G with the property of Baire is either open (and hence $[G : H] \leq \aleph_0$) or $[G : H] = 2^{\aleph_0}$. (See Corollary 6.5 of [77].) Our main result, Theorem 9.3.9, holds with the stronger definition.

The list of structures which are known to have the small index property includes: $(\mathbb{Q}, <)$, the countable atomless boolean algebra, every vector space of countable dimension over a finite or countable field, every countable ω-stable $\aleph_0$-categorical structure, and the random graph. In this section we prove that all countable arithmetically saturated models of arithmetic belong in this list. We start with two general lemmas.

Lemma 9.3.2 *Let H be an open subgroup of a Polish group G. If $K < G$ and $H \cap K$ is comeager in H, then K is open.*

Proof For each $a \in M$, the map $x \mapsto ax$ is a homeomorphism; hence, each coset of $H \cap K$ in H is also comeager. Since cosets are disjoint, there can be only one such coset; hence, H is contained in K, so K is open. □

In Section 9.1, we studied generic automorphisms, now we need to discuss generic tuples. If $G = \mathrm{Aut}(M)$, then for each $n < \omega$, we consider the group product G^n with the product topology. A tuple $(f_0, \ldots, f_{n-1}) \in G^n$ is *generic*, and its conjugacy class

$$[f_0, \ldots, f_{n-1}]_G = \{(f_0^g, \ldots, f_{n-1}^g) : g \in G\}$$

is comeager in G^n.

All of the results from Section 9.1 concerning Lascar generics can be generalized to generic tuples. The proofs are essentially the same. Here are the results. Each is equipped with the reference to the corresponding result for single automorphisms.

Definition 9.3.3 (9.1.2) Let $K \prec_{\mathsf{sm}} M$. We say that the tuple $(f_0, \ldots, f_n) \in (\mathrm{Aut}^M(K))^{n+1}$ is *existentially closed* if for every formula $\varphi(x, \bar{y})$ with parameters in K and for all $h_0, \ldots, h_n \in \mathrm{Aut}(M)$, if $f_i \subseteq h_i$ for $i \leq n$, and $M \models \exists x \varphi(x, h_0(x), \ldots, h_n(x))$, then $K \models \exists x \varphi(x, f_0(x), \ldots, f_n(x))$.

Lemma 9.3.4 (9.1.3) *Let M be countable and arithmetically saturated. Suppose that $a, a_0, \ldots, a_n \in M$ are such that $\mathrm{tp}(a) = \mathrm{tp}(a_0) = \cdots = \mathrm{tp}(a_n)$. Then there are $K \prec_{\mathsf{sm}} M$ and an existentially closed tuple $(f_0, \ldots, f_n) \in (\mathrm{Aut}^M(K))^{n+1}$ such that $a \in K$ and $f_i(a) = a_i$ for all $i \leq n$.* □

Definition 9.3.5 (9.1.4) Let M be a model and $G = \mathrm{Aut}(M)$. A tuple $(g_0, \ldots, g_n) \in G^{n+1}$ is *Lascar generic* if the following conditions are satisfied:

(1) For each finite $A \subseteq M$, there is $K \prec_{\mathsf{sm}} M$ such that $A \subseteq K$, $g_i(K) = K$ for $i \leq n$, and the tuple $(g_0 \restriction K, \ldots, g_n \restriction K)$ is existentially closed.

(2) If $K \prec_{\mathsf{sm}} M$, and $g_i(K) = K$ for $i \leq n$, and $(g_0 \restriction K, \ldots, g_n \restriction K)$ is existentially closed, $K \prec L \prec_{\mathsf{sm}} M$, and $(f_0, \ldots, f_n)$ is such that for all $i \leq n$ $f_i \in \mathrm{Aut}^M(L)$, and $g_i \restriction K \subseteq f_i$, then there is $h \in G_{(K)}$ such that for all $i \leq n$, $f_i \subseteq g_i^h$.

Proposition 9.3.6 (9.1.5) *Let $(f_0, \ldots, f_n)$ and $(g_0, \ldots, g_n)$ be two Lascar generic tuples for a model M, and let $G = \mathrm{Aut}(M)$. Suppose that $M_0 \prec_{\mathsf{sm}} M$, $f_i(M_0) = g_i(M_0) = M_0$, $f_i \restriction M_0 = g_i \restriction M_0$, for $i \leq n$, and $(f_0 \restriction M_0, \ldots, f_n \restriction M_0)$ is existentially closed. Then $[f_0, \ldots, f_n]_G = [g_0, \ldots, g_n]_G$.* □

Lemma 9.3.7 (9.1.6) *For each $n < \omega$, the set of Lascar generic n-tuples is comeager in G^n.* □

Finally another extension of Theorem 9.1.6:

Lemma 9.3.8 *Let $(g_0, \ldots, g_n)$ be a Lascar generic tuple. Then the set*

$$X = \{g \in G : (g_0, \ldots, g_n, g) \text{ is Lascar generic}\}$$

is comeager in G.

Proof It follows from Definition 9.3.5 that X is an intersection of countably many open sets. We will prove that X is dense. Let $\bar{a}, \bar{b} \in M$, be such that $\mathrm{tp}(\bar{a}) = \mathrm{tp}(\bar{b})$. We will find $g \in X$ such that $g(\bar{a}) = \bar{b}$. Let $K \prec_{\mathsf{sm}} M$ be such that $\bar{a}, \bar{b} \in K$, $g_i(K) = K$, $i \leq n$, and $(g_0 \restriction K, \ldots, g_n \restriction K)$ is existentially closed. By Theorem 9.1.6, there is a Lascar generic tuple $(g_0', \ldots, g_n', g')$ such that $g'(\bar{a}) = \bar{b}$ and for all $x \in K$ and all $i \leq n$, $g_i(x) = g_i'(x)$. Since $(g_0', \ldots, g_n')$ is also Lascar generic by Lemma 9.1.5, there is $h \in G_{(K)}$ such that $g_i' = g_i^h$ for $i \leq n$. Let $g = (g')^h$. Then $(g_0, \ldots, g_n, g)$ is Lascar generic and $g(\bar{a}) = \bar{b}$. □

Here is the main theorem of this section:

Theorem 9.3.9 *All countable arithmetically saturated models of* PA^* *have the small index property.*

Proof Let M be countable and arithmetically saturated, and let $G = \mathrm{Aut}(M)$. Suppose $H < G$ is non-open such that $[G : H] < 2^{\aleph_0}$.

We construct a family $\{M_\sigma : \sigma \in 2^{<\omega}\}$ of small elementary submodels of M and two families $\{g_\sigma : \sigma \in 2^{<\omega}\}$, $\{h_\sigma : \sigma \in 2^{<\omega} \setminus \{\emptyset\}\}$ of automorphisms of M such that:

(1) For all $\sigma, \tau \in 2^{<\omega}$, if $\sigma \subseteq \tau$, then $M_\sigma \prec M_\tau$;

(2) For all $F : \omega \longrightarrow \{0,1\}$, $\bigcup_{n<\omega} M_{F\restriction n} = M$;

(3) For all $\sigma \in 2^{<\omega}$, the tuple $(g_{\sigma\restriction 0}, g_{\sigma\restriction 1}, \ldots, g_{\sigma\restriction \ell(\sigma)-1}, g_\sigma)$ is Lascar generic;

(4) For all $\sigma, \sigma', \tau \in 2^{<\omega}$, if $\tau \subseteq \sigma$ and $\tau \subseteq \sigma'$, then $h_\sigma(x) = h_{\sigma'}(x)$ for all $x \in M_\tau$;

(5) For all $\sigma, \sigma', \tau \in 2^{<\omega}$, if $\tau \subset \sigma \subseteq \sigma'$, then $g_\tau^{h_\sigma} = g_\tau^{h_{\sigma'}}$;

(6) For all $\sigma \in 2^{<\omega}$, $g_\sigma^{h_{\sigma\frown 0}} \in H$ and $g_\sigma^{h_{\sigma\frown 1}} \notin H$.

Let $\langle a_n : n \in \omega \rangle$ be an enumeration of M. By Theorem 9.3.1, there is a Lascar generic $g_\emptyset \in H$. Let $M_\emptyset \prec_{\mathsf{sm}} M$ be such that $g_\emptyset(M_\emptyset) = M_\emptyset$ and $g_\emptyset \restriction M_\emptyset$ is existentially closed, and let h_0 be the identity on M. Now, by Lemma 9.3.2, there is a Lascar generic $f \in G_{(M_\emptyset)} \setminus H$. By Proposition 9.1.5, there is h_1 such that $g_\emptyset^{h_1} = f$.

Assume now that g_σ, M_σ, $h_{\sigma\frown 0}$, and $h_{\sigma\frown 1}$ have been defined. Let $e \in \{0,1\}$, and let $\tau = \sigma\frown e$. We will define g_τ, M_τ, $h_{\tau\frown 0}$, and $h_{\tau\frown 1}$.

Since $[G : H^{h_\tau^{-1}}] < 2^{\aleph_0}$, Theorem 9.3.1 and Lemma 9.3.8 imply that there is $g_\tau \in H^{h_\tau^{-1}}$ such that the tuple

$$(g_\emptyset, g_{\sigma\restriction 1}, \ldots, g_\sigma, g_\tau)$$

is Lascar generic. Let $h_{\tau\frown 0} = h_\tau$. Then $g_\tau^{h_{\tau\frown 0}} \in H$. The tuple

$$(g_\emptyset^{h_\tau}, g_{\sigma\restriction 1}^{h_\tau}, \ldots, g_\sigma^{h_\tau}, g_\tau^{h_\tau})$$

is also Lascar generic. Thus, there exists $M' \prec_{\mathsf{sm}} M$ such that $h_\tau(M_\sigma \cup \{a_{\ell(\sigma)}\}) \subseteq M'$ and

$$(g_\emptyset^{h_\tau} \restriction M', g_{\sigma\restriction 1}^{h_\tau} \restriction M', \ldots, g_\sigma^{h_\tau} \restriction M', g_\tau^{h_\tau} \restriction M')$$

is existentially closed. Let $M_\tau = h_\tau^{-1}(M')$. By Lemma 9.3.2, there is $f \notin H$ such that $f(x) = g_\tau^{h_\tau}(x)$ for all $x \in M'$, and the tuple

$$(g_\emptyset^{h_\tau}, g_{\sigma\restriction 1}^{h_\tau}, \ldots, g_\sigma^{h_\tau}, g_\tau^{h_\tau}, f)$$

is Lascar generic. By Proposition 9.3.6, there is $h \in G_{(M')}$ such that $g_\tau^{h_\tau h} = f$ and $g_\rho^{h_\tau h} = g_\rho^{h_\tau}$ for $\rho \subseteq \sigma$. Finally, we set $h_{\tau\frown 1} = hh_\tau$. One can check now that the conditions (1) through (6) are satisfied.

Consider an $F : \omega \longrightarrow \{0, 1\}$. The sequence $\langle h_{F\restriction n} : n < \omega \rangle$ is a Cauchy sequence (by conditions (2) and (4)). Let h_F be its limit. If $\sigma = F \restriction k$, then the sequence $\left\langle g_\sigma^{h_{F\restriction n}} : n < \omega \right\rangle$ is eventually constant (condition (5)) and its limit $h_\sigma^{h_F}$ is equal to $g_\sigma^{h_{\sigma\frown 0}}$ if $\sigma\frown 0 \subseteq F$, and it is equal to $g_\sigma^{h_{\sigma\frown 1}}$ if $\sigma\frown 1 \subseteq F$ (by continuity). Now, let $F' : \omega \longrightarrow \{0, 1\}$ be different from F, and let n be the largest initial segment of ω on which F and F' agree and let $\sigma = F \restriction n$. Assume, for example, that $\sigma\frown 0 \subseteq F$ and $\sigma\frown 1 \subseteq F'$. Then $g_\sigma^{h_F} \in H$ and $g_\sigma^{h_{F'}} \notin H$. It follows that $h_F h_{F'} \notin H$; hence, h_F and $h_{F'}$ are in different cosets of H in G. Thus $[G : H] = 2^{\aleph_0}$, which is a contradiction. □

9.3.1 *The cofinality of the automorphism group*

Let G be a group which is not finitely generated. The *cofinality* of G, denoted by $\mathrm{cf}(G)$, is the least cardinal number λ such that G is the union of an increasing chain of λ many proper subgroups. The group of all permutations of ω, the automorphism group of the random graph, and the automorphism groups of various $\aleph_0$-categorical structures have uncountable cofinality. In this section, we show that the cofinality of the automorphism group of a countable arithmetically saturated model is uncountable. As in proofs for other structures, the proof involves generic automorphisms. However, for models M of PA^* we can prove more: the cofinality of $\mathrm{Aut}(M)$ is uncountable iff M is arithmetically saturated. This is the main theorem of this subsection.

Theorem 9.3.10 *Let M be countable and recursively saturated. Then M is arithmetically saturated iff* $\mathrm{cf}(\mathrm{Aut}(M)) > \aleph_0$. *Moreover, if M is not arithmetically saturated, then* $\mathrm{Aut}(M)$ *is the union of a countable chain of its open subgroups.*

Proof Let $G = \mathrm{Aut}(M)$, and assume that M is arithmetically saturated. Suppose G is the union of a countable chain of proper subgroups $\langle H_n : n < \omega \rangle$. Notice first that, by Theorem 8.10.3, none of the H_n's can be open in G. By

Lemma 9.3.2, for all $a \in M$ and $n < \omega$, $G_a \cap H_n$ is not comeager in G_a. Also, since the union of countably many meager sets is meager, without loss of generality we can assume that no H_n is meager in G. Since for each a and each n, H_n is the union of countably many cosets of $H_n \cap G_a$ in H_n, it follows that none of the $H_n \cap G_a$ is meager.

Now we use the construction from the proof of Theorem 9.3.9. We define the automorphisms g_σ, and h_σ as before; however this time we require that for each $\sigma \in 2^n$, $g_{\sigma\frown 0} \in H_n$ and $g_{\sigma\frown 1} \notin H_n$. Then for any $\sigma \in 2^n$, and $F, F' \in 2^\omega$ such that $\sigma\frown 0 \subseteq F$ and $\sigma\frown 1 \subseteq F'$, as in the proof of Theorem 9.3.9, we have $h_F h_{F'} \notin H_n$.

Since $G = \bigcup_{n<\omega} H_n$, there are an $n < \omega$ and an uncountable $\Sigma \subseteq 2^\omega$ such that $h_F \in H_n$ for all $F \in \Sigma$. Since Σ is uncountable, there must be $m \geq n$, and $F, F' \in \Sigma$ such that $F \upharpoonright m = F' \upharpoonright m$ and $F \upharpoonright m+1 \neq F' \upharpoonright m+1$. But then $h_F h_{F'} \notin H_m$ while $H_n \subseteq H_m$, which is a contradiction.

Suppose now that M is not arithmetically saturated. It is shown in the proof of Theorem 8.10.3 that there is $a \in M$ such that:

(1) For each $d > \mathbb{N}$, there is a standard i such that $\mathbb{N} < (a)_i < d$;
(2) For each standard i, if $(a)_i > \mathbb{N}$, then $(a)_i \notin \mathrm{Scl}(0)$;

We will define a sequence of $\langle c_i : i < \omega \rangle$, where each c_i codes a sequence whose terms are among $\{(a)_j : j < \ell(a) \wedge (a)_j \leq (a)_i\}$. For $i < \omega$, let $\ell(c_i) = (a)_i$, and for $k < (a)_i$ let

$$(c_i)_k = \begin{cases} (a)_k & \text{if } k < \ell(a) \text{ and } (a)_k \leq (a)_i, \\ (a)_i & \text{otherwise.} \end{cases}$$

The sequence $\langle c_i : i < \omega \rangle$ is coded in M. Notice that for all $i < \omega$, $c_i \in \mathbb{N}$ iff $(a)_i \in \mathbb{N}$.

Suppose that $i, j < \omega$ and $(a)_i < (a)_j$. Then $\ell(c_i) = (a)_i = (c_j)_i$ and for all $k < \ell(c_i)$,

$$(c_i)_k = \begin{cases} (c_j)_k & \text{if } (c_j)_k \leq (c_j)_i, \\ (c_j)_i & \text{otherwise.} \end{cases}$$

Since $\ell(c_i) = (c_j)_i$, this proves that $c_i \in \mathrm{Scl}(c_j)$.

Let $\langle k_i : i < \omega \rangle$ be a sequence of standard numbers such that $(a)_{k_0} > (a)_{k_1} > \cdots > \mathbb{N}$ and $\inf\{(a)_{k_i} : i < \omega\} = \mathbb{N}$. Then $G_{c_{k_0}} \leq G_{c_{k_1}} \leq \cdots$ is a chain of open subgroups of G. Since for every $i < \omega$, $(a)_i \in \mathrm{Scl}(c_i)$, all these groups are proper subgroups of G. Since the sequence $\langle c_i : i < \omega \rangle$ is coded in M, the proof of Lemma 8.6.4 shows that for every $f \in G$, there is a nonstandard n such that for all standard k, if $c_k < n$, then $f(c_k) = c_k$. Since $\inf\{c_{k_i} : i < \omega\} = \mathbb{N}$, this proves that $G = \bigcup_{i<\omega} G_{c_{k_i}}$, and the result follows. □

Corollary 9.3.11 *Let M and N be countable recursively saturated models. If M is arithmetically saturated and* $\mathrm{Aut}(M) \cong \mathrm{Aut}(N)$, *then N is arithmetically saturated as well.*

Proof If M is arithmetically saturated, then $\mathrm{cf}(\mathrm{Aut}(M)) > \aleph_0$ and it follows that $\mathrm{cf}(\mathrm{Aut}(N)) > \aleph_0$, as the groups are isomorphic. Then N is arithmetically saturated. □

9.3.2 *Property* FA

A group is said to have property FA if, whenever it acts without inversion on a tree T, then there is a vertex fixed by all $g \in G$. The property has been defined by Serre, see [182] for all terms unexplained here. Theorem 18 of [182] states that a group G has property FA iff the following three conditions hold:

(1) G is not a free product with amalgamation.

(2) The infinite cyclic group is not a homomorphic image of G.

(3) G has uncountable cofinality.

Macpherson & Thomas [125] proved that if a Polish group G has a comeager conjugacy class, then it satisfies conditions (1) and (2) above. Hence, by Theorem 9.3.10, Proposition 9.2.1, and Theorem 9.2.3, we get the following corollary:

Corollary 9.3.12 *Let M be a countable recursively saturated model of* TA. *Then* $\mathrm{Aut}(M)$ *has property* FA *iff M is arithmetically saturated.* □

Of course, TA in the above corollary could be replaced by PA if we knew that the automorphism group of every countable arithmetically saturated model has a dense conjugacy class, which, in light of the results discussed in Section 9.2, seems plausible but most likely will be difficult to prove.

For every countable arithmetically saturated model, condition (2) above holds, and we have an alternative proof for it. Observe that, by Kaye's characterization of closed normal subgroups, no normal subgroup of the automorphism group of a countable recursively saturated model can be open. This fact can be given an independent proof. We ask the reader to do it in Exercise 9.6.4.

Proposition 9.3.13 *Let G be the automorphism group of a countable arithmetically saturated model M. Then no group of cardinality less than $2^{\aleph_0}$ is a homomorphic image of G.*

Proof Let G be as above and suppose that $f : G \longrightarrow H$ is a homomorphism with $|H| < 2^{\aleph_0}$. Then $\ker(f)$ is a normal subgroup of G and $[G : \ker(f)] < 2^{\aleph_0}$. Since M has the small index property, $\ker(f)$ is open. But, as we just observed, G does not have nontrivial open normal subgroups. □

9.4 Coding the standard system

Does the isomorphism type of the automorphism group of a countable recursively saturated model determine the isomorphism type of the model? This question might be too difficult since, in particular, one would need to analyze if and how the theory of a model is coded in the automorphism group. At present we do not seem to know enough to attempt this. Still, some information about the theory of a countable arithmetically saturated model can be recovered from its automorphism group.

Let M and N be countable arithmetically saturated models. If M is a model of TA and $\mathrm{Aut}(M) \cong \mathrm{Aut}(N)$, then N is also a model of TA. This follows from the theorem of Kaye [74] which we already mentioned in the previous section. The theorem completely characterizes closed normal subgroups of the automorphism group of a countable recursively saturated model: $H < \mathrm{Aut}(M)$ is closed and normal iff there is a $\mathrm{Aut}(M)$ invariant cut $I \subseteq_{\mathsf{end}} M$ such that H is the pointwise stabilizer of I. It is easy to see that every model having nonstandard definable elements has continuum many invariant cuts, while every model of TA has only one—$\mathbb{N}$. This information can be recovered from $\mathrm{Aut}(M)$ as topological group, and hence, due to the small index property, from $\mathrm{Aut}(M)$ as an abstract group as well. Notice that, by Corollary 9.3.11, in the above argument we could assume that M and N are recursively saturated and one of the models is arithmetically saturated. The same remark applies to the main result of this section, Theorem 9.4.1 below.

We show that the reconstruction problem restricted to the class of countable arithmetically saturated models of a given completion of PA has a positive solution. Once we fix the theory of an arithmetically saturated model, then the only isomorphism invariant left to determine is the standard system. This is accomplished by the following theorem.

Theorem 9.4.1 *Suppose M and N are countable arithmetically saturated models such that* $\mathrm{Aut}(M) \cong \mathrm{Aut}(N)$. *Then* $\mathrm{SSy}(M) = \mathrm{SSy}(N)$. □

To prove Theorem 9.4.1, we develop some coding machinery. *For the rest of this section, let M be a countable recursively saturated model, and let $G =$* $\mathrm{Aut}(M)$. We define several properties of G and its subgroups culminating in Definition 9.4.15, in which we say what it means for a group to encode a subset of $\mathbb{N}$. The definitions are formulated for G and M as above, but they all apply to arbitrary topological groups.

Definition 9.4.2 Given $H, K \leq G$ we say that H *precedes* K if $\bigcap_{h\in H} K^h$ is the trivial group.

Observe that if $H < G$, then G precedes H iff H does not contain any nontrivial normal subgroups of G. Also, H precedes H iff H is trivial. If H precedes K, $H \leq H'$, and $K' \leq K$, then H' precedes K'. Thus, H precedes K means, in some way, that H is "big" with respect to K. The term "precedes" is motivated by one of its important features. We will show that if I and J are distinct proper elementary cuts of M, then $I < J$ iff $G_{(I)}$ precedes $G_{(J)}$ iff $G_{\{I\}}$ precedes $G_{\{J\}}$.

Lemma 9.4.3 *Let I and J be elementary cuts in M. Then:*

(1) *If $I < J$, then $G_{(I)}$ precedes $G_{\{J\}}$.*
(2) *If $I \subseteq J$, then $G_{\{J\}}$ does not precede $G_{(I)}$.*

Proof For the proof of (1), consider some nonidentity $f \in G$. By the Moving Gaps Lemma (Theorem 8.3.2), there is $a > I$ such that $\mathrm{gap}(a) \neq \mathrm{gap}(f(a))$. By considering a larger cut I' such that $I < I' < \mathrm{gap}(\min\{a, f(a)\})$, if needed, we can assume that I is not downward ω coded. Then, by Corollary 8.4.6, there is $h \in G_{(I)}$ such that $h(a) \in J$ iff $h(f(a)) \notin J$. Therefore $f \notin h^{-1}G_{\{J\}}h$, completing the proof.

To prove (2), notice that if $f \in G_{(J)}$ and $k \in G_{\{J\}}$ then $kfk^{-1} \in G_{(I)}$. Hence, $G_{(J)} \subseteq k^{-1}G_{(I)}k$, but $G_{(J)}$ is nontrivial, so the proof is complete. □

Lemma 9.4.4 *Let H be a maximal open subgroup of G such that G precedes H. Then $H = G_{\{I\}}$ for some elementary cut I of M.*

Proof Let $J = J(H) = \inf\{a \in M : G_a \leq H\}$. Since H is open, it follows that $J \neq M$. Since G precedes H, it follows that $\mathrm{gap}(0) < J$. For suppose to the contrary that $J \subseteq \mathrm{gap}(0)$. If $J < \mathrm{gap}(0)$, then $G_{(\mathrm{gap}(0))}$ is a subgroup of each conjugate of H (Do IT!). If $J = \mathrm{gap}(0)$, then, for every $f \in G_{(>J)}$ (see the definition preceding Proposition 8.8.2) and every $g \in G$, there is $a > \mathrm{gap}(0)$ such that $G_{g(a)} \leq H$ and $f(a) = a$. Clearly $f \in g^{-1}Hg$; hence, $G_{(>J)}$ is a subgroup of each conjugate of H.

Let I be the largest elementary cut such that $I \subseteq J$. Clearly, $\mathrm{gap}(0) < I \neq M$ and $G_{\{J\}} \leq G_{\{I\}}$. Also, if $f \in G \setminus G_{\{J\}}$, then, without loss of generality, there is $a > J$ such that $f(a) \in J$ and $G_a \leq H$. Then $f \notin H$, since otherwise, by Proposition 8.8.1, we would have $G_{f(a)} \leq H$, which contradicts the definition of J. Thus we have established that $H \leq G_{\{J\}} \leq G_{\{I\}}$; hence, by maximality of H, $H = G_{\{I\}}$. □

Now we define two operations on subgroups of G. The meaning of these somewhat cryptic definitions is revealed in the following lemma and its corollaries. Our goal is to provide topological characterizations of the pointwise and setwise stabilizers of elementary cuts of M.

Definition 9.4.5 For $H \leq G$,

(1) $\Pi(H) = \bigcap\{H^g : g \in G \text{ and } H^g \text{ precedes } H\}$;

(2) $\Sigma(H)$ is the topological closure of the subgroup generated by

$$\bigcup\{\Pi(H^g) : g \in G \text{ and } H \text{ precedes } H^g\}.$$

Lemma 9.4.6 *Let J be a tall elementary cut of M. Then $\Pi(G_{\{J\}}) = G_{(J)}$.*

Proof Let $H = G_{\{J\}}$. Suppose first that $f \in G_{(J)}$ and K is a conjugate of H which precedes H. Then $K = G_{\{I\}}$, where I is an elementary cut and, according to Lemma 9.4.3 (2), $I < J$. Clearly, then $f \in K$. Thus $G_{(J)} \leq \Pi(H)$.

Next, suppose that $f \notin G_{(J)}$, so that $f(a) \neq a$ for some $a \in J$. Since J is tall, the Moving Gaps Lemma implies that there is $a \in J$ such that $f(\text{gap}(a)) \neq \text{gap}(a)$ (Do it!). Without loss of generality, we can assume that $\text{gap}(a) < \text{gap}(f(a))$. There is $h \in G_a$ such that $h(f(a)) > J$. Then $K = h^{-1}Hh = G_{\{I\}}$, where $I = h^{-1}(J) < J$, so that, by Lemma 9.4.3 (1), K precedes H. But $f \notin K$, so $f \notin \Pi(H)$. Therefore $\Pi(H) \leq G_{(J)}$. □

Corollary 9.4.7 (1) *Let D be a gap of M. Then $\Sigma(G_{\{D\}}) = G_{(\sup(D))} \lhd G_{(\inf(D))} = \Pi(G_{\{D\}})$.*

(2) *Let J be a tall elementary cut of M such that $J \neq \inf(D)$ for any gap D of M. Then $\Pi(G_{\{J\}}) = \Sigma(G_{\{J\}}) = G_{(J)}$.*

Proof Let D be a gap of M. Clearly, $G_{(\sup(D))} \lhd G_{(\inf(D))}$. Let $H = G_{\{D\}}$. Notice that, since $H = G_{\{\inf(D)\}}$, Lemma 9.4.6 implies that $\Pi(H) = G_{(\inf(D))}$. Also, Lemmas 9.4.6 and 9.4.3(1) imply that $\Sigma(H)$ is the closure of $G_{(>\sup(D))}$. Since $\sup(D)$ is an elementary cut, by Proposition 8.8.3, the closure of $G_{(>\sup(D))}$ is $G_{(\sup(D))}$, which finishes the proof of (1).

To prove (2), just as in (1) we get $\Pi(H) = G_{(J)}$ and $\Sigma(H)$ is the closure of $G_{(>J)}$, which is $G_{(J)}$. □

Suppose that D is a gap of M and $a \in D$ realizes a minimal type. Then G_a is an open maximal subgroup of G (Proposition 8.9.1) and $G_a = G_{\{D\}}$. For every elementary cut I and every unbounded type $p(x)$ realized in M, either $\sup(p^M \cap I) = I$ or $\inf(p^M \cap (M \setminus I)) = M \setminus I$. Let $p(x)$ be a minimal type realized in M and let I be an elementary cut of M.Then either I is tall and

$$I = \bigcup\{\sup(D) : D \cap p^M \neq \emptyset \wedge D < I\}$$

or I is short and then

$$I = \bigcap\{\sup(D) : D \cap p^M \neq \emptyset \wedge I < D\}.$$

This argument, combined with Corollary 9.4.7, proves the following corollary (Do it!).

Corollary 9.4.8 *Let H be a subgroup of G. The following are equivalent:*

(1) $H = G_{(I)}$ *for some elementary cut* I*;*

(2) *There is a nonempty set* $\mathcal{K}$ *of open maximal subgroups* $K < G$ *which are preceded by* G *such that* $H = \bigcap \{\Pi(K) : K \in \mathcal{K}\}$ *or* H *is the closure of* $\bigcup \{\Pi(K) : K \in \mathcal{K}\}$. □

Corollary 9.4.8 characterizes pointwise stabilizers of elementary cuts. The following simple lemma is needed to obtain a similar characterization of setwise stabilizers.

Lemma 9.4.9 *Let I be an elementary cut of M. Then $G_{\{I\}}$ is the normalizer of $G_{(I)}$; that is,* $G_{\{I\}} = \{f \in G : fG_{(I)} = G_{(I)}f\}$.

Proof For any $f \in G$, $fG_{(I)} = G_{(I)}f$ iff $G_{(I)} = G_{(f(I))}$. Since I is an elementary cut, it easily follows that the latter condition is equivalent to $f(I) = I$. Hence $\{f \in G : fG_{(I)} = G_{(I)}f\} = G_{\{I\}}$. □

Corollary 9.4.10 *Let H be a subgroup of G. The following are equivalent:*

(1) $H = G_{\{I\}}$ *for some elementary cut* I*;*

(2) *there is a nonempty set* $\mathcal{K}$ *of open maximal subgroups* $K < G$ *which are preceded by* G *such that* H *is the normalizer of* $\bigcap \{\Pi(K) : K \in \mathcal{K}\}$ *or* H *is the normalizer of the closure of* $\bigcup \{\Pi(K) : K \in \mathcal{K}\}$.

Proof Directly from Corollary 9.4.8 and Lemma 9.4.9. □

Next we define the notion of a gap stabilizer. The choice of terms is justified by the corollary that follows.

Definition 9.4.11 A subgroup H of G is a *gap stabilizer* if there is a nonempty set $\mathcal{K}$ of open maximal subgroups $K < G$ which are preceded by G such that H is the normalizer of $\bigcap \{\Pi(K) : K \in \mathcal{K}\}$ and $\Pi(H) \neq \Sigma(H)$.

Corollary 9.4.12 *A subgroup H of G is a gap stabilizer iff $H = G_{\{D\}}$ for some gap D of M.* □

Definition 9.4.13 Given subgroups $K, H_0, H_1, \dots$ of G, we say that K *supports* $\langle H_0, H_1, \dots \rangle$ if the following conditions hold:

(1) K and all H_i for $i < \omega$, are gap stabilizers;

(2) if $i < j$, then H_i precedes H_j;

(3) if $K \leq H < G$ and H is a gap stabilizer, then $H = K$ or $H = H_i$ for some $i < \omega$.

Lemma 9.4.14 *If $K, H_0, H_1, \ldots$ are subgroups of G and K supports $\langle H_0, H_1, \ldots \rangle$, then there are gaps $D_0 < D_1 < \cdots$ and $a \in M$ such that $H_i = G_{\{D_i\}}$ and $(a)_i \in D_i$ for each $i < \omega$.*

Proof The existence of gaps $D, D_0, D_1, \ldots$ such that $K = G_{\{D\}}$ and $H_i = G_{\{D_i\}}$ for $i < \omega$, follows directly from definitions and Corollary 9.4.12.

Let b be an element of D, and let $p(x)$ be the type of b. We claim that for any Skolem term $t(x)$ the following conditions are equivalent:

$$t(b) \in D \cup D_0 \cup D_1 \cup \ldots, \tag{1}$$

$$\forall b'[b' \in (D \cap p^M) \longrightarrow t(b') \in \text{gap}(t(b))]. \tag{2}$$

To prove that (1) implies (2), first notice that if $t(b) \in D$, then there is a Skolem term $t'(x)$ such that the formula $x < t'(t(x))$ is in $p(x)$. Hence, in this case we must have $t(b') \in D$. So now suppose that $t(b) \in D_i$ for some i. Let $f \in G$ be such that $f(b) = b'$. Then $f \in G_{\{D\}} = K \leq G_{\{D_i\}}$; hence, $t(b') = f(t(b)) \in D_i = \text{gap}(t(b))$.

To prove that (2) implies (1), suppose that $t(b') \in \text{gap}(t(b))$ whenever $b' \in D$ realizes $p(x)$. Therefore for any $f \in G_{\{D\}}$, we have $f(t(b)) \in \text{gap}(t(b))$; hence, $f \in G_{\{\text{gap}(t(b))\}}$. From Definition 9.4.13(3) it follows that $G_{\{\text{gap}(t(b))\}}$ must be either K or one of the H_i's. Therefore $t(b) \in D \cup D_0 \cup D_1 \ldots$.

Let $\langle t_n(x) : n < \omega \rangle$ be a recursive enumeration of all Skolem terms. Let $\varphi_n(x, y)$ be the formula $y \leq t_n(x) \wedge x \leq t_n(y)$. Directly from the definitions it follows that for any $a \in M$

$$\text{gap}(a) = \{c \in M : M \models \varphi_n(a, c), \text{ for some } n < \omega\}. \tag{3}$$

Now let $p(x) = \{\theta'_m(x) : m < \omega\}$, where $\langle \theta'_m(x) : m < \omega \rangle$ is recursive in $p(x)$, and then let $\theta_m(x) = \theta'_0(x) \wedge \cdots \wedge \theta'_m(x)$. We claim that for each Skolem term $t(x)$, the following is equivalent to (2) (and thus also to (1)):

$$\forall n < \omega \exists m < \omega \; M \models \forall x[\theta_m(x) \wedge \varphi_n(b, x) \longrightarrow \varphi_m(t(b), t(x))]. \tag{4}$$

Clearly, (4) implies (2) (Do it!). For the proof of (2)$\Longrightarrow$(4), suppose that (4) is false and let $n < \omega$ be a witness to this. By recursive saturation there is $b' \in M$ such that for all $m < \omega$,

$$M \models \theta_m(b') \wedge \varphi_n(b, b') \wedge \neg\varphi_m(t(b), t(b')).$$

Therefore, $b' \in \text{gap}(b)$, b' realizes $p(x)$, and $t(b') \notin \text{gap}(t(b))$, so (2) is false.

Let k_i be the ith element in the set

$$\{k < \omega : t_k(b) \in D \cup D_0 \cup D_1 \cup \ldots\}.$$

Since (1) is equivalent to (4), the sequence $\langle k_i : i < \omega \rangle$ is arithmetic in $p(x)$ and then so is the set of formulas $\{(x)_i = t_{k_i}(b) : i < \omega\}$. By arithmetic saturation, let $c \in M$ be such that $(c)_i = t_{k_i}(b)$ for each $i < \omega$.

We claim that for each $r < \omega$ there is a Skolem term $t(x)$ such that $t(b) \in D_r$. For suppose there was no such $t(x)$. Then for any $a \in D_r$ there is $a' \notin D_r$ such that $\text{tp}(a, b) = \text{tp}(a', b)$ (Do IT!). Therefore, by recursive saturation, there would be $f \in G_b$ such that $f(a) = a'$. Such an automorphism would contradict the fact that $K \leq H_r$.

Now let i_r be the least i such that $(c)_i \in D_r$; that is, i_r is the least i such that $(c)_i \notin \text{gap}((c)_{i_s})$ for each $s < r$ and for each $m < \omega$, either $(c)_m \in \text{gap}((c)_{i_s})$ for some $s < r$, $(c)_m \in \text{gap}((c)_i)$, or $(c)_i < (c)_m$. Because of (3), $\langle i_r : r < \omega \rangle$ is arithmetic in $\text{tp}(c)$. Therefore, by arithmetic saturation, there is $a \in M$ such that $(a)_r = (c)_{i_r}$ for all $r < \omega$. Thus $(a)_i \in D_i$ for all $i < \omega$. □

Now we come to our final definition.

Definition 9.4.15 Let X be subset of $\mathbb{N}$. We say that G encodes X if either X is finite or $X = \{i_0, i_1, \ldots\}$, where $i_0 < i_1 < \cdots$ and there are subgroups $K_1, K_2, H_0, H_1, H_2, \ldots$ of G such that K_1 supports $\langle H_0, H_1, H_2, \ldots \rangle$ and K_2 supports $\langle H_{i_0}, H_{i_1}, H_{i_2}, \ldots \rangle$.

Theorem 9.4.16 *For every $X \subseteq \mathbb{N}$, G encodes X iff $X \in \text{SSy}(M)$.*

Proof Suppose first that G encodes X. Since all finite sets are in $\text{SSy}(M)$,we can assume that $X = \{i_0, i_1, \ldots\}$, where $i_0 < i_1 < \cdots$ and $K_1, K_2, H_0, H_1, H_2, \ldots$ are as in Definition 9.4.15. By Lemma 9.4.14, there are distinct gaps $D_0, D_1 \ldots$ such that $H_i = G_{\{D_i\}}$ for $i < \omega$, and there are $a, b \in M$ such that $(a)_i \in D_i$ and $(b)_n \in D_{i_n}$, whenever $i, n < \omega$. By arithmetic saturation, there is $c \in M$ such that $(c)_n$ is the least i such that $(a)_i \in \text{gap}((b)_n)$ (Do IT!). But then $(c)_n = i_n$, so that $X \in \text{SSy}(M)$.

Assume now that $X = \{i_0, i_1, \ldots\}$, where $i_0 < i_1 < \cdots$ and $X \in \text{SSy}(M)$. Let $p(x)$ be a minimal type realized in M. Then, by recursive saturation, there is $a \in M$ which codes an increasing sequence $\langle (a)_i : i < \omega \rangle$ of elements realizing $p(x)$. Applying Lemma 2.1.10, we can get a minimal type $q(x)$ of $\text{Th}((M, (a)_0, (a)_1, \ldots))$ which is recursive in $\text{tp}(a)$, with the additional property that for each $i < \omega$, there is a Skolem term $t_i(x)$ such that the formula $t_i(x) = (a)_i$ is in $q(x)$. Let b realizes $q(x)$. Since $q(x)$ is unbounded and 2-indiscernible, by Lemma 3.1.18, $\text{tp}(b)$ is a rare type of $\text{Th}(M)$. Notice that $\text{tp}(b)$ is not a minimal, nor even a selective, type of $\text{Th}(M)$.

Let $H_i = G_{a_i}$ and $K_1 = G_b$. We shall show that K_1 supports $\langle H_0, H_1, \ldots \rangle$. Each $(a)_i$ realizes a minimal type; hence, each H_i is a gap stabilizer. Since the type of b in M is rare, by Corollary 9.4.12, K_1 is a gap stabilizer as well.

By Lemma 9.4.3(1), if $i < j$, then H_i precedes H_j. Finally, to show that Definition 9.4.13(3) holds, consider a gap D such that $K \leq G_{\{D\}} < G$. Then there is a Skolem term $t(x)$ such that $t(b) \in D$ (Do it!). Since b realizes a minimal type of $\mathrm{Th}((M, (a)_0, (a)_1, \dots))$, there is a least $k < \omega$ such that either

$$t'((a)_0, (a)_1, \dots, (a)_{k-1}, t(b)) = b$$

for some Skolem term $t'(\bar{x})$ or else

$$t'((a)_0, (a)_1, \dots, (a)_{k-1}) = t(b)$$

for some Skolem term $t'(\bar{x})$. In the first case, $t(b) \in \mathrm{gap}(b) = D$, so that $K = G_{\{D\}}$. In the second case, $k > 0$ (else $D = \mathrm{gap}(0)$ so that $G_{\{D\}} = G$) and then, because $(a)_0, (a)_1, \dots, (a)_{k-1}$ realize the minimal type $p(x)$, $t(b) \in \mathrm{gap}(a)_j$ for some $j \leq k-1$, so $G_{\{D\}} = H_j$. Therefore, K_1 supports $\langle H_0, H_1, \dots \rangle$.

Using the fact that $X \in \mathrm{SSy}(M)$, we can find in the same way an element $c \in M$ such that if $K_2 = G_c$ then K_2 supports $\langle H_{i_0}, H_{i_1}, \dots \rangle$. Therefore G encodes X. □

Now we are ready to prove Theorem 9.4.1. Suppose M_1 and M_2 are countable arithmetically saturated models, G_1 and G_2 are their automorphism groups, respectively, and $F : G_1 \longrightarrow G_2$ is an isomorphism. Since M_1 has the small index property, F is also an isomorphism of G_1 and G_2 qua topological groups. Suppose $X \in \mathrm{SSy}(M_1)$. By Theorem 9.4.16, G_1 encodes X; hence, G_2 encodes X as well, and, by the same theorem, $X \in \mathrm{SSy}(M_2)$. Thus $\mathrm{SSy}(M_1) = \mathrm{SSy}(M_2)$.

9.5 The spectrum of automorphism groups

What have we learned so far about the automorphism groups of countable recursively saturated models? By Theorem 9.4.1, there are continuum many nonisomorphic automorphism groups of arithmetically saturated models for any completion of PA. By Theorems 9.3.9 and 9.3.10, no automorphism group of an arithmetically saturated model can be isomorphic to an automorphism group of a recursively saturated model which is not arithmetically saturated. By Kaye's characterization of closed normal subgroups as the pointwise stabilizers of invariant cuts, if M and N are recursively saturated, only one of the models is a model of TA, and at least one is arithmetically saturated, then $\mathrm{Aut}(M) \not\cong \mathrm{Aut}(N)$ (Do it!). There is also a related recent Theorem by Nurkhaidarov on which we comment at the end of the Remarks and References section. In all other cases the question of which automorphism groups are isomorphic has not been settled.

Automorphism groups of countable recursively saturated models are large not only because of their size. By Corollary 5.5.2, if M is countable and recursively saturated model and $(I, <)$ a countable linearly ordered set, then $\mathrm{Aut}(I, <)$ embeds in $\mathrm{Aut}(M)$. In particular, we have:

Proposition 9.5.1 *Let M be countable and recursively saturated model. Then* $\mathrm{Aut}(\mathbb{Q},<)$ *embeds into* $\mathrm{Aut}(M)$. □

Proposition 9.5.2 *Let M be countable and recursively saturated. Then there is an embedding of* $\mathrm{Aut}(M)$ *onto a dense subgroup of* $\mathrm{Aut}(\mathbb{Q},<)$.

Proof If $p(x)$ is a minimal type realized in M, then $(p^M, <_M) \cong (\mathbb{Q},<)$, and it follows from Corollary 8.6.8 that the function $f \mapsto f \upharpoonright p^M$ induces a faithful embedding of $\mathrm{Aut}(M)$ onto a dense subgroup of $\mathrm{Aut}(\mathbb{Q},<)$ (Do it!). □

So $\mathrm{Aut}(M)$ and $\mathrm{Aut}(\mathbb{Q},<)$ are very much alike. Still $\mathrm{Aut}(M) \not\cong \mathrm{Aut}(\mathbb{Q},<)$. In fact, we have a more general result.

Theorem 9.5.3 *Let M be a countable arithmetically saturated model and let $\mathfrak{A}$ be a countable $\aleph_0$-categorical first-order structure. Then* $\mathrm{Aut}(M) \not\cong \mathrm{Aut}(\mathfrak{A})$.

Proof Every increasing chain of open subgroups of the automorphism group of any $\aleph_0$-categorical structure is finite. (See, for example, Fact 3.5 of [57]). This is not the case for $\mathrm{Aut}(M)$. To see this, consider the recursive type $p(v)$:

$$\{\exists x[(v)_n = \langle (v)_{n+1}, x\rangle] : n < \omega\} \cup \{t((v)_{n+1}) < (v)_n : n < \omega \text{ and } t \in \mathsf{Term}\}.$$

Clearly, $p(v)$ is finitely realizable. If $a \in M$ realizes $p(v)$, then for each $n < \omega$, $(a)_{n+1} \in \mathrm{Scl}((a)_n)$ and $(a)_n \notin \mathrm{Scl}((a)_{n+1})$. Hence $G_{(a)_0} < G_{(a)_1} < \cdots$ is an infinite increasing chain of open subgroups of $\mathrm{Aut}(M)$. This argument shows that $\mathrm{Aut}(M)$ and $\mathrm{Aut}(\mathfrak{A})$ are not isomorphic as topological groups. It follows that $\mathrm{Aut}(M)$ and $\mathrm{Aut}(\mathfrak{A})$ are not isomorphic as abstract groups. For suppose to the contrary that $F : \mathrm{Aut}(M) \longrightarrow \mathrm{Aut}(\mathfrak{A})$ is a group isomorphism. Then for every open $H < \mathrm{Aut}(\mathfrak{A})$, $F^{-1}(H)$ has at most countable index in $\mathrm{Aut}(M)$; hence, by the small index property $F^{-1}(H)$ is open in $\mathrm{Aut}(M)$, which proves that F is a homeomorphism. □

In the argument above we used the small index property in a substantial way. We do not know if the result is true for recursively saturated but not arithmetically saturated models.

By Theorem 9.4.1, the automorphism group of a countable arithmetically saturated model determines its standard system. We finish this section with another result relating standard systems to automorphism groups. The following lemma is due to B. H. Neumann & H. Neumann [138].

Lemma 9.5.4 *There is a recursive sequence of group terms $t_n(x,y)$, $n < \omega$ such that for every $A \subseteq \mathbb{N}$ there are $f, g \in \mathrm{Aut}(\mathbb{Q},<)$ such that*

$$t_n(f,g) = \mathrm{id} \iff n \in A.$$ □

Theorem 9.5.5 *Let M and N be countable recursively saturated models. Suppose there is an embedding $F : \mathrm{Aut}(M) \longrightarrow \mathrm{Aut}(N)$ such that for all $f, g \in \mathrm{Aut}(M)$, if (M, f, g) is recursively saturated, then $(N, F(f), F(g))$ is recursively saturated. Then $\mathrm{SSy}(M) \subseteq \mathrm{SSy}(N)$.*

Proof Suppose that there is an embedding $F : \mathrm{Aut}(M) \longrightarrow \mathrm{Aut}(N)$ which maps recursively saturated pairs of automorphisms to recursively saturated pairs of automorphisms. We will show that $\mathrm{SSy}(M) \subseteq \mathrm{SSy}(N)$. So let $A \in \mathrm{SSy}(M)$ be given. By Lemma 9.5.4 and Corollary 9.5.1, there are $f, g \in \mathrm{Aut}(M)$ such that

$$A = \{n \in \mathbb{N} : t_n(f, g) = \mathrm{id}\}.$$

Hence the theory

$$T = \{f \text{ and } g \text{ are automorphisms}\} \cup$$
$$\{t_n(f, g) = \mathrm{id} : n \in A\} \cup \{t_n(f, g) \neq \mathrm{id} : n \notin A\}$$

is coded in $\mathrm{SSy}(M)$ and is consistent. By chronic resplendency, we get such f and g so that $(M, f, g) \models T$ and (M, f, g) is recursively saturated. Then

$$A = \{n \in \mathbb{N} : (N, F(f), F(g)) \models t_n(F(f), F(g)) = \mathrm{id}\},$$

and, since $(N, F(f), F(g))$ is recursively saturated, this implies that $A \in \mathrm{SSy}(N)$. □

Corollary 9.5.6 *Let M and N be countable recursively saturated models such that $M \equiv N$ and there is an isomorphism $F : \mathrm{Aut}(M) \longrightarrow \mathrm{Aut}(N)$ such that for all $f, g \in \mathrm{Aut}(M)$, (M, f, g) is recursively saturated iff $(N, F(f), F(g))$ is recursively saturated. Then $M \cong N$.* □

9.6 Exercises

♣**9.6.1** Every recursively saturated model has small elementary submodels which are not finitely generated.

♣**9.6.2** If f is an *automorphism* of countable recursively saturated model, then the conjugacy class of f is not open. Some conjugacy classes are not closed. Every countable arithmetically saturated model has an *automorphism* whose conjugacy class is the intersection of a countable family of open sets.

♦**9.6.3** If M is a countable recursively saturated model of TA and $G = \mathrm{Aut}(M)$, then G precedes each of its open maximal subgroups. Consequently, each open maximal subgroup of G is a setwise stabilizer of an elementary cut of M.

♣**9.6.4** If M is a countable, arithmetically saturated model and $H < \mathrm{Aut}(M)$ is normal, then H is not an open subgroup of $\mathrm{Aut}(M)$. This follows from Kaye's theorem on closed normal subgroups. An independent proof can be given (♥).

♦**9.6.5** If M is a countable and arithmetically saturated model and $H < \mathrm{Aut}(M)$, then $[\mathrm{Aut}(M) : H] \geq \aleph_0$.

♦**9.6.6** An alternative proof of Theorem 9.5.3 can be given, by showing that the automorphism group of a countable recursively saturated model has at least continuum many open subgroups.

9.7 Remarks & References

The proof of the small index property for countable arithmetically saturated models of PA follows closely Lascar's proof in [120] with some minor modifications. In particular, Lemmas 9.1.3 and 9.3.4 have a new, more direct proof. For background material on generic sequences in $\aleph_0$-categorical structures see [77] and the paper [59] by Hodges, Hodkinson, Lascar, and Shelah. In the definiton of the small index property we use the condition "$[G : H]$ is at most countable." As Hodges notes in [58]: "Some people say '$<2^\omega$' in place of 'at most ω'; in practice it makes little difference." The proof of Theorem 9.3.9 shows that for countable arithmetically saturated models indeed there is no difference.

The connection between dense conjugacy classes and graph coloring is exploited in Schmerl [178]. Let $H^*(k,n)$ be the same statement as $H(k,n)$ but applied to (undirected) graphs rather than digraphs. It is easy to see that $\forall n \neg H^*(n, n+1)$. The statement $\forall n H^*(n,n)$ is known as Hedetniemi's Conjecture. See the survey [159]. A weaker form of Hedetniemi's Conjecture, which is still open, is $\forall n \exists k H^*(k,n)$. It is known that $\forall n \exists k H^*(k,n) \iff \exists k H^*(k,10) \iff \exists k H(k,4) \iff \forall n \exists k H(k,n)$. The middle equivalence appears in [204].

John Truss proved (personal communication) that if an automorphism group G of a countable structure has an automorphism whose conjugacy class is dense, then every local generic of G is generic.

The result on cofinality of arithmetically saturated models was proved in Kossak & Schmerl [107]. For a survey of the group cofinality problem see Simon Thomas' [206].

Coding the standard system of an arithmetically saturated model in its automorphism group is from [108]. The proof owes much to earlier work of Kaye and Kotlarski on the same problem from [114]. In a recent paper, Nurkhaidarov [139] modified some arguments from [108] and combined them with a result of Seetapun [181] on Ramsey's theorem for pairs to show the following:

> Theorem: There are countable arithmetically saturated models M_0, M_1, M_2, M_3 which share the same standard system such that $\mathrm{Aut}(M_i) \not\cong \mathrm{Aut}(M_j)$ for $i \neq j$.

Theorem 9.5.3 is from [75].

Theorem 9.5.5 is a slightly simplified special case of a theorem of Kaye [73]. Corollary 9.5.6 holds for arbitrary countable recursively saturated first-order structures. The proof is similar. Kaye's question whether pairs in Theorem 9.5.5 can be replaced by single recursively saturated automorphisms is open.

10

ω_1-LIKE MODELS

We already have discussed some ω_1-like models in Chapter 2. Constructions of Jónsson models and rather classless recursively saturated models show how the model theory of uncountable structures differs from the model theory of countable ones. As evidenced throughout this book, countable nonstandard models form a complex class. Set theory involved in the study of uncountable models brings vast new layers of complexity. Many interesting structures can be built. Many natural problems remain open.

Recall that a linearly ordered structure $(A, <)$ is *κ-like* for an infinite cardinal κ if $|A| = \kappa$ and every proper initial segment of $(A, <)$ is of cardinality smaller than κ. The ω_1-like models of PA are worth studying for (at least) two reasons. First: ω_1 iterations of end extensions of countable models generate ω_1-like models with interesting second-order properties. Second: the study of ω_1-like models is really the study of their countable elementary substructures. It has often been the case that questions about ω_1-like models translate into open problems concerning countable structures providing valuable insights.

Results on ω_1-like models can be sometimes generalized to κ-like models for all or some uncountable cardinals κ. The question of which of the results generalize and in what form leads to nontrivial set-theoretic and model-theoretic questions. We say more about it in the Remarks and References section of this chapter.

10.1 ω_1-Like recursively saturated models

To begin, we need to know that ω_1-like recursively saturated models exist. For an easy example, start with $M = (\mathbb{N}, S)$, where S is the satisfaction relation of $\mathbb{N}$. Let κ be a cardinal and let N be a κ-canonical extension of M. Then N is κ-like and its reduct to $\mathcal{L}_{\mathsf{PA}}$ is recursively saturated. In this way we get recursively saturated κ-like models of TA, but the same argument works if instead of the standard model we begin with (M, S), where M is a countable recursively saturated models and S is a partial inductive satisfaction class.

There are many other constructions. The starting point is the following proposition. Recall that if M and N are countable and recursively saturated, then

$$M \cong N \text{ iff } M \equiv N \quad \text{and} \quad \mathrm{SSy}(M) = \mathrm{SSy}(N).$$

In particular, for countable and recursively saturated M and N, if $M \prec_{\text{end}} N$, then $M \cong N$.

Proposition 10.1.1 *Every countable recursively saturated model M has a recursively saturated elementary end extension. Moreover, every such model M has an elementary end extension N such that (N, M) is recursively saturated.*

Proof We will prove the moreover part. Let T be the theory in $\mathcal{L}(K)$, where K is a new predicate symbol expressing that K is an elementary cut of M. Clearly T is consistent with $\text{Th}(M)$ (because there are models of $\text{Th}(M)$ which have proper elementary cuts). By chronic resplendency there is $K \prec_{\text{end}} M$ such that (M, K) is recursively saturated. Then K is recursively saturated and $\text{SSy}(M) = \text{SSy}(K)$; hence $M \cong K$, and the result follows. □

In Corollary 10.1.6, we show that Proposition 10.1.1 does not generalize to uncountable models. In other words, the recursively saturated version of the MacDowell–Specker Theorem is false.

Whereas not every model has a nonconservative elementary end extension, it is easy to show that every resplendent model has one (DO IT!). In fact, more is true:

Proposition 10.1.2 *If $M \prec N$ and N is recursively saturated, then N is not a conservative extension of M.*

Proof We only need to consider the case when N is an elementary end extension of M. Suppose N is recursively saturated. By recursive saturation, for every $a \in N$, there is $s_a \in N$ such that for all formulas $\varphi(v)$,

$$N \models \forall v < a(\langle \varphi, v \rangle \in s_a \longleftrightarrow \varphi(v)).$$

Pick an $a > M$ and let $S = s_a \cap M$. By Corollary 1.9.5, S is undefinable in M. Since S is coded in N, N is not a conservative extension. □

Thus, if $M \prec_{\text{end}} N$ and N is recursively saturated, then there is an undefinable $X \subseteq M$ which is coded in N. By contrast, the next proposition shows that for every such X, there is a recursively saturated elementary end extension of M in which X is not coded. The proposition is an immediate consequence of Corollary 8.2.5.

Proposition 10.1.3 *If X is an undefinable subset of a countable recursively saturated model M, then M has a recursively saturated elementary end extension in which X is not coded.* □

Another proof of Proposition 10.1.3 using Q_e-classes is sketched in Exercise 10.7.2.

A digression: in connection with Proposition 10.1.3, one could ask whether for every undefinable subset X of recursively saturated model M, there is an elementary end extension N of M such that there is no $Y \subseteq M$ coded in N such that $(M, X) \equiv (M, Y)$. The negative answer is a consequence of the result on perfect generics in Exercise 6.6.7. (DO IT!).

Theorem 10.1.4 *Every countable recursively saturated model M has an undefinable inductive subset X such that for all $I \prec_{\text{end}} M$, $(I, X \cap I) \prec (M, X)$.* □

Recall that a model M is rather classless if every class of M is definable. We will use Proposition 10.1.3 to show that there exist recursively saturated rather classless models. This was first proved by Kaufmann [67] using Jensen's set-theoretic principle $\diamondsuit$ which states that there exists a sequence of sets $\langle S_\alpha : \alpha < \omega_1 \rangle$ such that for every $X \subseteq \omega_1$, the set $\{\alpha < \omega_1 : X \cap \alpha = S_\alpha\}$ is a stationary subset of ω_1. $\diamondsuit$ is true in the constructible universe, and it implies (in ZFC) that $\aleph_1 = 2^{\aleph_0}$.

We say that M is a *Kaufmann* model if M is ω_1-like, recursively saturated and rather classless. We will show that Kaufmann models exist. The proof below uses $\diamondsuit$; however, as shown by Shelah in [184], the theorem is provable in ZFC.

Theorem 10.1.5 *Every countable recursively saturated model has an elementary end extension to a Kaufmann model.*

Proof Let M_0 be a countable recursively saturated model. We define a continuous chain of recursively saturated elementary end extensions $\langle M_\alpha : \alpha < \omega_1 \rangle$. For a successor $\alpha < \omega_1$, let $M_{\alpha+1}$ be a countable recursively saturated elementary end extension of M_α. For limit $\lambda < \omega_1$, let $M_\lambda = \bigcup_{\alpha<\lambda} M_\alpha$. We can assume that the universe of M_λ is λ. To define $M_{\lambda+1}$, consider S_λ from the $\diamondsuit$ sequence. If S_λ is definable in M_λ, then let $M_{\lambda+1}$ be any countable recursively saturated elementary end extension of M_λ. Otherwise, let $M_{\lambda+1}$ be a countable recursively saturated elementary end extension of M_λ in which S_λ is not coded, given by Proposition 10.1.3.

The model $N = \bigcup_{\alpha<\omega_1} M_\alpha$ is a recursively saturated ω_1-like elementary end extension of M_0. Let X be a class of N and let $X_\alpha = X \cap M_\alpha$. The set $A = \{\alpha < \omega_1 : (M_\alpha, X_\alpha) \prec (N, X)\}$ is closed and unbounded in ω_1. Hence, there is a limit $\lambda \in A$ such that $X_\lambda = S_\lambda$. Since X is a class of N, S_λ is coded in $M_{\lambda+1}$. From the construction it follows that $S_\lambda \in \text{Def}(M_\lambda)$. But $(M_\lambda, S_\lambda) \prec (N, X)$; hence, $X \in \text{Def}(N)$, and this finishes the proof. □

It is easy to see that all elementary end extensions of a rather classless model must be conservative. Combining this with Proposition 10.1.2 we get.

Corollary 10.1.6 *Kaufmann models have no recursively saturated elementary end extensions.* □

Recall that $X \subseteq M \models \mathsf{PA}^*(\mathcal{L})$ is a class of M iff $(M, X) \models I\Sigma_0$. If M is κ-like, for a regular $\kappa > \aleph_0$, then, for set-theoretic reasons, for every $X \subseteq M$,

(M, X) satisfies the collection principle for all formulas of $\mathcal{L} \cup \{X\}$. Hence, every class of a κ-like model, for regular $\kappa > \aleph_0$, is inductive.

Kaufmann models are recursively saturated and do not have partial inductive satisfaction classes; however to get a model with these properties, a simpler construction can be given. Let M be such a model. It follows from the discussion above that every class of M is inductive. This implies M does not have recursively saturated elementary end extensions; thus we get a proof of Corollary 10.1.6 without constructing Kaufmann models.

Despite the fact that Theorem 10.1.5 is provable in ZFC, it is still an open problem to give a direct construction of a Kaufmann model. This is one of the open questions in the list of Twenty Questions in the last chapter. Below we give another construction which uses a weaker combinatorial principle, weak $\diamondsuit$ for ω_1. The principle, denoted by Φ in the literature, says that: given a "partition" function $P : 2^{<\omega_1} \longrightarrow 2$, there is a function $\rho : \omega_1 \longrightarrow 2$ such that for all $\sigma \in 2^{\omega_1}$ the set $\{\alpha < \omega_1 : P(\sigma \restriction \alpha) = \rho(\alpha)\}$ is stationary in ω_1.

As shown by Devlin & Shelah [25], Φ is equivalent to $2^{\aleph_0} < 2^{\aleph_1}$.

To build a Kaufmann model , we will use a slightly modified equivalent version of Φ.

> WEAK $\diamondsuit$: For every function $P : 2^{<\omega_1} \times 2^{<\omega_1} \longrightarrow 2$ there is $\rho : \omega_1 \longrightarrow 2$ such that for all $(\sigma, \tau) \in 2^{\omega_1} \times 2^{\omega_1}$ the set
>
> $$\{\alpha < \omega_1 : P(\sigma \restriction \alpha, \tau \restriction \alpha) = \rho(\alpha)\}$$
>
> is stationary in ω_1.

We will utilize Corollary 8.2.5, which states that every countable recursively saturated model M has two countable recursively saturated elementary end extensions M^0 and M^1 such that

$$\mathrm{SSy}(M^0/M) \cap \mathrm{SSy}(M^1/M) = \mathrm{Def}(M).$$

Let M be a countable recursively saturated model. For every $s \in 2^{<\omega_1}$ we define a model M_s by induction on the length of s. Let $M_\emptyset = M$. For nonempty s, we have the usual two cases to consider.

Successor step. If M_s is defined, then we let $M_{s^\frown 0}$ and $M_{s^\frown 1}$ be countable recursively saturated elementary end extensions of M_s such that

$$\mathrm{SSy}(M_{s^\frown 0}/M_s) \cap \mathrm{SSy}(M_{s^\frown 1}/M_s) = \mathrm{Def}(M_s).$$

Limit step. If $\ell(s)$ is a limit ordinal and $M_{s'}$ is defined for all $s' \lhd s$, then $M_s = \bigcup_{s' \lhd s} M_{s'}$.

We can assume that if $\ell(s) = \lambda$ and λ is a limit ordinal, then the universe of M_s is λ.

Now we define the partition function P. For notational convenience we identify subsets of ordinals $\alpha < \omega_1$ with their characteristic functions. For $X \subseteq \alpha$ and $s \in 2^\alpha$, let

$$P(X, s) = \begin{cases} 0, & \text{if } X \in \mathrm{SSy}(M_{s^\frown 1}/M_s) \setminus \mathrm{Def}(M_s), \\ 1, & \text{if } X \in \mathrm{SSy}(M_{s^\frown 0}/M_s) \setminus \mathrm{Def}(M_s). \end{cases}$$

Notice that P is defined only for some pairs (X, s). It is irrelevant how P is defined for other pairs.

Let ρ be the guessing function for the P given by Φ and let $N = \bigcup_{\alpha<\omega_1} M_{\rho\restriction\alpha}$.

Clearly N is ω_1-like and recursively saturated. We will prove that it is also rather classless. Suppose X is a class of N. Since the set

$$E = \{\alpha < \omega_1 : P(X \cap \alpha, \rho \restriction \alpha) = \rho(\alpha)\}$$

is stationary, there is a limit ordinal $\lambda \in E$ such that for $s = \rho \restriction \lambda$ and $X_s = M_s \cap X$,

$$(M_s, X_s) \prec (N, X).$$

We claim that X_s is definable in M_s, which implies that X is definable in N and this will finish the proof. So suppose X_s is not definable in M_s. Then for an $e \in \{0, 1\}$, X_s is in $\mathrm{SSy}(M_{s^\frown e}/M_s) \setminus \mathrm{Def}(M_s)$. Since X_s is coded in N, we must have $\rho(\lambda) = e$. By the definition of P, we also have $P(X \cap M_s, \rho \restriction \lambda) = 1 - e$, hence $\rho(\lambda) = 1 - e$, and we get a contradiction.

10.2 Similar nonisomorphic models

A countable recursively saturated model is characterized up to isomorphism by its complete theory and its standard system. For models of arbitrary cardinality, this is generalized to the following proposition.

Proposition 10.2.1 *Let M and N be recursively saturated models and suppose that $M \equiv N$. Then, $\mathrm{SSy}(M) = \mathrm{SSy}(N)$ iff $M \equiv_{\infty\omega} N$.* □

$\mathcal{L}_{\infty\omega}$-equivalence has a characterization in terms of partial isomorphisms. Let $\mathfrak{A}$ and $\mathfrak{B}$ be first-order structures. Then, $\mathfrak{A} \equiv_{\infty\omega} \mathfrak{B}$ iff there exists a *back-and-forth* system for $\mathfrak{A}$ and $\mathfrak{B}$, that is, a nonempty set I of pairs $(\bar{a}, \bar{b})$ of finite tuples of elements of $\mathfrak{A}$ and $\mathfrak{B}$, respectively such that:

(1) if $(\bar{a}, \bar{b})$ is in I, then $\mathrm{tp}(\bar{a}) = \mathrm{tp}(\bar{b})$;

(2) for every pair $(\bar{a}, \bar{b})$ in I and every $c \in \mathfrak{A}$, there is $d \in \mathfrak{B}$ such that $(\bar{a}c, \bar{b}d)$ is in I;

(3) for every pair $(\bar{a}, \bar{b})$ in I and every $d \in \mathfrak{B}$, there is $c \in \mathfrak{A}$ such that $(\bar{a}c, \bar{b}d)$ is in I.

If M and N are recursively saturated elementarily equivalent models and $\mathrm{SSy}(M) = \mathrm{SSy}(N)$, then the back-and-forth system for M and N is provided by the set $I = \{(\bar{a}, \bar{b}) : \bar{a} \in M, \bar{b} \in N, \mathrm{tp}(\bar{a}) = \mathrm{tp}(\bar{b})\}$. We will show that if, in addition, M and N are ω_1-like, then M and N have back-and-forth system consisting of partial countable isomorphisms. This is made precise in the following characterization of elementary equivalence of structures of cardinality $\aleph_1$ in $\mathcal{L}_{\infty\omega_1}$.

Theorem 10.2.2 *Let $\mathfrak{A}$ and $\mathfrak{B}$ be first-order structures of cardinality $\aleph_1$ for the same countable language. Then, $\mathfrak{A} \equiv_{\infty\omega_1} \mathfrak{B}$ iff A and B can be represented as unions of chains $\langle \mathfrak{A}_\alpha : \alpha < \omega_1 \rangle$, $\langle \mathfrak{B}_\alpha : \alpha < \omega_1 \rangle$ of countable submodels of $\mathfrak{A}$ and $\mathfrak{B}$ respectively, and there is $\langle G_\alpha : \alpha < \omega_1 \rangle$ such that each G_α is nonempty set of isomorphisms $f : \mathfrak{A}_\alpha \longrightarrow \mathfrak{B}_\alpha$ and for all $\alpha < \beta < \omega_1$, every isomorphism in G_α extends to one in G_β.* □

Note that the chains in Theorem 10.2.2 are not required to be continuous.

Recall that $a \in M$ codes an *ascending sequence of gaps*, abbreviated $a \in \mathrm{ASG}(M)$, if $\ell(a)$ is nonstandard and for each $i < \ell(a)$, $\mathrm{gap}((a)_i) < (a)_{i+1}$.

For $a \in \mathrm{ASG}(M)$ and a cut $I < \ell(a)$, let

$$M(I, a) = \sup\{(a)_i : i \in I\}.$$

Proposition 10.2.3 *Let M be recursively saturated. Then:*

(1) *For each $c \in M$ there is $a \in \mathrm{ASG}(M)$ such that $c = (a)_0$.*

(2) *If $p(v)$ is an unbounded type realized in M, then for each $a \in \mathrm{ASG}(M)$ there is $a' \in \mathrm{ASG}(M)$ such that $M(\omega, a) = M(\omega, a')$ and for every $n < \omega$, $(a')_n$ realizes $p(v)$.*

Proof Part (1) is an easy exercise. The proof of (2) is not difficult either. Let $p(v)$ be a type realized in M. For $a \in \mathrm{ASG}(M)$, consider the type $p'(v, a)$

$$\{\varphi((v)_n) : \varphi(v) \in p(v) \text{ and } n < \omega\} \cup \{(a)_n < (v)_n < (a)_{n+1} : n < \omega\}.$$

This type is recursive in $p(v)$, and it is easy to see that it is finitely realizable in M (Do it!). Any a' realizing $p'(v)$ in M has the required property. □

The following technical proposition and its corollary will be used later.

Proposition 10.2.4 *Let $p(v)$ be a minimal type realized in a countable recursively saturated model M. If $\langle a_n : n < \omega \rangle$, $\langle b_n : n < \omega \rangle$ are increasing unbounded sequences of p^M, then there is an automorphism f of M such that for each $n < \omega$, $f(a_n) = b_n$.*

Proof The proof is based on the fact that elements realizing a minimal type are strongly indiscernible (see Lemma 3.2.9). Since p is an unbounded indiscernible type, $\bar{a} = \langle a_0, \dots, a_{n-1} \rangle$, $\bar{b} = \langle b_0, \dots, b_{n-1} \rangle$ are increasing sequences of elements of p^M then, for all $c < \operatorname{gap}(\min\{a_0, b_0\})$ and all formulas φ

$$M \models \varphi(\bar{a}, c) \longleftrightarrow \varphi(\bar{b}, c).$$

In fact, we have a stronger property: if $c, c' < \operatorname{gap}(\min\{a_0, b_0\})$ and $\operatorname{tp}(c) = \operatorname{tp}(c')$, then for all formulas φ,

$$M \models \varphi(\bar{a}, c) \longleftrightarrow \varphi(\bar{b}, c').$$

See the proof of Lemma 3.2.9. Now (1) can be proved by the usual back-and-forth construction begun by enumerating M in such a way that no element c is considered in the construction before the image of a_n is declared to be b_n, where n is the smallest such that $c < \operatorname{gap}(\min\{a_n, b_n\})$ (DO IT!). □

Corollary 10.2.5 *Let $p(v)$ be a minimal type realized in a countable recursively saturated model M. Then every increasing unbounded sequence $\langle a_n : n < \omega \rangle$ of elements realizing p in M is coded in some recursively saturated elementary end extension N of M.*

Proof Consider the type $q(v)$

$$\{\varphi((v)_n) : \varphi(v) \in p(v) \text{ and } n < \omega\} \cup \{(v)_n < (v)_{n+1} : n < \omega\}.$$

The type $q(v)$ is recursive in $p(v)$ and finitely realizable, hence it is realized in M. Let b realizes $q(v)$. Then $M(\omega, b)$ is isomorphic to M. By Proposition 10.2.4, there is an isomorphism $f : M(\omega, b) \cong M$ such that for all $n < \omega$, $f((b)_n) = a_n$, and the result follows. □

Let $p(v)$ be a minimal type realized in M. Proposition 10.2.3 (2) implies that for every $a \in \operatorname{ASG}(M)$, there is $a' \in \operatorname{ASG}(M)$ such that $M(\omega, a) = M(\omega, a')$ and for each $n < \omega$, $(a')_n$ realizes $p(v)$. Since $p(v)$ is indiscernible, the above observation has further consequences, which are summarized in the next proposition. Part (2) below is formulated in a rather mysterious way due to a particular application we have in mind.

Proposition 10.2.6 *Let M be countable and recursively saturated.*

(1) *For all $a, b \in \operatorname{ASG}(M)$, $(M, M(\omega, a)) \cong (M, M(\omega, b))$, moreover:*
(2) *If $a_0, \dots, a_n$, $b_0, \dots, b_n \in \operatorname{ASG}(M)$ are such that $M(\omega, a_0) < \cdots < M(\omega, a_n)$, $M(\omega, b_0) < \cdots < M(\omega, b_n)$, $k \leq n$ is such that for all $i < k$, $a_i = b_i$ and $c \in M(\omega, a_k) \cap M(\omega, b_k)$, then*

$$(M, M(\omega, a_0), \dots, M(\omega, b_n), c) \cong (M, M(\omega, b_0), \dots, M(\omega, b_n), c).$$

(3) *Suppose* $a, b \in \mathrm{ASG}(M)$, $K \prec_{\mathsf{end}} M$, K *is not downward* ω*-coded and* $(a)_0, (b)_0 > K$. *Then every* $f \in \mathrm{Aut}(K, \mathrm{Cod}(M/K))$ *can be extended to an automorphism* g *of* M *such that* $g(M(\omega, a)) = M(\omega, b)$.

Proof For the proof of (1) notice that, by recursive saturation and indiscernibility, if for $a, b \in \mathrm{ASG}(M)$, $(a)_n$ and $(b)_n$ for $n < \omega$, realize $p(v)$, then there are $a', b' \in M$ such that $(a)_n = (a')_n$ and $(b)_n = (b')_n$ for all $n < \omega$, and $\mathrm{tp}(a') = \mathrm{tp}(b')$.

The proof of (2) is essentially the same using finite tuples of elements of $\mathrm{ASG}(M)$ and strong indiscernibility (Do IT!).

To prove (3) notice that, by previous arguments, we can assume that,

$$(M, a, c)_{c \in K} \equiv (M, b, c)_{c \in K},$$

because if a and b do not satisfy this condition, we can replace them by elements a', b' which code ω-sequences of indiscernibles which are cofinal in $M(\omega, a)$ and $M(\omega, b)$, respectively, and for which the above condition holds. Now the result follows from Theorem 8.5.1 (Do IT!). □

Theorem 10.2.7 *Let* M *and* N *be* ω_1*-like and recursively saturated. Then*

$$M \equiv_{\infty\omega_1} N \iff [M \equiv N \text{ and } \mathrm{SSy}(M) = \mathrm{SSy}(N)].$$

Proof Let $\langle a_\alpha : \alpha < \omega_1 \rangle, \langle b_\alpha : \alpha < \omega_1 \rangle$ be such that for all $\alpha < \beta < \omega_1$, $a_\alpha \in \mathrm{ASG}(M)$, $b_\alpha \in \mathrm{ASG}(M)$, $M(\omega, a_\alpha) \prec M(\omega, a_\beta)$, and $N(\omega, b_\alpha) \prec N(\omega, b_\beta)$. For each $\alpha < \omega_1$, let G_α be the set of all isomorphisms $f : M(\omega, a_\alpha) \longrightarrow N(\omega, b_\alpha)$ which satisfy the condition: for all $X \subseteq M(\omega, a_\alpha)$, $Y \subseteq N(\omega, b_\alpha)$, X is coded in M iff $f(X)$ is coded in N. Proposition 10.2.6 shows that the conditions of Theorem 10.2.2 are satisfied, and this proves the nontrivial part of the theorem. □

Every countable recursively saturated model has $2^{\aleph_1}$ nonisomorphic ω_1-like recursively saturated elementary end extensions. (There are many ways to prove this, and we approach this topic more systematically in the next section.) Then, Theorem 10.2.7 implies that there are nonisomorphic ω_1-like recursively saturated models which are $\mathcal{L}_{\infty\omega_1}$-elementarily equivalent. Theorem 10.2.9 is an even stronger version of this result.

Recall that a chain $\langle M_\alpha : \alpha < \kappa \rangle$ is *continuous* if for all limit ordinals $\lambda < \kappa$, $M_\lambda = \bigcup_{\alpha < \lambda} M_\alpha$. A *filtration* of M is a continuous chain of elementary extensions $\langle M_\alpha : \alpha < \kappa \rangle$ such that $|M_\alpha| < \kappa$ for all $\alpha < \kappa$ and $\bigcup_{\alpha < \kappa} M_\alpha = M$.

Let L and K be countable and recursively saturated with $K \prec_{\mathsf{end}} L$. We say that a model M is (L, K)-*filtrated* if it has a filtration $\langle M_\alpha : \alpha < \omega_1 \rangle$ such that the set

$$\{\alpha < \omega_1 : (M_{\alpha+1}, M_\alpha) \cong (L, K)\}$$

is closed and unbounded in ω_1. Of course, any (L, K)-filtrated model of cardinality $\aleph_1$ is ω_1-like and recursively saturated.

Proposition 10.2.8 *If a model M is (L, K)-filtrated, then for every filtration $\langle M_\alpha : \alpha < \omega_1 \rangle$ of M, the set*

$$\{\alpha : \forall \beta < \omega_1 \ [\alpha < \beta \longrightarrow (M_\beta, M_\alpha) \cong (L, K)]\}$$

is closed and unbounded in ω_1.

Proof First notice that since M is (L, K)-filtrated, for any filtration $\langle M_\alpha : \alpha < \omega_1 \rangle$, the set

$$\{\alpha : \ [(M_{\alpha+1}, M_\alpha) \cong (L, K)]\}$$

is closed and unbounded in ω_1 (Do it!). By Theorem 8.5.2, if $K \prec_{\mathsf{end}} L \prec_{\mathsf{end}} M$, all models are countable and recursively saturated, then the identity on K extends to an isomorphism $f : L \longrightarrow M$; hence $(M, K) \cong (L, K)$. The assumption in Theorem 8.5.2 that K is not downward ω-coded can be satisfied by replacing K with a K' such that $K \prec_{\mathsf{end}} K' \prec_{\mathsf{end}} L$ and K' is not downward ω-coded, if needed. Hence, for $\alpha < \omega_1$, if $(M_{\alpha+1}, M_\alpha) \cong (L, K)$, then for every countable $\beta \geq \alpha$, $(M_\beta, M_\alpha) \cong (L, K)$. □

There are ω_1-like recursively saturated models which are not (L, K)-filtrated for any $K \prec_{\mathsf{end}} L$. For example, by Theorem 1.10.2, there is a recursively saturated ω_1-like model M with a filtration $\langle M_\alpha : \alpha < \omega_1 \rangle$ such that for all $\alpha < \beta < \omega_1$, $(M_{\alpha+1}, M_\alpha)$ and $(M_{\beta+1}, M_\beta)$ are elementarily inequivalent.

If a model M is (L, K)-filtrated, is M determined up to isomorphism by (L, K)? The next theorem provides a negative answer.

Theorem 10.2.9 *Let $K = L(\omega, a)$, where L is countable recursively saturated and $a \in \mathrm{ASG}(L)$. Then L has $2^{\aleph_1}$ nonisomorphic, ω_1-like, (L, K)-filtrated elementary end extensions.*

Proof Let $f : \omega \to \omega$ eventually dominate every function $g \in \mathrm{SSy}(L)$. That is, for every $g \in \mathrm{SSy}(L)$, there is $n < \omega$ such that for all $k > n$, $f(k) > g(k)$.

We will construct nonisomorphic (L, K)-filtrated models M and N. The construction can be generalized to obtain $2^{\aleph_1}$ pairwise nonisomorphic such models.

We will build $M = \bigcup\{M_\alpha : \alpha < \omega_1\}$ and $N = \bigcup\{N_\alpha : \alpha < \omega_1\}$ by transfinite induction. Let $M_0 = N_0 = L$. If α is a successor ordinal, then let $M_{\alpha+1}$ and $N_{\alpha+1}$ be any countable recursively saturated models such that $M_\alpha \prec_{\mathsf{end}} M_{\alpha+1}$ and $N_\alpha \prec_{\mathsf{end}} N_{\alpha+1}$. Now, suppose M_α and N_α have been defined for all $\alpha < \lambda$ and λ is a limit ordinal. We let $M_\lambda = \bigcup_{\alpha\in\lambda} M_\alpha$ and $N_\lambda = \bigcup_{\alpha\in\lambda} N_\alpha$. To define $M_{\lambda+1}$ and $N_{\lambda+1}$, let $\langle\delta_n : n < \omega\rangle$ be an increasing sequence of ordinals whose limit is λ. We let $M_{\lambda+1}$ be such that $(M_{\lambda+1}, M_\lambda) \cong (L(\omega, a), K)$ and such that for some $a_\lambda \in M_{\lambda+1}$ and all $n < \omega$, $(a_\lambda)_n \in M_{\delta_{n+1}} \setminus M_{\delta_n}$ (possible by Proposition 10.2.5). We let $N_{\lambda+1}$ be such that $(N_{\lambda+1}, N_\lambda) \cong (M, K)$ and for some $b_\lambda \in N_{\lambda+1}$ and all $n < \omega$, $(b_\lambda)_n \in N_{\delta_{f(n)+1}} \setminus N_{\delta_{f(n)}}$ (possible by the same proposition).

We claim that $M \not\cong N$. For suppose $F : M \to N$ is an isomorphism. Let $C = \{\alpha < \omega_1 : F(M_\alpha) = N_\alpha\}$, and let λ be a limit point of C. Let $a = a_\lambda, b = b_\lambda$, and $c = F(a)$. Let $g : \omega \to \omega$ be defined by

$$g(i) = \min\{m : (c)_i \leq (b)_m\}.$$

For arbitrarily large $n < \omega$, there is $\gamma \in C$ such that $\delta_{f(n)} < \gamma \leq \delta_{f(n+1)}$. For every such γ we have, for all $i \leq f(n)$,

$$(c)_i = (F(a))_i = F((a)_i) \in N_\gamma,$$

hence for all i,

$$i \leq f(n) \to g(i) \leq n + 1.$$

Now, since g is coded in N (and hence in M as well), so is the function $h(n) = \mathrm{card}^N \{i : g(i) \leq n + 1\}$. But we have shown that for arbitrarily large n, $h(n) \geq f(n)$, which is a contradiction. □

The above proof relies heavily on the special type of the filtrations used. This leaves open the question:

Problem 10.2.10 *Are there countable recursively saturated $K \prec_{\mathsf{end}} L$ such that any two (L, K)-filtrated models are isomorphic?*

10.3 Finitely determinate structures and PA(aa)

By Theorem 10.2.7, any two ω_1-like elementarily equivalent models with the same standard system are $\mathcal{L}_{\infty\omega_1}$-equivalent. As we have seen in the previous section, there are many nonisomorphic such models. For another example, let $K_0 \prec_{\mathsf{end}} L_0$, $K_1 \prec_{\mathsf{end}} L_1$, be countable recursively saturated models and suppose $(L_0, K_0) \not\cong (L_1, K_1)$ and $K_0 \cong K_1$. If M_0 is (L_0, K_0)-filtrated and M_1 is (L_1, K_1)-filtrated, then $M_0 \not\cong M_1$ and, by Theorem 10.2.7, $M_0 \equiv_{\infty\omega_1} M_1$. There is another logic in which the difference between M_0 and M_1 can be expressed. It is the stationary logic $\mathcal{L}(\mathsf{aa})$.

Let $\mathcal{P}_{\aleph_1}(M)$ be the set of countable subsets of M. A collection of $\mathcal{A} \subseteq \mathcal{P}_{\aleph_1}(M)$ is *unbounded* if for every $X \in \mathcal{P}_{\aleph_1}(M)$, there is $Y \in \mathcal{A}$ such that $X \subseteq Y$. $\mathcal{A}$ is *closed* if whenever $X_0 \subseteq \cdots \subseteq X_n \subseteq \cdots$ is a countable chain of sets in $\mathcal{A}$, then $\bigcup\{X_n : n < \omega\} \in \mathcal{A}$.

For a language $\mathcal{L}$, $\mathcal{L}(\mathsf{aa})$ is $\mathcal{L}$ augmented with the membership symbol $\in$, a new set of set variables $s, t, \ldots$, and a quantifier aa to bind them. Semantics for $\mathcal{L}(\mathsf{aa})$ is defined by adding the following to Tarski's definition of satisfaction. For an $\mathcal{L}$-structure M: $M \models \mathsf{aa}s\varphi(s)$ iff the set

$$\{X \in \mathcal{P}_{\aleph_1}(M) : (M, X) \models \varphi(X)\}$$

is closed and unbounded in $\mathcal{P}_{\aleph_1}(M)$.

The sentence $\mathsf{aa}s\exists x(x \notin s)$ is considered an axiom of $\mathcal{L}(\mathsf{aa})$; hence, all models of stationary logic are uncountable. All models of $\mathcal{L}(\mathsf{aa})$ satisfy the *Diagonal Intersection Scheme*:

$$\forall x\ \mathsf{aa}s\varphi(x, s) \longrightarrow \mathsf{aa}s\ \forall x \in s\varphi(x, x).$$

The Diagonal Intersection Scheme is a formal expression of the Diagonal Intersection Lemma: for any regular uncountable cardinal κ, the diagonal intersection of a κ-sequence of closed unbounded subsets of κ is closed and unbounded in κ. The diagonal intersection of a sequence $\langle X_\alpha : \alpha < \kappa\rangle$ is $\{\alpha < \kappa : \alpha \in \bigcup_{\nu<\alpha} X_\nu\}$.

An $\mathcal{L}$-structure M is *finitely determinate* if it satisfies the scheme Det:

$$\mathsf{aa}s_0\mathsf{aa}s_1 \ldots \mathsf{aa}s_{n_1}\forall\bar{x}[\mathsf{aa}t\varphi(\bar{x}, \bar{s}, t) \vee \mathsf{aa}t\neg\varphi(\bar{x}, \bar{s}, t)].$$

By a direct proof by induction on complexity of formulas, one can show the following proposition on normal forms for formulas of $\mathcal{L}(\mathsf{aa})$:

Proposition 10.3.1 *Let $\mathcal{L}$ be a first-order language, and let $\mathfrak{A}$ be a finitely determinate $\mathcal{L}$-structure. Then every formula of $\mathcal{L}(\mathsf{aa})$ is equivalent in $\mathfrak{A}$ to one of the form $\mathsf{aa}s_0 \ldots \mathsf{aa}s_{n-1}\varphi(\bar{s}, \bar{t}, \bar{x})$, where φ has no second-order quantifiers.* □

The following Eklof–Mekler Criterion characterizes finitely determinate structures of cardinality $\aleph_1$.

Theorem 10.3.2 *Let $\mathfrak{A}$ be a structure of cardinality $\aleph_1$. Then $\mathfrak{A}$ is finitely determinate iff it has a filtration $\langle \mathfrak{A}_\alpha : \alpha < \omega_1\rangle$ such that whenever $k \leq n < \omega$, $\langle \alpha_0, \ldots, \alpha_n\rangle$, $\langle \beta_0, \ldots, \beta_n\rangle \in [\omega_1]^{n+1}$ are such that $\alpha_j = \beta_j$ for $j < k$; and $c_0, \ldots, c_r \in \mathfrak{A}_{\alpha_k} \cap \mathfrak{A}_{\beta_k}$ for some $r < \omega$, then*

$$(\mathfrak{A}, \mathfrak{A}_{\alpha_0}, \ldots, \mathfrak{A}_{\alpha_n}, c_0, \ldots, c_r) \equiv (\mathfrak{A}, \mathfrak{A}_{\beta_0}, \ldots, \mathfrak{A}_{\beta_n}, c_0, \ldots, c_r).$$ □

By Proposition 10.2.6 and Theorem 10.2.9, we get the following corollary.

Corollary 10.3.3 *Every countable recursively saturated model has $2^{\aleph_1}$ pairwise nonisomorphic finitely determinate recursively saturated ω_1-like elementary end extensions.* □

By a theorem of Shelah (see Lemma 4.1 in [93]), the models constructed in the proof of Theorem 10.2.9 have an additional property: they are all $\mathcal{L}_{\mathsf{PA}\,\infty\omega_1}(\mathsf{aa})$-equivalent.

Let $\mathsf{PA}(\mathsf{aa})$ be Peano Arithmetic in which the induction schema ranges over all formulas of $\mathcal{L}_{\mathsf{PA}}(\mathsf{aa})$. The results below apply also to the similarly defined $\mathsf{PA}^*(\mathsf{aa})$, but for notational simplicity, we will consider PA only.

Is $\mathsf{PA}(\mathsf{aa})$ consistent? What do its models look like? Let us prove an easy fact first.

Proposition 10.3.4 *If $N \models \mathsf{PA}(\mathsf{aa})$, then N is ω_1-like.*

Proof Suppose $N \models \mathsf{PA}(\mathsf{aa})$ and consider the formula

$$\varphi(x) = \mathsf{aa}s\forall y\ (y < x \longrightarrow y \in s).$$

Clearly

$$N \models \varphi(0) \wedge \forall x(\varphi(x) \longrightarrow \varphi(x+1)),$$

so that $N \models \forall x\ \varphi(x)$. Thus, for every $a \in N$, a_N is countable. Since all $\mathcal{L}(\mathsf{aa})$ structures are uncountable, N is ω_1-like. □

The rest of the section is devoted the construction of a model of $\mathsf{PA}(\mathsf{aa})$.

We will need some results concerning CA, the second-order theory in $\mathcal{L}_{\mathsf{PA}}$ consisting of the basic axioms of PA, the induction axiom, and the full comprehension schema

$$\exists X\forall x\ (x \in X \longleftrightarrow \varphi(x)),$$

where $\varphi(x)$ is a second-order formula which can have undisplayed first- and second-order variables other than X. We will consider $\mathsf{CA}+\mathsf{AC}$, where AC is the following version of the axiom of choice:

$$\forall x\exists X\varphi(X,x) \longrightarrow \exists X\forall x\ \varphi((X)_x,x),$$

where $\varphi(X,x)$ can have additional undisplayed free variables.

Theorem 10.3.5 *Let $(M,\mathfrak{X}) \models \mathsf{CA}+\mathsf{AC}$ be countable. Let $\mathfrak{X} = \{A_0, A_1, \dots\}$, let p be a minimal type of* $\mathrm{Th}(M, A_0, A_1, \dots)$*, and let $(I,<)$ be an ω_1-like ordering.*

For $M^* = (M, A_0, A_1, \dots)$, let $M^*(I)$ be the canonical I-extension of M^*. Then the reduct of $M^*(I)$ to $\mathcal{L}_{\mathsf{PA}}$ is a model of $\mathsf{PA}(\mathsf{aa}) + \mathsf{Det}$. □

The proof of Theorem 10.3.5 is based on the following two technical lemmas proved in [171]. We will show how the theorem follows from the lemmas. The proofs of the lemmas are rather involved, and we will not give them here. For a linearly ordered set $(I, <)$ and $J \subseteq I$, we write $J < I$ if J is an a proper initial segment of I.

Lemma 10.3.6 *Suppose $(I, <)$ is a linearly ordered set, $n < \omega$, and let $I_0 < \cdots < I_{n-1} < I$. Let $J \subseteq I$ be such that if we set $J_i = I_i$ for $i < n$, then $J_0 < \cdots < J_{n-1} < J$. Then*

$$(M^*(J), M^*(J_0), \dots, M^*(J_{n-1})) \prec (M^*(I), M^*(I_0), \dots, M^*(I_{n-1})). \quad \square$$

Lemma 10.3.7 *Suppose $(I, <)$ is a linearly ordered set, $n < \omega$, and $I_0 < \cdots < I_{n-1} < I$. Suppose $D \in \mathrm{Def}(M^*(I), M^*(I_0), \dots, M^*(I_{n-1}))$. Then $D \cap M^*(I_0) \in \mathrm{Def}(M^*(I_0))$.*

Proof of Theorem 10.3.5 Since $(I, <)$ is ω_1-like, it has a filtration $\langle I_\alpha : \alpha < \omega_1 \rangle$. Then $\langle M^*(I_\alpha) : \alpha < \omega_1 \rangle$ is a filtration of $M^*(I)$. We will show that this filtration satisfies the Eklof–Mekler Criterion.

Let $k, n, r, \alpha_0, \dots, \alpha_n, \beta_0, \dots, \beta_n, c_0, \dots, c_r$ be as in the criterion. For $j = k, k+1, \dots, n-1$, pick some $a_j \in I_{\alpha_{j+1}} \setminus I_{\alpha_j}$ and $b_j \in I_{\beta_{j+1}} \setminus I_{\beta_j}$ and pick $a_n \in I \setminus I_{\alpha_n}$ and $b_n \in I \setminus I_{\beta_n}$. Let

$$J = (I_{\alpha_k} \cap I_{\beta_k}) \cup \{a_k, a_{k+1}, \dots, a_n\}$$

and

$$K = (I_{\alpha_k} \cap I_{\beta_k}) \cup \{b_k, b_{k+1}, \dots, b_n\}.$$

By Theorem 3.3.11, there is an isomorphism $f : M^*(J) \longrightarrow M^*(K)$ which is the identity on $M^*(I_{\alpha_k} \cap I_{\beta_k})$ and which maps $M^*(J) \cap I_{\alpha_i}$ onto $M^*(K) \cap I_{\beta_i}$ for $i \leq n$ and a_i to b_i whenever $k \leq i < n$. By Lemma 10.3.6 both

$$(M^*(J), M^*(J \cap I_{\alpha_0}), \dots, M^*(J \cap I_{\alpha_n}))$$

and

$$(M^*(K), M^*(K \cap I_{\alpha_0}), \dots, M^*(K \cap I_{\alpha_n}))$$

are elementary substructures of

$$(M^*(I), M^*(I_{\alpha_0}), \dots, M^*(I_{\alpha_n})).$$

Therefore

$$(M^*(J), M^*(J \cap I_{\alpha_0}), \ldots, M^*(J \cap I_{\alpha_n}), c_0, \ldots, c_r)$$

is elementarily equivalent to

$$(M^*(K), M^*(K \cap I_{\alpha_0}), \ldots, M^*(K \cap I_{\alpha_n}), c_0, \ldots, c_r),$$

so the Eklof–Mekler Criterion is verified. Thus, $M^*(I)$ is finitely determinate so its reduct to $\mathcal{L}_{\mathsf{PA}}$ also is.

Let N be the reduct of $M^*(I)$ to $\mathcal{L}_{\mathsf{PA}}$. We will prove that $N \models \mathsf{PA}(\mathsf{aa})$. Suppose $a \in N$ and $\varphi(x, y)$ is an $\mathcal{L}_{\mathsf{PA}}(\mathsf{aa})$ formula in which the only free variables are the first-order variables x and y. Suppose $D = \{x \in N : N \models \varphi(x, a)\}$ is such that $0 \in D$ and $x+1 \in D$ whenever $x \in D$. By Proposition 10.3.1, we can assume that $\varphi(x, y)$ is of the form $\mathsf{aa}s_0 \ldots \mathsf{aa}s_{n-1}\psi(\bar{s}, x, y)$. By Lemma 10.3.6, whenever $I_0 < \cdots < I_{n-1} < I$, there is a formula in the language of $M^*(I)$ with parameters from $M^*(I_0)$ defining

$$D_{I_0} = \{x \in M^*(I_0) : N \models \psi(M^*(I_0), \ldots, M^*(I_{n-1}), x, a)\}.$$

By Lemma 10.3.7, this defining formula depends on I_0 only. For $J < I$, define $f(J)$ to be some $K < J$ such that D_J is definable in $M^*(J)$ with parameters from $M^*(K)$, if there is such K, and let $f(J) = J$ otherwise. By Fodor's lemma f is constant on a stationary set. It follows from Lemma 10.3.6 that there is a single formula $\theta(x)$ of the language of $M^*(I)$ such that the collection

$$\{I_0 : D_{I_0} = \{x \in M^*(I_0) : M^*(I_0) \models \theta(x)\}\}$$

is stationary. Then it easily follows that $\theta(x)$ defines D in $M^*(I)$. But $M^*(I) \models \mathsf{PA}^*$, so $D = M^*(I) = N$. □

10.4 Ramsey quantifiers and $\mathsf{PA}(Q^2)$

The theory $\mathsf{PA}(Q^2)$ is an extension of PA in $\mathcal{L}_{\mathsf{PA}}$ augmented by a Ramsey quantifier Q^2 which binds two free variables. The intended interpretation is: $M \models Q^2x_0, x_1\varphi(x_0, x_1)$ iff there is an unbounded $X \subseteq M$ such that $M \models \varphi(a_0, a_1)$ for all distinct a_0, a_1 in X. A set X with this property is a *witness* for $Q^2x_0, x_1\varphi(x_0, x_1)$. Under this interpretation, every $\mathsf{PA}(Q^2)$ formula has an equivalent second-order form in which every occurrence of $Q^2x_0, x_1\varphi(x_0, x_1)$ is replaced by

$$\exists X[\forall y \exists z(z > y \wedge z \in X) \wedge \forall x_0, x_1 \in X(x_0 \neq x_1 \longrightarrow \varphi(x_0, x_1))].$$

$\mathsf{PA}(Q^2)$ is Peano Arithmetic formulated in $\mathcal{L}_{\mathsf{PA}}(Q^2)$. A *weak model* of $\mathsf{PA}(Q^2)$ is the second-order structure $(M, \mathfrak{X})$, where $\mathfrak{X} \subseteq \mathcal{P}(M)$, with the interpretation

$(M,\mathfrak{X}) \models Q^2 x_0, x_1 \varphi(x_0,x_1)$ iff $Q^2 x_0, x_1 \varphi(x_0,x_1)$ has a witness in $\mathfrak{X}$. A *strong model* M is identified with the weak model $(M, \mathcal{P}(M))$.

We begin with a proposition that justifies the presence of $\mathsf{PA}(Q^2)$ in this chapter.

Proposition 10.4.1 *If M is a strong model of $\mathsf{PA}(Q^2)$, then M is κ-like for some regular cardinal κ.*

Proof Suppose that for some $a \in M$ there is a function f whose range is unbounded in M and whose domain is contained in a_M. Then the set of codes of (standard) finite restrictions of f witnesses $M \models \varphi(a)$, where $\varphi(z)$ is

$$Q^2 x, y[x \text{ and } y \text{ code functions with domains contained in } [0,z] \\ \wedge \forall i(i \in \mathrm{dom}(x) \cap \mathrm{dom}(y) \longrightarrow x(i) = y(i))].$$

Since $M \models \mathsf{PA}(Q^2)$, there must be a least b such that $\varphi(b)$ holds. Clearly $b \neq 0$. But if X is a witness to $\varphi(b)$, then the set $\{\sigma \restriction [0, b-1] : \sigma \in X\}$ is a witness to $\varphi(b-1)$, which is a contradiction. So, for each a, $|a_M| < |M|$. □

The rest of this section is devoted to the proof that $\mathsf{PA}(Q^2)$ has κ-like models for every regular cardinal κ. Of course, the standard model is a strong model of $\mathsf{PA}(Q^2)$. Proposition 10.4.1 shows that it is the only countable strong model.

Now we prove that if a formula of $\mathcal{L}_{\mathsf{PA}}(Q^2)$ has a witness, then it has a definable one. More precisely:

Lemma 10.4.2 *For any formula $\varphi(x_0,x_1)$ of $\mathcal{L}_{\mathsf{PA}}(Q^2)$ there is a formula $W_\varphi(x)$ $\mathcal{L}_{\mathsf{PA}}(Q^2)$ such that for every weak model $(M,\mathfrak{X})$ of $\mathsf{PA}(Q^2)$,*

$$(M,\mathfrak{X}) \models \forall x_0, x_1 (W_\varphi(x_0) \wedge W_\varphi(x_1) \wedge x_0 \neq x_1 \longrightarrow \varphi(x_0,x_1)),$$

and

$$(M,\mathfrak{X}) \models Q^2 x_0, x_1 \varphi(x_0,x_1) \longleftrightarrow \forall x \exists y > x\ W_\varphi(x).$$

Proof Let $(M,\mathfrak{X})$ be a weak model of $\mathsf{PA}(Q^2)$. We define a sequence of elements of M by formal induction as follows. Let

$$(s)_0 = \min\left\{w : Q^2 x_0, x_1[\varphi(x_0,x_1) \wedge w \neq x_0 \wedge \varphi(w,x_0) \wedge \varphi(x_0,w)]\right\},$$

if this set is nonempty, and $w_0 = 0$ otherwise. For $n \in M$, let $(s)_{n+1}$ be the smallest w such that for all $i \leq n$, $w \neq (s)_i$ and in $(M,\mathfrak{X})$,

$$Q^2 x_0, x_1[\varphi(x_0,x_1) \wedge \forall i \leq n(\varphi((s)_i,x_0) \wedge \varphi(x_0,(s)_i) \wedge \varphi((s)_i,w) \wedge \varphi(w,(s)_i))],$$

if there is such a w, and $(s)_{n+1} = 0$ otherwise.

There is a formula $\psi(s)$ such that $(M, \mathfrak{X}) \models \psi(s)$ iff s is an initial segment of the sequence $\langle (s)_n : n \in M \rangle$. Then define $W_\varphi(x)$ to be

$$x > 0 \wedge \exists s[\psi(s) \wedge \exists i < \ell(s)(x = (s)_i). \qquad \square$$

As an immediate consequence we have:

Corollary 10.4.3 *Let $(M, \mathfrak{X})$ be a weak model of $\mathsf{PA}(Q^2)$, and let $\mathfrak{X}'$ be the collection of subsets of M definable in $(M, \mathfrak{X})$ by formulas of $\mathcal{L}_{\mathsf{PA}}(Q^2)$ with parameters from M. Then $(M, \mathfrak{X}')$ is a weak model of $\mathsf{PA}(Q^2)$, and $(M, \mathfrak{X})$ and $(M, \mathfrak{X}')$ satisfy the same $\mathcal{L}_{\mathsf{PA}}(Q^2)$ sentences with parameters from M.* □

To prove the existence of κ-like strong models of $\mathsf{PA}(Q^2)$, we need one more lemma.

Lemma 10.4.4 *Let κ be an uncountable regular cardinal. Suppose $|M| < \kappa$ and N is a κ-canonical elementary end extension of M. If $N \models Q^2 x_0, x_1 \varphi(x_0, x_1)$ for a formula $\varphi(x_0, x_1)$ with parameters from N, then there is a witness for this fact which is definable in N.*

Proof Let $\langle M_\alpha : \alpha < \kappa \rangle$ be filtration of N, and let p be a minimal type such that for each α, $M_{\alpha+1}$ is generated over M_α by some c_α realizing p. Then for each $\alpha < \kappa$, $N = \mathrm{Scl}(M_\alpha \cup \{c_\xi : \alpha \leq \xi < \kappa\})$.

Assume $N \models Q^2 x_0, x_1 \varphi(x_0, x_1)$. Let $X \subseteq M$ be an unbounded witness for this fact. Then X is a collection of elements of the form $f(a, c_{\xi_1}, \ldots, c_{\xi_n})$ for some $a \in M_\alpha$, $n < \omega$, and a Skolem term f. Let $\alpha < \kappa$ be such that the parameters in $\varphi(x_0, x_1)$ are in M_α. Since $|X| = \kappa$ and κ is regular, a counting argument shows that there are a Skolem term f, an $a \in M_\alpha$, an $n < \omega$, and an increasing sequence of n-tuples of $\langle \bar{c}^\alpha : \alpha < \kappa \rangle$ of elements of $\langle c_\alpha : \alpha < \kappa \rangle$ such that $f(a, \bar{c}^\alpha) \in X$ and $f(\bar{c}^\alpha) < f(\bar{c}^\beta)$, for all $\alpha < \beta < \kappa$. We can assume that the parameters in $\varphi(x_0, x_1)$ are in $\mathrm{Scl}(a)$.

By strong indiscernibility of $\langle c_\alpha : \alpha < \kappa \rangle$, there is $b \in N$ and a formula $\theta(x)$ defining an unbounded subset of N such that:

$$\begin{aligned} N \models \forall x_1, \ldots, x_{2n}[b < x_1 < \cdots < x_{2n} \wedge \theta(x_1) \wedge \cdots \wedge \theta(x_{2n}) \\ \longrightarrow [\varphi(f(a, x_1, \ldots, x_n), f(a, x_{n+1}, \ldots, x_{2n})) \\ \wedge \varphi(f(a, x_{n+1}, \\ \ldots, x_{2n}), f(a, x_1, \ldots, x_n))]. \end{aligned}$$

Let $g(b, z)$ be the z-th element of $\{x : b < x \wedge \theta(x)\}$, and let

$$Y = \{f(a, g(b, nz + 1), \ldots, g(b, nz + n)) : z \in N\}.$$

Then Y is a definable witness for $Q^2 x_0, x_1 \varphi(x_0, x_1)$. □

Theorem 10.4.5 *Let κ be a regular uncountable cardinal, and let $(M, \mathfrak{X})$ be a countable weak model of* $\mathsf{PA}(Q^2)$*. Then there exists a κ-like strong model N which is an $\mathcal{L}_{\mathsf{PA}}(Q^2)$ elementary end extension of $(M, \mathfrak{X})$.*

Proof Let $\langle U_i : i < \omega \rangle$ be an enumeration of the subsets of M which are 0-definable in $(M, \mathfrak{X})$. Let M^* be a model of PA^* obtained from M by adjoining the U_i's as extra predicates. Let N^* be a κ-canonical extension of M^*. Let N be the reduct of N^* to the language of PA.

Clearly $(M^*, \mathfrak{X})$ is a weak model of $\mathsf{PA}^*(Q^2)$. To prove that N is as desired, we will show that for each formula $\varphi(\bar{u})$ in $\mathcal{L}_{\mathsf{PA}^*}(Q^2)$ with the only free variables shown, there exists a formula $\chi(\bar{u})$ of $\mathcal{L}_{\mathsf{PA}^*}$ with the same free variables such that both $(M^*, \mathfrak{X})$ and N^* satisfy $\forall \bar{u}\ (\varphi(\bar{u}) \longleftrightarrow \chi(\bar{u}))$.

The proof is by induction on complexity of $\varphi(\bar{u})$. The only nontrivial case is when $\varphi(\bar{u})$ is of the form $Q^2 x_0, x_1 \psi(\bar{u}, x_0, x_1)$, where ψ is in $\mathcal{L}_{\mathsf{PA}^*}$. By Lemma 10.4.2 applied to $\mathsf{PA}^*(Q^2)$, there exists a formula $W_\psi(\bar{u}, x)$ such that:

$$(x_0 \neq x_1 \wedge W_\psi(\bar{u}, x_0) \wedge W_\psi(\bar{u}, x_1)) \longrightarrow \psi(\bar{u}, x_0, x_1)$$

and

$$Q^2 x_0, x_1 \psi(\bar{u}, x_0, x_1) \longleftrightarrow \forall y \exists x > y W_\psi(\bar{u}, x)$$

hold in any weak model of $\mathsf{PA}(Q^2)$.

Let i be such that

$$U_i = \{(\bar{a}, d) \in M : (M^*, \mathfrak{X}) \models W_\psi(\bar{a}, d)\}.$$

Let $\chi(\bar{u})$ be the formula $\forall y \exists x > y\ U_i(\bar{u}, x)$. Trivially

$$(M^*, \mathfrak{X}) \models \forall \bar{u}(\varphi(\bar{u}) \longleftrightarrow \chi(\bar{u})).$$

We will show that the same sentence holds in N^*. Directly from the definition of i it follows that

$$M^* \models \forall \bar{u} \forall x_0, x_1 [x_0 \neq x_1 \wedge U_i(\bar{u}, x_0) \wedge U_i(\bar{u}, x_1) \longrightarrow \psi(\bar{u}, x_0, x_1)].$$

Since $M^* \prec N^*$, the same sentence is true in N^*; hence,

$$N^* \models \forall \bar{u}(\chi(\bar{u}) \longrightarrow \varphi(\bar{u})).$$

To finish the proof suppose that $N^* \models \varphi(\bar{a})$, that is, $N^* \models Q^2 x_0, x_1 \psi(\bar{a}, x_0, x_1)$. By Lemma 10.4.4, there is an $\mathcal{L}_{\mathsf{PA}^*}$ formula $\theta(\bar{v}, x)$ such that

$N^* \models \eta(\bar{a})$, where $\eta(\bar{u})$ is the formula

$$\exists \bar{\nu}\{\forall y \exists x > y\theta(\bar{\nu}, x) \wedge \forall x_0, x_1[x_0 \neq x_1 \wedge \theta(\bar{\nu}, x_0) \wedge \theta(\bar{\nu}, x_1) \longrightarrow \psi(\bar{u}, x_0, x_1)]\}.$$

Clearly $(M^*, \mathfrak{X}) \models \forall \bar{u}(\eta(\bar{u}) \longrightarrow \varphi(\bar{u}))$; hence, $M^* \models \forall \bar{u}(\eta(\bar{u}) \longrightarrow \chi(\bar{u}))$, so the same sentence is true in N^*. Hence $N^* \models \chi(\bar{a})$. □

10.5 Rigid recursively saturated models

In contrast with the numerous results on automorphism groups of countable models discussed in previous chapters, in this section we prove that there is a recursively saturated ω_1-like model which is rigid, that is, it has no nontrivial automorphisms. We prove an even stronger result: there is a recursively saturated ω_1-like model which has no elementary endomorphisms. In fact, as the next proposition shows, such models have no endomorphisms at all.

Proposition 10.5.1 *Let M be a κ-like model of* PA *for an uncountable cardinal κ. Then every embedding of M into itself is elementary.*

Proof Notice that, since M is κ-like, for every embedding $f : M \longrightarrow M$, $f(M) \subseteq_{\mathsf{cof}} M$. Hence the result follows from Gaifman's theorem on cofinal extensions (Theorem 1.3.3). □

To construct models without endomorphisms, we use prime satisfaction classes. A partial inductive satisfaction class S of a model M is *prime* if (M, S) is prime. We say that $S_1, S_2 \subseteq M$ are elementarily equivalent (isomorphic, etc.) if (M, S_1) and (M, S_2) are elementarily equivalent (isomorphic, etc.). Every countable recursively saturated model has many elementarily inequivalent prime classes. This, as we prove below, is a consequence of Scott's Theorem 1.5.3.

Theorem 10.5.2 *If a countable model M has an Q_e-class, where either $e \in M$ is nonstandard or $e = \infty$, then M has continuum many pairwise elementarily inequivalent prime Q_e-classes.*

Proof We consider the case of $e \in M$, the other case being similar. Let $\mathcal{L} = \mathcal{L}_{\mathsf{PA}} \cup \{S, e\}$, and let $T_0 = \mathrm{Th}(M, e) +$ "S is an Q_e-class". Since for each formula φ of $\mathcal{L}_{\mathsf{PA}}$,

$$M \models \varphi(e) \longleftrightarrow T_0 \vdash (\varphi(e) \in S),$$

and the axioms of Q_e-classes form a recursive set, T_0 represents itself. Moreover, since M is recursively saturated, $T_0 \in \mathrm{SSy}(M)$; hence, by Theorem 1.5.3, there are continuum many completions T of T_0 such that $\mathrm{Rep}(T) = \mathrm{SSy}(M)$. For each such T, let (M_T, e_T, S_T) be the prime model of T. Now, since M_T is recursively saturated, $\mathrm{SSy}(M_T) = \mathrm{SSy}(M)$, and $(M_T, e_T) \equiv (M, e)$, there is an isomorphism $f : M_T \longrightarrow M$ such that $f(e_T) = e$. Then $f(S_T)$ is a prime Q_e-class for M, and the result follows. □

If $M \prec_{\mathsf{cof}} N$ and $(M, X) \models \mathsf{PA}^*$, then by $\overline{X}$ we denote the unique $Y \subseteq N$ such that $(M, X) \prec (N, Y)$. (See Theorem 1.3.7)

Lemma 10.5.3 *Suppose that $N \prec_{\mathsf{cof}} M$, S is a prime Q_e-class for N, where $e \in N$ is nonstandard or $e = \infty$, and $f : N \longrightarrow M$ is an elementary embedding such that $f(N) \prec_{\mathsf{cof}} M$, either $f(e) \in N$ or $e = \infty$, and $\overline{f(S)} \in \mathrm{Def}((M, \overline{S}))$. Then f is the identity function.*

Proof Notice that, by the assumptions, both N and M are recursively saturated. First, we consider the case that $e \in N$ and let $d = f(e)$. We will show that $d = e$. If not, then first we will show that we can assume that $d < e$. If $d > e$, then $f(S)$ is a prime Q_d-class for $f(N)$; hence $\overline{f(S)}$ is a Q_d-class for M. Since $\overline{f(S)} \in \mathrm{Def}((M, \overline{S}))$ and $\overline{S}$ is a Q_e class, by Lemma 1.9.8, $\overline{f(S)}|_e = \overline{S}$. This argument shows that if $d > e$, then we can consider $f^{-1} : f(N) \longrightarrow M$ instead of f in the lemma we are proving.

So now we assume that $d < e$. Since $\overline{f(S)} \in \mathrm{Def}((M, \overline{S}))$ and $\overline{S}$ is an Q_e-class for M, we have $\overline{f(S)} = \overline{S}|_d$, hence $\mathrm{Th}((N, S)) = \mathrm{Th}((f(N), f(S))) = \mathrm{Th}((M, \overline{f(S)})) = \mathrm{Th}((M, \overline{S}|_d)) = \mathrm{Th}((N, S|_d))$. Since $\mathrm{tp}(e) = \mathrm{tp}(d)$, we get that $d + \mathbb{N} < e$; hence, $(N, S|_d)$ is recursively saturated, and, consequently, $\mathrm{Th}((N, S|_d)) \in \mathrm{SSy}(N)$. But (N, S) is prime; hence, by Corollary 1.5.2, $\mathrm{Th}(N, S) \notin \mathrm{SSy}(N)$, which is a contradiction. Hence $d = e$.

Thus, in either case ($d = e$ or $e = \infty$) we get $\overline{f(S)} = \overline{S}$ (DO IT!), so that $(N, S) \prec (M, \overline{S})$ and $(f(N), f(S)) \prec (M, \overline{S})$. But both these substructures of $(M, \overline{S})$ are prime, consequently $(N, S) = (f(N), f(S))$, and f being an automorphism of a prime model must be the identity. □

Corollary 10.5.4 *If S is a prime Q_e-class for M, where either $e \in M$ is nonstandard or $e = \infty$, then each cofinal embedding $f : M \longrightarrow M$ for which $\overline{f(S)} \in \mathrm{Def}((M, S))$ is the identity function.* □

Corollary 10.5.5 *Let M be countable and recursively saturated. Then there is a countable recursively saturated N such that $M \prec_{\mathsf{end}} N$ and for every end extension N' of N and every embedding $f : N' \longrightarrow N'$ such that $f(M)$ is cofinal in M, $f \restriction M$ is the identity function.*

Proof Let S be a prime partial inductive satisfaction class for M, and let (N, S') be a conservative elementary end extension of (M, S). Corollary 10.5.4 implies that N has the required property. □

Theorem 10.5.6 *Every countable recursively saturated model has an ω_1-like recursively saturated elementary end extension which has no nontrivial embeddings into itself.*

Proof Let M_0 be countable and recursively saturated. For $\alpha < \omega_1$, let $M_{\alpha+1}$ be the extension of M_α given by Corollary 10.5.5. For limit λ, let $M_\lambda = \bigcup\{M_\alpha : \alpha < \lambda\}$. Let $M = \bigcup\{M_\alpha : \alpha < \omega_1\}$. Suppose $f : M \longrightarrow M$ is an embedding. Since there are arbitrarily large $\alpha < \omega_1$ for which $f(M_\alpha)$ is cofinal in M_α, Corollary 10.5.5 implies that f is the identity function on M. □

10.6 Isomorphic + nonisomorphic ×

In this section we use a slightly illogical convention: for a model M, $(M, +)$ denotes the reduct of M to $+$, and $(M, \times)$ is the reduct of M to $\times$. The basic facts about these reducts are: for every model model M, $(M, +) \equiv (\mathbb{N}, +)$, $(M, \times) \equiv (\mathbb{N}, \times)$, and, if M is nonstandard, both $(M, +)$ and $(M, \times)$ are recursively saturated. These results follow from quantifier elimination for Presburger Arithmetic ($= \mathrm{Th}(\mathbb{N}, +)$) and Skolem Arithmetic ($= \mathrm{Th}(\mathbb{N}, \times)$).

Presburger Arithmetic and Skolem Arithmetic are rich theories. (See Remarks and References section in Chapter 1 and (Do it!).) As a corollary we have:

Proposition 10.6.1 *If M and N are countable, nonstandard, and* $\mathrm{SSy}(M) = \mathrm{SSy}(N)$, *then $(M, +) \cong (N, +)$ and $(M, \times) \cong (N, \times)$. Consequently, for all countable M and N, $(M, +) \cong (N, +)$ iff $(M, \times) \cong (N, \times)$.* □

For models of arbitrary cardinality we also know that if $(M, \times) \cong (N, \times)$, then $(M, +) \cong (N, +)$. To see this, note that primes are definable in $(M, \times)$ and the only prime dividing a power of 2 is 2 and conversely, the powers of 2 are just the numbers which are not divisible by any prime other than 2. Hence, if $f : (M, \times) \longrightarrow (N, \times)$ is an isomorphism, then for each $a \in M$, there is a unique $a' \in N$ such that $f(2^a) = 2^{a'}$. The mapping $a \mapsto a'$ is an isomorphism of the additive reducts of M and N (Do it!). We will show that in general the additive structure of a model does not determine its multiplicative structure. The examples are suitably chosen ω_1-like models. The part of Proposition 10.6.1 concerning additive reducts has the following generalization to ω_1-like models. We will not prove it here.

Theorem 10.6.2 *If M and N are ω_1-like models and* $\mathrm{SSy}(M) = \mathrm{SSy}(N)$, *then $(M, +) \cong (N, +)$.* □

Now it remains to show that there are ω_1-like models with the same standard systems whose reducts to $\times$ are nonisomorphic.

Theorem 10.6.3 *Every countable model M has $2^{\aleph_1}$ ω_1-like elementary end extensions with pairwise nonisomorphic multiplicative reducts.*

Proof Let K be an ω_1-like model and let X be a set of primes of K. Then, X is a class of K iff for every countable $A \subseteq M$ there is $a \in K$ such that $\forall x \in A(x \in X \longleftrightarrow x|a)$ (Do it!). Consequently, if X is a set of primes of K

and $f : (K, \times) \longrightarrow (L, \times)$ is an isomorphism, then $f(X)$ is a set of primes of L and if X is a class, then $f(X)$ is a class as well.

Now let M be a countable model, let I be stationary subset of ω_1, and let M_I be an ω_1-like elementary end extension of M with a filtration $\langle M_\alpha : \alpha < \omega_1 \rangle$ such that $M_\alpha \prec_{\mathsf{end}} M_\beta$, for all $\alpha < \beta < \omega_1$ with the property that the extension $M_\alpha \prec_{\mathsf{end}} M_{\alpha+1}$ is conservative iff $\alpha \in I$. We will show that I can be recovered from $(M_I, \times)$ modulo a closed unbounded subset of ω_1, which will finish the proof.

By Theorem 2.2.14, each M_I is rather classless. Let $\langle C_\nu : \nu < \omega_1 \rangle$ be a nonrepeating enumeration of all those classes (= definable sets) of M_I which consist of primes only. By the observation above, the set $\{C_\nu : \nu < \omega_1\}$ is determined by $(M_I, \times)$.

Let P_α be the collection of sets of primes of M_α which are coded in $M_{\alpha+1}$. Let $J \subseteq \omega_1$ be the set of those $\alpha < \omega_1$ for which $\{C_\nu \cap M_\alpha : \nu < \alpha\}{=}P_\alpha$. To see that J is nonempty, consider the function $\nu \mapsto \beta_\nu$, where β_ν is the least β such that C_ν is definable using parameters from M_β. The set $\{\alpha : \forall \nu(\nu < \alpha \longleftrightarrow \beta_\nu < \alpha)\}$ is closed and unbounded in ω_1 (Do it!).

Now consider the set

$$Z = \{\alpha < \omega_1 : \alpha \in \mathsf{Lim} \wedge \forall \nu(\nu < \alpha \longleftrightarrow \beta_\nu < \alpha)\}\,.$$

We will show that $I \cap Z = J \cap Z$.

Consider some $\alpha \in I \cap Z$. Since $M_{\alpha+1}$ is a conservative extension of M_α, the set of coded subsets of M_α is $\mathrm{Def}(M_\alpha)$. To prove that $\alpha \in J$, it suffices to show that the set $\{C_\nu \cap M_\alpha : \nu < \alpha\}$ is the set of all definable set of primes of M_α. If $C \subseteq M_\alpha$ is a definable sets of primes, then $C = C_\nu \cap M_\alpha$ for some ν for which $\beta_\nu \leq \alpha$. But since α is a limit ordinal and β_ν is not, $\beta_\nu < \alpha$. Then $\nu < \alpha$ because $\alpha \in Z$. Conversely, if $C = C_\nu \cap M_\alpha$ for some $\nu < \alpha$, then $\beta_\nu < \alpha$, so that $C \in \mathrm{Def}(M_\alpha)$.

Now, let $\alpha \in J \cap Z$, intending to show that $M_{\alpha+1}$ is a conservative extension of M_α. Since an extension in which all coded set of primes are definable in the ground model must be conservative, it is enough to show that each set of primes $C \subseteq M_\alpha$ which is coded in $M_{\alpha+1}$ is definable in M_α. Let C be such a set. Then there is $\nu < \alpha$ such that $C = C_\nu \cap M_\alpha$. But then $\beta_\nu < \alpha$, so that $C \in \mathrm{Def}(M_\alpha)$. □

10.7 Exercises

♣**10.7.1** If M is a countable recursively saturated model, S is a Q_e-class for M for some nonstandard e, and X is an undefinable subset of M, then there is a nonstandard e' such that X is undefinable in $(M, S|_{e'})$.

♣**10.7.2** Use previous exercise to prove Proposition 10.1.3.

♣10.7.3 If $\mathrm{cf}(\kappa) > \aleph_0$, then every model M such that $|M| < \kappa$ has a κ-like, rather classless elementary end extension. If $\kappa > \aleph_0$, then every model M such that $|M| \leq \kappa$ has a rather classless elementary end extension of cardinality κ.

♦10.7.4 Prove without using $\diamondsuit$ in any form that there is an ω_1-like recursively saturated model without partial inductive satisfaction class.

♣10.7.5 If M is a recursively saturated ω_1-like model without inductive satisfaction classes, then M has no recursively saturated elementary end extension.

♦10.7.6 Every countable recursively saturated model has $2^{\aleph_1}$ nonisomorphic recursively saturated rigid elementary end extensions.

♦10.7.7 Every countable recursively saturated model has a recursively saturated rigid elementary end extension without partial inductive satisfaction class.

♦10.7.8 Every countable recursively saturated model has a recursively saturated rigid elementary end extension with a partial inductive satisfaction class.

♦10.7.9 If M and N are countable and recursively saturated models and $M \prec_{\mathsf{end}} N$, then there is an automorphism of M which cannot be extended to an automorphism of N.

♦10.7.10 If K is countable and recursively saturated, then there are continuum many pairwise nonisomorphic structures of the form (K, L), where L is an elementary cut of K.

♣10.7.11 If $p(v)$ is an unbounded type realized in a recursively saturated model M, then for any $c \in M$, there is an $a \in M$ such that $\ell(a) = c$ and for all $i < c$, $(a)_i$ realizes p.

Let M be a recursively saturated model, and let I be a cut of M. Let $\mathrm{ASG}(M, I)$ be the set of those $a \in M$ such that $\ell(a) > I$ and for all $i \in I$, $\mathrm{gap}((a)_i) < (a)_{i+1}$. For $a \in \mathrm{ASG}(M, I)$, let $M(I, a) = \sup\{(a)_i : i \in I\}$.

♣10.7.12 For any cut I and $a \in \mathrm{ASG}(M, I)$, $M(I, a)$ is a recursively saturated elementary cut of M. If I is semiregular in M and $a \in \mathrm{ASG}(M, I)$, then I is definable in $(M, M(I, a))$.

♥10.7.13 If M is countable and recursively saturated, then there are a semiregular I and $a, b \in \mathrm{ASG}(M, I)$ such that $(M, M(I, a)) \not\equiv (M, M(I, b))$.

♦10.7.14 Let K be countable and recursively saturated. There are continuum many pairwise elementarily inequivalent structures of the form (K, L), where L is an elementary cut of K. (HINT: use the previous exercise and the fact that ever nonstandard model has continuum many pairwise elementarily inequivalent semiregular cuts.)

♣**10.7.15** If M is countable and arithmetically saturated, $K \prec_{\mathsf{end}} M$ is recursively saturated, and for some $c \in M$, $K = \sup\{(c)_n : n < \omega\}$, then $K = M(\omega, a)$ for some $a \in \mathrm{ASG}(M)$.

♥**10.7.16** If M is countable and recursively saturated and every elementary cut K of M which is of the form $\sup\{(c)_n : n < \omega\}$, is also of the form $K = M(\omega, a)$ for some $a \in \mathrm{ASG}(M)$, then M is arithmetically saturated.

♣**10.7.17** If M be countable and recursively saturated, then M has countable recursively saturated elementary end extensions K and L such that for some $a \in K$, $M = \sup\{(a)_n : n < \omega\}$ and there is no such a in L.

♣**10.7.18** Every countable recursively saturated model M has $2^{\aleph_1}$ recursively saturated elementary end extensions, whose reducts to $\times$ are pairwise nonisomorphic. (HINT: use the previous exercise.)

♣**10.7.19** If K and L are countable and recursively saturated, and $K \prec_{\mathsf{end}} L$, then K has an ω_1-like recursively saturated elementary end extension M which is (L, K)-filtrated.

A model M is *e-lofty*, if for every recursive, finitely realizable type $p(x, \bar{b})$, $\bar{b} \in M$, there is $s \in M$ such that $\mathrm{card}^M(s) = e$ and the type $p(v, \bar{b}) \cup \{x \in s\}$ is finitely realizable as well. A model is *lofty* if it is e-lofty for some $e \in M$. It is shown in [68] that a countable M is lofty iff it has a simple elementary extension which is recursively saturated.

♣**10.7.20** If N is a recursively saturated simple extension of M, then $M \prec_{\mathsf{cof}} N$.

♣**10.7.21** Every simple cofinal extension of a κ-like model is κ-like.

♥**10.7.22** If M is lofty and has no recursively saturated simple extensions, then M is κ-like for some regular uncountable cardinal κ.

10.8 Remarks & References

Bovykin & Kaye [18] give a survey and some new results on Friedman's 14th problem concerning the spectrum of order types of nonstandard models.

Kaufmann used $\diamondsuit$ to construct a Kaufmann model in [67]. The crucial Lemma 10.1.3 has many proofs. Kaufmann's proof in [67] uses the theory of admissible structures. The proof in Exercise 10.7.2 is from [164]. Much is known about κ-like, recursively saturated, rather classless models for $\kappa > \omega_1$. The following theorems are proved in [164]:

(1) If $\kappa > \mathrm{cf}(\kappa) > \aleph_0$, then every consistent extension of PA^* has a κ-like, recursively saturated, rather classless model.

(2) If κ is a regular cardinal and PA has a κ-like, recursively saturated, rather classless model, then there is an Aronszajn κ-tree.

(3) Assume V=L. Let κ be an inifinite cardinal, and let T be a consistent extension of PA*. Then the following are equivalent:

  (a) there is a κ-like, recursively saturated, rather classless model of T;

  (b) $\mathrm{cf}(\kappa) > \aleph_0$ and κ is not weakly compact.

(4) If $\kappa > \aleph_0$, then every consistent extension of PA* has a recursively saturated, rather classless model of cardinality κ.

The question: how highly saturated can a rather classless model be? is considered by Schmerl in [172] and [176]. In particular, Theorem 1.2 of [176] says:

> If κ is a singular cardinal, λ is a regular cardinal, $\mathrm{cf}(\kappa) > \lambda$, and M is a λ-saturated κ-like model of PA*. Then there is a rather classless, λ-saturated, κ-like $N \equiv M$.

Every model M of PA* has a naturally defined extension to a real closed field. By constructing special uncountable models, one obtains fields with interesting properties. This is explored by Schmerl [165] and Keisler & Schmerl [79] in connection with the question of Sikorski concerning ordered fields with the Bolzano–Weierstrass property.

While Corollary 10.1.6 shows that the MacDowell–Specker theorem is not valid when restricted to the class of recursively saturated models, some special cases of it hold. It is shown in [91] that every short recursively saturated model and every recursively saturated model with cofinality ω has a recursively saturated elementary end extension.

Theorems 10.5.2 and 10.5.6 are from Kossak & Schmerl [106].

Theorem 10.2.2, characterizing $\mathcal{L}_{\infty\omega_1}$-equivalence for models of cardinality $\aleph_1$ is due to David Kueker [118].

Cuts of the form $M(\omega, a)$, where $a \in \mathrm{ASG}(M)$, were introduced and studied by Smoryński in [192] and [195]. The results of Section 10.2, with the exception of Theorem 10.6.2, are from Kossak [92] and [93]. An example giving a solution to the problem in Exercise 10.7.13 is given in [97].

Ramsey quantifiers were introduced and studied in the general model-theoretic setting by Magidor & Malitz [126]. Macintyre [124] pointed out the importance of PA(Q^2) as a natural extension of PA which eliminates Paris–Harrington type independence phenomena and he proved several relevant results. Proposition 10.4.1 and Lemma 10.4.2 are due to him. Macintyre has also shown that PA(Q^2) has a definable truth definition for $\mathcal{L}_{\mathsf{PA}}$ formulas; hence, the first-order reducts of nonstandard models of PA(Q^2) are recursively saturated. This and some results concerning provability of some combinatorial principles independent of PA were obtained independently by Morgenstern [133]. Theorem 10.4.5 was proved in Schmerl & Simpson [179] as a part of the completeness theorem for PA(Q^2). It is shown that PA(Q^2) and $\Pi^1_1 - \mathsf{CA}_0$ have the same first-order consequences. A weak model $(M, \mathfrak{X})$ of PA(Q^2) is called *reduced* if

$\mathfrak{X} = \mathrm{Def}(M, \mathfrak{X})$. A model $(M, \mathfrak{X})$ of $\Pi^1_1 - \mathsf{CA}_0$ is called reduced if for each $A \in \mathfrak{X}$ there is $k < \omega$ such that A is Δ_1 in $0^{(k)}$ of $(M, \mathfrak{X})$. It is proved in [179] that the reduced weak models of $\mathsf{PA}(Q^2)$ are the same as the reduced models of $\Pi^1_1 - \mathsf{CA}_0$. The results of Macintyre and Morgenstern become easy corollaries of this statement.

The material on $\mathsf{PA}(\mathsf{aa})$ is from [171]. The main theorem, Theorem 10.3.5 constitutes a half of the proof that $\mathsf{PA}(\mathsf{aa})+\mathsf{Det}$ and CA have the same first-order consequences.

The theorem on existence of recursively saturated ω_1-like rigid models is from Kossak & Schmerl [106].

Victor Harnik [53, 54] developed a theory countable and ω_1-like recursively saturated models of Presburger arithmetic. Theorem 10.6.2 is due to him [53]. Theorem 10.6.3 is from Kossak et al. [104].

The fact that additive and multiplicative reducts of models of PA are recursively saturated is based on quantifier elimination for the theories of addition and multiplication. Quantifier elimination for addition is the well-known theorem of Presburger, see [198] for the result and its history. While the decidability of multiplication of integers was proved already by Skolem in 1930, the result on elimination of quantifiers for multiplication, needed in the proof of Propositiom 10.6.1, was proved independently by Cegielski [19] and Nadel [136] in 1981.

Kaufman & Schmerl [68, 69] present a theory of loftiness. Among many other results, it is shown that the converse to the statement in Exercise 10.7.22 holds: If M is κ-like for some regular κ and it is not recursively saturated, then M has no recursively saturated simple extension. It is an interesting open problem whether there is an ω_1-like lofty model which is not recursively saturated.

11

ORDER TYPES

Every model M of Peano Arithmetic has a reduct $(M, <)$ that is a linearly ordered set. There is only one possible order type of $(M, <)$ if M is countable and nonstandard and that order type is very well understood. By contrast, very little is known about $(M, <)$ for uncountable M. One especially vexing question is: *is it true that whenever M, N are uncountable models, then there is $M' \equiv M$ such that $(M', <) \cong (N <)$?* This chapter contributes nothing to this question. But it does contain some results on order types of models of Peano Arithmetic.

The common thread of the main results of this chapter is the following notion of a (κ, λ)-cut.

Definition 11.0.1 Let $I \subseteq_{\mathsf{end}} M$ be a cut of a model of PA^*, and let κ, λ be infinite cardinals. Then I is a (κ, λ)-*cut* if $\mathrm{cf}(I) = \kappa$ and $\mathrm{dcf}(I) = \lambda$.

In this definition, $\mathrm{cf}(I)$ is the "external" cofinality, which is the least cardinal κ for which there exists $X \subseteq I$ that is unbounded in I and $|X| = \kappa$. Similarly, $\mathrm{dcf}(I)$ is the least cardinal λ for which there is $Y \subseteq M \backslash I$ such that $|Y| = \lambda$ and, for each $x > I$ there is $y \in Y$ with $y < x$.

11.1 On (κ, κ)-cuts

Not very much that is nontrivial can be said, in general, about the existence of (κ, λ)-cuts other than the following theorem.

Theorem 11.1.1 *For every nonstandard M there are an infinite cardinal κ and a (κ, κ)-cut $I \subseteq_{\mathsf{end}} M$.*

Proof Before starting the proof, let us see what goes wrong with a naive approach. Pick some nonstandard elements $a_0, b_0 \in M$ with b_0 very much bigger than a_0. Inductively, and slowly, construct a transfinite increasing sequence $a_0 < a_1 < \cdots < a_\alpha < \cdots$ and simultaneously a transfinite decreasing sequence $b_0 > b_1 > \cdots > b_\alpha > \cdots$, always making sure that $a_\alpha < b_\alpha$. Of course, this cannot go on forever. Suppose there is a limit ordinal γ_0 at which it stops; that is, there is no c such that $a_\alpha < c < b_\alpha$ for all $\alpha < \gamma_0$. Then, letting $I = \sup\{a_\alpha : \alpha < \gamma_0\} = \{x \in M : x \leq a_\alpha \text{ for some } \alpha < \gamma_0\}$, we have a cut I for which $\mathrm{cf}(I) = \mathrm{dcf}(I) = \mathrm{cf}(\gamma_0)$. The problem is that this construction may end, not at a limit ordinal but at a successor ordinal. The construction in the proof is designed to circumvent this problem.

Instead of constructing just one pair of increasing and decreasing sequences, we will construct infinitely many, and we will not construct these sequences by adjoining one point at each stage but ω points. Thus, for some limit ordinal γ_0 and for each $n < \omega$ we will have, for $\alpha < \beta < \gamma_0$, that

$$a_{n,\alpha} < a_{n,\beta} < b_{n,\beta} < b_{n,\alpha} \,. \qquad (\star)$$

It will also be required that the sequence of a_n's grows fast, the growth rate being determined by the sequence of b_{n+1}'s. We will make use of the function $E(x) = x^x$. Specifically, we require that

$$a_{n,\alpha} + E(b_{n+1,\alpha}) < a_{n,\alpha+1} \,. \qquad (\star)$$

We start by obtaining $a_{n,\alpha}$ and $b_{n,\alpha}$ for $n, \alpha < \omega$, satisfying both $(\star)$'s for $n < \omega$ and $\alpha < \beta < \omega$. One way to get these is first, by overspill, get a descending sequence $d_0 > d_1 > d_2 > \cdots$, where $E(d_{n+1}) = d_n$, and then set $a_{n,i} = i \cdot d_{2n}$ and $b_{n,i} = 2d_{2n} - i$, for $n, i < \omega$. It is easily checked (DO IT!) that these definitions do the job.

Now suppose that, for the limit ordinal γ, we have managed to get $a_{n,\alpha}$ and $b_{n,\alpha}$ satisfying both $(\star)$'s for $n < \omega$ and $\alpha < \gamma$. We try to get, for each $n < \omega$, some c_n such that $a_{n,\alpha} < c_n < b_{n,\alpha}$ for every $\alpha < \gamma$. This leads to two cases.

Case 1: All the c_n exist. In this case we just continue with the two sequences by letting

$$a_{n,\gamma+i} = c_n - E(c_{n+1}) + i \cdot E(c_{n+1} - 1)$$

and

$$b_{n,\gamma+i} = c_n - E(c_{n+1}) + c_{n+1} \cdot E(c_{n+1} - 1) - i$$

whenever $n, i < \omega$. It is necessary to check that these extended sequences still satisfy both $(\star)$'s, a task that is left as an exercise for the reader to do. One helpful observation, however, is that $b_{n,\gamma+i} < c_n$ for $n, i < \omega$.

Case 2: Some c_n does not exist. In this case, we let $\kappa = \mathrm{cf}(\gamma)$ and $I = \sup\{a_{n,\alpha} : \alpha < \gamma\}$. Then $\mathrm{cf}(I) = \mathrm{dcf}(I) = \kappa$, so we are done.
Since Case 2 must eventually occur, the proof is complete. □

11.2 Saturation of the order reduct

There are connections between the extent of saturation of a model M and the existence of (κ, λ)-cuts. The following proposition is the simplest.

Proposition 11.2.1 *If M is a κ-saturated model of PA^*, where κ is an uncountable cardinal, then M has no (μ, λ)-cuts, where $\mu, \lambda < \kappa$.* □

The main result of this section, Theorem 11.2.4, shows that the converse also is true.

11.2.1 *Canonical codes*

There are lots of different sorts of objects that can be coded in PA and lots of different ways that they can be coded. Recall that if $a \in M \models \mathsf{PA}$ and $a_0, a_1, a_2, \ldots$ is an ω-sequence from M, then a codes this sequence if $M \models (a)_n = a_n$ for every standard $n \in M$. Any coded ω-sequence is necessarily bounded.

Every bounded definable subset $A \subseteq M$ (that is, A is M-finite) can be coded in a canonical way. We say that $a \in M$ is the *canonical code* for A if $a = \sum_{x \in A} 2^x$. For example, 0 is the canonical code for $\emptyset$, 1 is the canonical code for $\{0\}$, and 75 is the canonical code for $\{0, 1, 3, 6\}$. Every M-finite set has a unique canonical code and, conversely, every element of M is the canonical code of a unique M-finite set. We will write $x \in y$ if x is in the set canonically coded by y, and $x \subseteq y$ if the set canonically coded by x is a subset of the one canonically coded by y. For example, $3 \in 75$ and $1 \subseteq 75$.

Lemma 11.2.2 *Suppose $M \models \mathsf{PA}^*$, $p \in M$, $B, C, D \subseteq M$ are M-finite, $p_M \cap B = \emptyset$, and $D = B \cup p_M$. Let $b, c, d \in M$ be the canonical codes for B, C, D, respectively. Then $b \leq c \leq d$ iff $B \subseteq C \subseteq D$.*

Proof If $B \subseteq C \subseteq D$, then it is clear that $b \leq c \leq d$ (Do it!).

For the converse direction, we proceed by induction. For $i \in M$, let $S(i)$ be the following statement: *if p, B, C, D, b, c, d are as in the Lemma, $b \leq c \leq d$ and x is the ith largest element of B, then x is the ith largest element of C.* It suffices to prove that $S(i)$ holds for all i, and we will do this by induction on i.

Basis step. Here $i = 0$ so that $x = \max(B)$. Then $b > 0$ since $B \neq \emptyset$, from which it follows that $c > 0$ and then $C \neq \emptyset$. Let $y = \max(C)$. Clearly, $d < 2^x$, and as $c \leq d$, then $y \leq x$. Also, $2^{x-1} \leq b$, and as $b \leq c$, then $y \geq x$. Thus, $y = x$, so $x = \max(C)$.

Inductive step. Assume $S(i)$, and let x be the $(i+1)$th largest element of B. Let $B' = B \cap [0, x], C' = C \cap [0, x]$ and $D' = D \cap [0, x]$. Now the basis step can be applied to B', C', D' (Do it!) to show that $x = \max(C')$. Thus, x is the $(i+1)$th largest element of C. □

11.2.2 $\aleph_1$*-saturation*

The following theorem gives two characterizations of $\aleph_1$-saturation.

Theorem 11.2.3 *For each $M \models \mathsf{PA}^*$, the following are equivalent:*

(1) *M is $\aleph_1$-saturated.*
(2) *The reduct $(M, <)$ is $\aleph_1$-saturated.*
(3) *Every ω-sequence from M is coded.*

Proof The implication (1) $\Longrightarrow$ (2) is trivial. We will prove (2) $\Longrightarrow$ (3) $\Longrightarrow$ (1).

(3) $\Longrightarrow$ (1): Assume (3) holds. Then $\mathrm{cf}(M) \geq \aleph_1$, as otherwise there would be a coded, cofinal ω-sequence, which is impossible. Now let $\{\varphi_0(x), \varphi_1(x), \varphi_2(x), \ldots\}$ be a finitely satisfiable set of unary formulas, possibly involving some parameters from M. Our goal is get some $z \in M$ such that $M \models \varphi_n(z)$ for each $n < \omega$.

Let $b_n \in M$ be such that $M \models \varphi_0(b_n) \wedge \varphi_1(b_n) \wedge \cdots \wedge \varphi_n(b_n)$, and then choose $b \in M$ so that $b > b_n$ for each $n < \omega$. Let c_n be the canonical code for the set $\{x \in M : M \models x < b \wedge \bigwedge_{i \leq n} \varphi_i(x)\}$, and then let c be a code for the ω-sequence $c_0, c_1, c_2, \ldots$. Observe that for all standard $e \in M$

$$M \models \forall n < e[(c)_n \supseteq (c)_e > 0],$$

so, by overspill, there is a nonstandard $e \in M$ for which this is also true. Pick any $z \in (c)_e$. Then $z \in (c)_n$ for each $n < \omega$, so that $M \models \varphi_n(z)$ for each $n < \omega$, completing the proof that (1) holds.

(2) $\Longrightarrow$ (3): Assume (2) holds. Let $a_0, a_1, a_2, \ldots \in M$ be any ω-sequence from M, intending to show that it is coded. We define two other sequences $a'_0, a'_1, a'_2, \ldots$ and $a''_0, a''_1, a''_2, \ldots$ as follows. Let the first of these sequences be $a_0, 1+a_0+a_1, 2+a_0+a_1+a_2, \ldots, n+a_0+a_1+\cdots+a_n, \ldots$, so $a'_0 < a'_1 < a'_2 < \cdots$. Clearly, $\mathrm{cf}(M) \geq \aleph_1$, so we can let s be an upper bound for this sequence, and then let $a''_n = s - a'_n$, resulting in a decreasing sequence $a''_0 > a''_1 > a''_2 > \cdots$. It is evident that if any one of these sequences is coded, then all three are. Thus, without loss of generality, we can assume that the given sequence $a_0, a_1, a_2, \ldots$ is decreasing.

Let b_n be the canonical code for the set $\{a_0, a_1, \ldots, a_n\}$, and let c_n be the canonical code for the set $\{a_0, a_1, \ldots, a_n\} \cup [0, a_{n+1}]$. Then, $b_0 < b_1 < b_2 < \cdots < c_2 < c_1 < c_0$. By the $\aleph_1$-saturation of $(M, <)$, there is $d \in M$ such that $b_0 < b_1 < b_2 < \cdots < d < \cdots < c_2 < c_1 < c_0$. By Lemma 11.2.2, d is the canonical code of a set whose nth largest element is a_n. Clearly, there is $a \in M$ which codes a sequence whose nth term is the nth largest $x \in d$. Thus, $(a)_n = a_n$ for each $n < \omega$. □

11.2.3 κ-saturation for $\kappa > \aleph_1$

The equivalence of (1) and (2) of the previous theorem generalizes to all uncountable cardinals.

Theorem 11.2.4 *Let $M \models \mathsf{PA}^*$ and let κ be an uncountable cardinal. The following are equivalent:*

(1) *M is κ-saturated.*

(2) *The reduct $(M, <)$ is κ-saturated.*

Proof Fix $M \models \mathsf{PA}^*$. The implication (1) $\Longrightarrow$ (2) is trivial. The reverse implication is proved by induction on the uncountable cardinal κ. The basis step $\kappa = \aleph_1$ is Theorem 11.2.3.

For κ a limit cardinal, there is nothing to prove since in this case a structure is κ-saturated iff it is λ-saturated for all $\lambda < \kappa$.

So we will consider the case that κ is a successor cardinal, namely $\kappa = \lambda^+$. (For the rest of this proof, we will eschew κ in favor of λ^+.) We assume that $(M, <)$ is λ^+-saturated and, by the inductive hypothesis, that M is λ-saturated. Our goal is to prove that M is λ^+-saturated.

Let $\{\varphi_\alpha(x) : \alpha < \lambda\}$ be a finitely satisfiable set of λ unary formulas, possibly involving some parameters from M. The object is to get some $z \in M$ such that, for each $\alpha < \lambda$, $M \models \varphi_\alpha(z)$. For each finite subset $S \subseteq \lambda$, let $x_S \in M$ be such that $M \models \bigwedge_{\alpha \in S} \varphi_\alpha(x_S)$. Since there are at most λ of these witnesses and $\mathrm{cf}(M) \geq \lambda^+$ by the λ^+-saturation of $(M, <)$, there is $r \in M$ that is bigger than all of these witnesses. Letting a'_α be the canonical code of $\{x \in M : M \models x \leq r \wedge \varphi_\alpha(x)\}$, we see that $\varphi_\alpha(x)$ is equivalent to $x \in a'_\alpha$. Using the λ-saturation of M, we can find a strictly increasing sequence $\langle a_\alpha : \alpha < \lambda \rangle$ such that for any $u \in M$, $2u \in a_\alpha$ iff $u = a'_\alpha$. Let $\varphi(v, x)$ be the formula $\exists u[2u \in v \wedge x \in u]$. The point of doing this is that $\varphi(a_\alpha, x)$ is equivalent to $\varphi_\alpha(x)$. Henceforth, we will use $\varphi(a_\alpha, x)$ instead of $\varphi_\alpha(x)$.

Since M is λ-saturated, there is a sequence $\langle z_\alpha : \alpha < \lambda \rangle$ such that $\alpha \leq \beta < \lambda$, then $M \models \varphi(a_\alpha, z_\beta)$.

Using the λ-saturation of M, we can get a sequence $\langle c_\alpha : \alpha < \lambda \rangle$ (where we are thinking of each c_α as a canonical code) such that whenever $\alpha \leq \beta < \lambda$, the following sentences are true in M:

(1) $a_\alpha \in c_\beta$;

(2) $\forall u[u \in c_\alpha \rightarrow u \leq a_\alpha]$;

(3) $\forall u[u \in c_\alpha \rightarrow \varphi(u, z_\alpha)]$;

(4) $\forall u < a_\alpha[u \in c_\beta \rightarrow u \in c_\alpha]$.

Using the λ-saturation of M, we will be able to construct such a sequence inductively. From (1) and (2), there is no choice but to let $c_0 = 2^{a_0}$. Now suppose that $0 < \gamma < \lambda$, and that we already have $\langle c_\alpha : \alpha < \gamma \rangle$ such that (1)–(4) are true whenever $\alpha \leq \beta < \gamma$. We want $c = c_\gamma$ such that the sentences:

(1′) $a_\alpha \in c \wedge a_\gamma \in c$;

(2′) $\forall u[u \in c \rightarrow u \leq a_\gamma]$;

(3′) $\forall u[u \in c \rightarrow \varphi(u, z_\gamma)]$;

(4′) $\forall u < a_\alpha[u \in c \rightarrow u \in c_\alpha]$

are true whenever $\alpha < \gamma$. Since M is λ-saturated, it suffices that for each finite $S \subseteq \gamma$, there is $c \in M$ that makes these sentences true for each $\alpha \in S$. For finite $S \subseteq \gamma$, let $\delta = \max(S)$ (or $\delta = 0$ if $S = \emptyset$), and then let c be the canonical code for $\{u \in c_\delta : M \models \varphi(u, z_\gamma)\} \cup \{a_\gamma\}$. This c works (DO IT!), thereby proving the existence of the sequence $\langle c_\alpha : \alpha < \lambda \rangle$.

Our next goal is to obtain a strictly decreasing sequence $\langle d_\alpha : \alpha < \lambda \rangle$, with d_α encoding (in a way to be seen) both c_α and z_α. Start with a strictly increasing sequence $\langle s_\alpha : \alpha < \lambda \rangle$ such that, for every $\alpha < \lambda$, $(s_\alpha)_0 = c_\alpha$ and $(s_\alpha)_1 = z_\alpha$. Such a sequence can be constructed using the λ-saturation of M (DO IT!). Since $(M, <)$ is λ^+-saturated, this sequence has an upper bound $s \in M$. Now let $d_\alpha = s - s_\alpha$. Define the functions $C, Z : [0, s] \longrightarrow M$ by $C(y) = (s - y)_0$ and $Z(y) = (s - y)_1$. In particular, $C(d_\alpha) = c_\alpha$ and $Z(d_\alpha) = z_\alpha$.

The next sequence we want is $\langle b_\alpha : \alpha < \lambda \rangle$ (where we are thinking of each b_α as a canonical code) such that whenever $\alpha \leq \beta < \lambda$, the following sentences are true in M:

(5) $d_\alpha \in b_\beta$;
(6) $\forall y[y \in b_\alpha \to d_\alpha \leq y \leq d_0]$;
(7) $\forall y[y \in b_\alpha \to \forall u(u \in C(y) \to \varphi(u, Z(y)))]$;
(8) $\forall y \geq d_\alpha[y \in b_\alpha \leftrightarrow y \in b_\beta]$;
(9) $\forall y[y \in b_\beta \wedge y \leq d_\alpha \to a_\alpha \in C(y)]$.

Using the λ-saturation of M, we will be able to construct such a sequence inductively. From (5) and (6), there is no choice but to let $b_0 = 2^{d_0}$. Now suppose that $0 < \gamma < \lambda$, and that we already have $\langle b_\alpha : \alpha < \gamma \rangle$ such that (5)–(9) are true whenever $\alpha \leq \beta < \gamma$. We want $b = b_\gamma$ such that the sentences:

(5′) $d_\alpha \in b \wedge d_\gamma \in b$;
(6′) $\forall y[y \in b \to d_\gamma \leq y \leq d_0]$;
(7′) $\forall y[y \in b \to \forall u(u \in C(y) \to \varphi(u, Z(y)))]$;
(8′) $\forall y \geq d_\alpha[y \in b_\alpha \leftrightarrow y \in b]$;
(9′) $\forall y[y \in b \wedge y \leq d_\alpha \to a_\alpha \in C(y)]$;

are true whenever $\alpha < \gamma$. Since M is λ-saturated, it suffices that for each finite $S \subseteq \gamma$, there is $b \in M$ that makes these sentences true for each $\alpha \in S$. For finite $S \subseteq \gamma$, let $\delta = \max(S)$ (or $\delta = 0$ if $S = \emptyset$), and then let b be the canonical code for $B_\delta \cup \{d_\gamma\}$, where B_δ is the set canonically coded by b_δ. This b works (DO IT!), thereby proving the existence of the sequence $\langle b_\alpha : \alpha < \lambda \rangle$.

Let B_α be the set canonically coded by b_α. Then (5), (6), (8) imply that $d_\alpha \in B_\alpha \subseteq [d_\alpha, d_0]$ and $B_\alpha = B_\beta \cap [d_\alpha, d_0]$ whenever $\alpha \leq \beta < \lambda$. Then $b_\alpha + 2^{d_\alpha} - 1$ is the canonical code of $B_\alpha \cup [0, d_\alpha - 1]$, and

$$b_\alpha < b_\beta < b_\beta + 2^{d_\beta} - 1 < b_\alpha + 2^{d_\alpha} - 1$$

whenever $\alpha < \beta < \lambda$.

By the λ^+-saturation of $(M, <)$, there is $b_\infty \in M$ such that $b_\alpha < b_\infty < b_\alpha + 2^{d_\alpha} - 1$ for all $\alpha < \lambda$. Let B_∞ be the set canonically coded by b_∞. By Lemma 11.2.2, $B_\infty \cap [d_\alpha, d_0] = B_\alpha$; therefore, $d_\alpha \in B_\infty$ for each $\alpha < \lambda$.

For $\alpha < \lambda$, define w_α to be the least $w \in B_\infty$ such that $M \models \forall y[y \in B_\infty \wedge w \leq y \leq d_\alpha \rightarrow a_\alpha \in C(y)]$. From (9), $w_\alpha < d_\beta$ for all $\alpha, \beta < \lambda$. Therefore, by the λ^+-saturation of $(M, <)$, there is $d_\infty \in B_\infty$ such that $w_\alpha < d_\infty < d_\alpha$ for each $\alpha < \lambda$ (DO IT!). Let $z_\infty = Z(d_\infty)$.

Since $w_\alpha < d_\infty$, the defining property of w_α gives us that $a_\alpha \in C(d_\infty)$, and then from (7), (5), (1) that $M \models \varphi(a_\alpha, z_\infty)$. □

Suppose $(A, <)$ is a linearly ordered set. If $X, Y \subseteq A$ are such that $x < y$ whenever $x \in X$ and $y \in Y$, then we write $X < Y$. We say that $(A, <)$ is *κ-dense* if, whenever $X, Y \subseteq A$ are such that $|X|, |Y| < \kappa$ and $X < Y$, then there is $c \in A$ such that $x \leq c$ for $x \in X$ and $c \leq y$ for $y \in Y$. Every κ-saturated linearly ordered set is κ-dense, and there are κ-dense linearly ordered sets that are not κ-saturated. In the proofs of Theorems 11.2.3 and 11.2.4, the κ-density of $(M, <)$ was the only consequence of κ-saturation that was used. Therefore, if κ is an uncountable cardinal and $M \models \mathsf{PA}^*$ is such that $(M, <)$ is κ-dense, then $(M, <)$-saturated. Observe that that for an infinite cardinal κ, $(M, <)$ is κ-dense iff whenever there is a (μ, λ)-cut, then $\kappa \leq \max(\mu, \lambda)$. Thus, there is the following corollary.

Corollary 11.2.5 *Let $M \models \mathsf{PA}^*$ and let κ be an uncountable cardinal. The following are equivalent:*

(1) *M is κ-saturated.*

(2) *If $I \subseteq_{\mathsf{end}} M$ is a (μ, λ)-cut, then $\kappa \leq \max(\mu, \lambda)$.* □

But more is true and is more easily proved. See Exercise 11.3.4.

11.3 Exercises

♣**11.3.1** The order type of a nonstandard model is of the form $\omega + (\omega^* + \omega)\rho$, where ρ is the order type of a dense linear ordering without endpoints.

♦**11.3.2** Show that the order type ρ in the previous exercise cannot be the order type of the real numbers.

♣**11.3.3** Every ω_1-like model of PA^* has order type $\omega + (\omega^* + \omega) \cdot (\mathbb{Q} \cdot \omega_1)$.

♦**11.3.4** Let $(A, <)$ be a linearly ordered set having a first element such that every element has an immediate successor and every element but the first has an immediate predecessor. If κ is an uncountable cardinal, then $(A, <)$ is κ-dense iff it is κ-saturated.

♥**11.3.5** Suppose κ is an uncountable cardinal, M a model and $\lhd$ a definable linear ordering of M. If the ordered set $(M, \lhd)$ is κ-saturated, then the model M is κ-saturated.

Let $I \subseteq_{\mathsf{end}} N$ be a cut of the model N, and then define $M(I) = \{y \in N : yz \in I \text{ for all } z \in I\}$. For $x \in N$, let $xI = \{y \in N : y < xz \text{ for some } z \in I\}$ and $x/I = \{y \in N : x > yz \text{ for all } z \in I\}$. The cut I is a *type one* cut if either $I = xM(I)$ for some $x \in I$ or $I = x/M(I)$ for some $x \in N \backslash M(I)$. If I is not type one, then it is *type two*.

♦**11.3.6** If I is a type two cut, then I is a (κ, κ)-cut for some infinite cardinal κ. (HINT: $\kappa = \mathrm{dcf}(M(I))$.)

♥**11.3.7** Every nonstandard model has a type two cut.

11.4 Remarks & References

The still unanswered question stated at the beginning of this chapter is the 14th in H. Friedman's list of 102 problems in mathematical logic [42]. More on this question can be found in Bovykin & Kaye [18].

Theorem 11.1.1 is from Shelah [185]. The proof presented there is a model of clarity to which the proof given here is nearly identical. Theorem 11.2.4 is due to J.-F. Pabion [142]. The proof of Theorem 11.2.4 presented here, while following closely the one in [142], contains some simplifications. Theorem 11.2.3 also is from [142], where a good deal of the credit for this theorem is given to Denis Richard. Exercise 11.3.5, which generalizes Theorem 11.2.4, is a little later result of Pabion [141].

Exercise 11.3.6 comes from Keisler & Leth [78], where the connections with nonstandard analysis are investigated. Exercise 11.3.7 is a theorem of Renling Jin [61]; its proof uses ideas from the proof of Theorem 11.1.1.

12

TWENTY QUESTIONS

Needless to say, there is still work to be done. There are many open questions. We would like to conclude with our personal selection of twenty of them. Some have already been noted in previous chapters, some not. Most of them have been published elsewhere. We have reasons to believe that none of these problems is easy. If there is a question here that admits an easy answer, then most likely we are to blame, and it should be also easy to restate the question to restore its open status.

No list of open problems concerning models of PA is complete without the venerable Scott set Problem. As noted in Chapter 1, under CH every Scott set is the standard system of a nonstandard model of PA.

Question 1 Assume CH is false. Is every Scott set the standard system of a nonstandard model of PA?

Every model M has a minimal elementary end extension, and every nonstandard *countable* model a minimal cofinal extension. If Ramsey ultrafilters exist (which will be the case if CH holds, but is not always the case), then every nonstandard model whose standard system is $\mathcal{P}(\omega)$ has a minimal cofinal extension. What happens in general?

Question 2 Does every nonstandard model M have a minimal cofinal extension?

There are many questions that can be asked about substructure lattices. The following question would have a positive answer if we knew that every finite lattice has a finite congruence representation. Even if this turns out not to be the case, this question still could have a positive answer.

Question 3 Is every finite lattice isomorphic to some $\mathrm{Lt}(M)$?

For every finite distributive lattice L, every M has an elementary end extension N such that $\mathrm{Lt}(N/M) \cong L$. There are also nondistributive lattices L with this property; however, both $\mathbf{M}_3$ and $\mathbf{N}_5$ fail to have this property, although their failures are of different degrees. Every countable M has an elementary end extension N such that $\mathrm{Lt}(N/M) \cong \mathbf{N}_5$, but no M at all has an elementary end extension such that $\mathrm{Lt}(N/M) \cong \mathbf{M}_3$.

This situation suggests the following two questions, which we list together.

Question 4 What finite lattices L are such that every M has an elementary end extension N such that $\mathrm{Lt}(N/M) \cong L$? What finite lattices L are such that every *countable* M has an elementary end extension N such that $\mathrm{Lt}(N/M) \cong L$?

We prove in Chapter 7 that every countable model M has an elementary end extension N such that $(M, \mathrm{Cod}(N/M)) \models I\Sigma_1 + \neg B\Sigma_2$. A conjecture in [105] states that this statement generalizes to higher levels of the arithmetic hierarchy. Kanovei [65] proved the conjecture for countable recursively saturated M.

Question 5 Let M be a countable model. Is it true that for every $n > 0$, M has an elementary end extension N such that $(M, \mathrm{Cod}(N/M)) \models I\Sigma_n + \neg B\Sigma_{n+1}$?

All Jónsson models constructed in Corollary 2.1.13 realize uncountably many complete types.

Question 6 Is there a Jónsson model of PA which realizes only countably many complete types?

In Mills' counterexample to the MacDowell–Specker for uncountable languages, the model constructed is nonstandard. This leaves open the following question, posed in [131].

Question 7 Is there an expansion of the standard model having no conservative elementary extension?

Many important results concerning countable recursively saturated models rely on the existence of Lascar generic automorphisms and sequences. Since we do not know if non-arithmetically saturated recursively saturated models have Lascar generics, many questions concerning such models remain open. In Chapters 8 and 9, we have proved that for countable arithmetically saturated models all four questions below have positive answers (well, we have not proved everything: (3) below is our Exercise 9.6.5).

Question 8 Let M be a countable recursively, but not arithmetically, saturated model.

(1) Does M have a Lascar generic automorphism?
(2) Does M have the small index property?
(3) Does $\mathrm{Aut}(M)$ have subgroups of finite index?
(4) Is every open subgroup of $\mathrm{Aut}(M)$ contained in a maximal one?

As shown in [98], if M is countable recursively saturated and not arithmetically saturated, then for every $f \in \mathrm{Aut}(M)$, $\mathrm{fix}(f) \cong M$. If M is arithmetically saturated, then M has continuum many nonisomorphic elementary submodels which can be fixed point sets.

Question 9 Let M be countable and arithmetically saturated and let K be an elementary submodel of M. Is there $f \in \mathrm{Aut}(M)$ such that $\mathrm{fix}(f) \cong K$?

In Chapter 8, we have seen that for all countable recursively saturated M and N such that $M \prec_{\mathsf{end}} N$, there are $2^{\aleph_0}$ automorphisms of M which do not extend to automorphisms of N. In extreme cases only the identity on M extends.

Question 10 Let M be countable and recursively saturated. Is there a nontrivial automorphism of M which does not extend to any recursively saturated elementary end extension of M?

If $S \subseteq M$ is a partial inductive satisfaction class and f is an automorphism of (M, S), then f can be extended to a recursively saturated elementary end extension of M. Hence, Question 10 is related to the following question: let f be an automorphism of a countable recursively saturated model M. Is there a partial inductive satisfaction class S such that f is an automorphism of (M, S)?

Question 11 Are there countable, arithmetically saturated models M_1 and M_2 such that $\mathrm{Th}(\mathrm{Aut}(M_1)) \neq \mathrm{Th}(\mathrm{Aut}(M_2))$?

The next question is due to Richard Kaye [74], who proved that if M is countable and recursively saturated, then the closed normal subgroups of $G = \mathrm{Aut}(M)$ are exactly the pointwise stabilizers of invariant cuts of M. If I is such a cut and it is closed under exponentiation, then $G_{(>I)}$, which is the union of pointwise stabilizers of all cuts $J > I$, is also normal, and its topological closure is $G_{(I)}$.

Question 12 Is every normal subgroup of G either of the form $G_{(>I)}$ or $G_{(I)}$ for an invariant cut I? In particular, is $G_{(>\mathbb{N})}$ generated by automorphisms f such that (M, f) is recursively saturated?

It is shown in [174] that if M is a countable, recursively saturated model of PA, then $\mathrm{Th}(\mathrm{Aut}(M))$ is undecidable. How undecidable is it? The next question asks about the Turing degree.

Question 13 If M is countable and recursively saturated, is $\deg(\mathrm{Th}(\mathrm{Aut}(M))) \geq \mathbf{0}^{(\omega)}$?

Friedman's 14th was mentioned at the beginning of Chapter 11. We repeat it here.

Question 14 If M is uncountable and $T \supseteq \mathsf{PA}$ is a completion, does T have a model N such that the order reducts $(M, <)$ and $(N, <)$ are isomorphic?

Nurkhaidarov proved in [139] that there are a countable Scott set $\mathfrak{X}$ and four recursively saturated models whose standard system is $\mathfrak{X}$ with pairwise nonisomorphic automorphism groups.

Question 15 For a given countable Scott set $\mathfrak{X}$, what is the cardinality of the set of isomorphism types of automorphism groups of countable recursively saturated models whose standard system is $\mathfrak{X}$?

Loftiness was introduced and studied in [68] and [69]. If M is a countable model of PA, then M is lofty iff M has a simple cofinal extension that is recursively saturated. The right-to-left implication holds even if M is uncountable, but it is unknown if the converse implication holds for uncountable M. If it fails for some M, then M must be κ-like for some uncountable regular cardinal κ, and then there is also a counterexample with $\kappa = \omega_1$. It is known that if M is κ-like and has a simple cofinal extension that is recursively saturated, then M is already recursively saturated. Thus:

Question 16 Is there an ω_1-like model that is lofty but not recursively saturated?

The usual proof of the existence of a Kaufmann model proceeds by first showing, assuming $\diamondsuit$, that there is a Kaufmann model, and then eliminating $\diamondsuit$ by Shelah's Absoluteness Theorem from [184]. We are hoping for a still more direct proof, avoiding $\diamondsuit$. This problem is included in the list of open problems in Hodges' [56] in the form: "Prove the existence of rather classless recursively saturated models of Peano arithmetic in cardinality ω_1 without assuming diamond at any stage of the proof." A positive answer to the following question would do it.

Question 17 Suppose M be countable and $X \subseteq M$ is such that $\mathrm{Th}(M, X) \in \mathrm{SSy}(M)$. Does M always have an elementary end extension N that is recursively saturated such that if $Y \subseteq M$ is coded in N, then $(M, Y) \not\cong (M, X)$?

The next question is one we heard from Kanovei. The question has some cousins such as: does $\mathsf{ZF} \vdash \exists M[\mathrm{SSy}(M) = \mathcal{P}(\omega)]$?

Question 18 Does PA have a model M whose universe is the set $\mathbb{R}$ and whose $+$ and $\times$ are Borel subsets of $\mathbb{R}^3$ and $\mathrm{SSy}(M) = \mathcal{P}(\omega)$?

It is shown in [97] that no ω_1-like recursively saturated model is Jónsson. It is also shown there that if such an M has a partial inductive satisfaction class, then it has no recursively saturated elementary proper submodels of cardinality $\aleph_1$.

Question 19 Is there an ω_1-like recursively saturated model which has no recursively saturated proper elementary submodels of cardinality $\aleph_1$?

In the proof of Theorem 10.6.3 we constructed two ω_1-like models whose additive reducts are isomorphic and multiplicative reducts are not.

Question 20 Let M and N be ω_1-like models of the same completion of PA, whose both additive and multiplicative reducts are isomorphic. Are M and N isomorphic?

Added in proof After we circulated our list of 20 questions among our colleagues, Ali Enayat wasted no time and promptly answered two of them. He gave an elegant ultrapower construction to give the positive answer to Question 9 [30]. He also constructed a standard model with no conservative elementary extension, answering Question 7 [31]. His solution leaves open the following variant of Question 7.

Question 7' Is there a standard model $\mathfrak{A}$ such that some nonstandard model of $\mathrm{Th}(\mathfrak{A})$ has no (conservative) elementary end extension.

REFERENCES

The numbers in brackets following a reference indicate page numbers where the reference is cited.

[1] Abramson, Fred G. & Harrington, Leo A. (1978). Models without indiscernibles. *J. Symbolic Logic*, **43**(3), 572–600. [157]

[2] Argyros, S. A., Felouzis, V., & Kanellopoulos, V. (2002). A proof of Halpern-Läuchli partition theorem. *Eur. J. Combin.*, **23**(1), 1–10. [179]

[3] Balbes, Raymond & Dwinger, Philip (1974). *Distributive lattices.* University of Missouri Press, Columbia, Mo. [132]

[4] Bamber, Nicholas & Kotlarski, Henryk (1997). On interstices of countable arithmetically saturated models of Peano arithmetic. *Math. Logic Quart.*, **43**(4), 525–540. [227]

[5] Barwise, Jon & Schlipf, John (1975). *On recursively saturated models of arithmetic*, Volume 498 of *Lecture Notes in Mathematics*, pp. 42–55. Springer, Berlin. [23]

[6] Barwise, Jon & Schlipf, John (1976). An introduction to recursively saturated and resplendent models. *J. Symbolic Logic*, **41**(2), 531–536. [23]

[7] Baumgartner, James E. & Taylor, Alan D. (1978). Partition theorems and ultrafilters. *Trans. Amer. Math. Soc.*, **241**, 283–309. [88]

[8] Bigorajska, Teresa (2003). Strongly maximal subgroups determined by elements in interstices. *MLQ Math. Logic. Quart.*, **49**(1), 101–108. [229]

[9] Bigorajska, Teresa, Kotlarski, Henryk, & Schmerl, James H. (1998). On regular interstices and selective types in countable arithmetically saturated models of Peano arithmetic. *Fund. Math.*, **158**(2), 125–146. [227], [229]

[10] Birkhoff, Garrett (1933). On the combination of subalgebras. *Proc. Cambridge Philos. Soc.*, **29**, 441–464. [133]

[11] Birkhoff, Garrett (1967). *Lattice theory.* Third edition. American Mathematical Society Colloquium Publications, Vol. XXV. American Mathematical Society, Providence, RI. [132]

[12] Blass, Andreas (1972). The intersection of nonstandard models of arithmetic. *J. Symbolic Logic*, **37**, 103–106. [23]

[13] Blass, Andreas (1974). On certain types and models for arithmetic. *J. Symbolic Logic*, **39**, 151–162. [133]

[14] Blass, Andreas (1977). Amalgamation of nonstandard models of arithmetic. *J. Symbolic Logic*, **42**(3), 372–386.

[15] Blass, Andreas (1977). End extensions, conservative extensions, and the Rudin-Frolík ordering. *Trans. Amer. Math. Soc.*, **225**, 325–340.

[16] Blass, Andreas (1978). A model-theoretic view of some special ultrafilters. In *Logic Colloquium '77 (Proc. Conf., Wrocław, 1977),*

Volume 96 of *Stud. Logic Foundations Math.*, pp. 79–90. North-Holland, Amsterdam. [88]

[17] Blass, Andreas (1980). Conservative extensions of models of arithmetic. *Arch. Math. Logik Grundlag.*, **20**(3–4), 85–94. [48]

[18] Bovykin, Andrey & Kaye, Richard (2002). *Order-types of models of Peano arithmetic*, Volume 302 of *Contemporary Mathematics*, pp. 275–285. American Mathematical Society, Providence, RI. [278], [288]

[19] Cegielski, Patrick (1981). Théorie élémentaire de la multiplication des entiers naturels. In *Model theory and arithmetic (Paris, 1979–1980)*, Volume 890 of *Lecture Notes in Mathematics*, pp. 44–89. Springer, Berlin. [280]

[20] Chang, C. C. (1965). A note on the two cardinal problem. *Proc. Amer. Math. Soc.*, **16**, 1148–1155. [179]

[21] Cholak, Peter A., Jockusch, Carl G., & Slaman, Theodore A. (2001). On the strength of Ramsey's theorem for pairs. *J. Symbolic Logic*, **66**(1), 1–55. [201]

[22] Clote, P. (1985). *Partition relations in arithmetic*, Volume 1130 of *Lecture Notes in Mathematics*, pp. 32–68. Springer, Berlin. [201]

[23] Cohn, P. M. (1957). Groups of order automorphisms of ordered sets. *Mathematika*, **4**, 41–50. [157]

[24] Conrad, Paul (1959). Right-ordered groups. *Michigan Math. J.*, **6**, 267–275. [157]

[25] Devlin, Keith J. & Shelah, Saharon (1978). A weak version of $\diamondsuit$ which follows from $2^{\aleph_0} < 2^{\aleph_1}$. *Israel J. Math.*, **29**(2–3), 239–247. [259]

[26] Dimitracopoulos, C. & Paris, J. (1988). A note on a theorem of H. Friedman. *Z. Math. Logik Grundlag. Math.*, **34**(1), 13–17. [24]

[27] Duby, Grégory (2003). Automorphisms with only infinite orbits on non-algebraic elements. *Arch. Math. Logic*, **42**(5), 435–447. [228]

[28] Ehrenfeucht, Andrzej (1973). Discernible elements in models for Peano arithmetic. *J. Symbolic Logic*, **38**, 291–292. [23]

[29] Ehrenfeucht, A. & Mostowski, A. (1956). Models of axiomatic theories admitting automorphisms. *Fund. Math.*, **43**, 50–68. [157]

[30] Enayat, Ali (1986). Conservative extensions of models of set theory and generalizations. *J. Symbolic Logic*, **51**(4), 1005–1021. [47]

[31] Enayat, Ali (1988). Undefinable classes and definable elements in models of set theory and arithmetic. *Proc. Amer. Math. Soc.*, **103**(4), 1216–1220. [48]

[32] Enayat, Ali (1990). Minimal elementary extensions of models of set theory and arithmetic. *Arch. Math. Logic*, **30**(3), 181–192. [47]

[33] Enayat, Ali (1999). *Analogues of the MacDowell-Specker theorem for set theory*, Volume 203 of *Lecture Notes in Pure and Applied Mathematics*, pp. 25–50. Dekker, New York. [47]

[34] Enayat, Ali. Automorphisms of models of arithmetic: a unified view. To appear in Annals of Pure and Applied Logic.

[35] Enayat, Ali. A standard model of Peano Arithmetic with no conservative elementary extension. To appear.

[36] Engström, Fredrik (2004). Expansions, omitting types, and standard systems. Ph.D. thesis, Götenborg University. [228]

[37] Erdős, P., Hajnal, A., & Rado, R. (1965). Partition relations for cardinal numbers. *Acta Math. Acad. Sci. Hungar.*, **16**, 93–196. [157]

[38] Erdős, P. & Rado, R. (1956). A partition calculus in set theory. *Bull. Amer. Math. Soc.*, **62**, 427–489. [157]

[39] Feit, W. (1983). An interval in the subgroup lattice of a finite group which is isomorphic to M_7. *Algebra Universalis*, **17**, 220–221. [134]

[40] Felgner, Ulrich (1971). Comparison of the axioms of local and universal choice. *Fund. Math.*, **71**(1), 43–62. (errata insert). [179]

[41] Friedman, Harvey (1973). *Countable models of set theories*, Volume 337 of *Lecture Notes in Mathematics*, pp. 539–573. Springer, Berlin. [23], [24]

[42] Friedman, Harvey (1975). One hundred and two problems in mathematical logic. *J. Symbolic Logic*, **40**, 113–129. [288]

[43] Gaifman, Haim (1970). *On local arithmetical functions and their application for constructing types of Peano's arithmetic*, pp. 105–121. North-Holland, Amsterdam. [88], [179]

[44] Gaifman, Haim (1972). *A note on models and submodels of arithmetic*, Volume 255 of *Lecture Notes in Mathematics*, pp. 128–144. Springer, Berlin. [22], [48]

[45] Gaifman, H. (1976). Models and types of Peano's arithmetic. *Ann. Math. Logic*, **9**, 223–306. [vi], [23], [33], [48], [133]

[46] Gaifman, Haim & Dimitracopoulos, Constantine (1982). *Fragments of Peano's arithmetic and the MRDP theorem*, Volume 30 of *Monograph. Enseign. Math.*, pp. 187–206. University of Genève, Geneva. [22]

[47] Geiser, James R. (1968). Nonstandard logic. *J. Symbolic Logic*, **33**, 236–250. [24]

[48] Grätzer, G. & Schmidt, E. T. (1963). Characterizations of congruence lattices of abstract algebras. *Acta Sci. Math. (Szeged)*, **24**, 34–59. [132]

[49] Hájek, Petr (1981). Completion closed algebras and models of Peano arithmetic. *Comment. Math. Univ. Carolin.*, **22**(3), 585–594. [21]

[50] Hájek, Petr & Pudlák, Pavel (1993). *Metamathematics of first-order arithmetic.* Perspectives in Mathematical Logic. Springer-Verlag, Berlin. [viii], [7], [201]

[51] Hajnal, A. (1985). The chromatic number of the product of two $\aleph_1$-chromatic graphs can be countable. *Combinatorica*, **5**(2), 137–139.[235]

[52] Halpern, J. D. & Läuchli, H. (1966). A partition theorem. *Trans. Amer. Math. Soc.*, **124**, 360–367. [179]

[53] Harnik, Victor (1986). ω_1-like recursively saturated models of Presburger's arithmetic. *J. Symbolic Logic*, **51**(2), 421–429. [280]

[54] Harnik, Victor (1987). *Countable or ω_1-like models of Presburger's arithmetic*, pp. 89–106. Springer, Berlin. [280]

[55] Hirst, Jeffry, L. (1987). Combinatorics in subsystems of second order arithmetic. Ph.D. thesis, The Pennsylvania State University. [201]

[56] Hodges, Wilfrid (1985). *Building models by games*, Volume 2 of *London Mathematical Society Student Texts.* Cambridge University Press, Cambridge. [292]

[57] Hodges, Wilfrid (1989). *Categoricity and permutation groups*, Volume 129 of *Studies in Logic and the Foundations of Mathematics*, pp. 53–72. North-Holland, Amsterdam. [252]

[58] Hodges, Wilfrid (1993). *Model theory*, Volume 42 of *Encyclopedia of Mathematics and its Applications.* Cambridge University Press, Cambridge. [254]

[59] Hodges, Wilfrid, Hodkinson, Ian, Lascar, Daniel, & Shelah, Saharon (1993). The small index property for ω-stable ω-categorical structures and for the random graph. *J. London Math. Soc. (2)*, **48**(2), 204–218. [254]

[60] Jensen, D. & Ehrenfeucht, A. (1976). Some problem in elementary arithmetics. *Fund. Math.*, **92**(3), 223–245. [21], [23], [24]

[61] Jin, Renling (1997). Type two cuts, bad cuts and very bad cuts. *J. Symbolic Logic*, **62**(4), 1241–1252. [288]

[62] Jockusch, Jr., Carl G. (1972). Ramsey's theorem and recursion theory. *J. Symbolic Logic*, **37**, 268–280. [201]

[63] Jockusch, Jr., Carl G. & Soare, Robert I. (1972). Π_1^0 classes and degrees of theories. *Trans. Amer. Math. Soc.*, **173**, 33–56. [179]

[64] Kanamori, Akihiro (2003). *The higher infinite*. Second edition. Springer Monographs in Mathematics. Springer-Verlag, Berlin. Large cardinals in set theory from their beginnings. [157]

[65] Kanovei, Vladimir (1998). *On "star" schemata of Kossak & Paris*, Volume 12 of *Lecture Notes Logic*, pp. 101–114. Springer, Berlin. [201], [290]

[66] Kanoveĭ, V. (1995). Uniqueness, collection, and external collapse of cardinals in IST and models of Peano arithmetic. *J. Symbolic Logic*, **60**(1), 318–324. [227]

[67] Kaufmann, Matt (1977). A rather classless model. *Proc. Amer. Math. Soc.*, **62**(2), 330–333. [vii], [258], [278]

[68] Kaufmann, Matt & Schmerl, James H. (1984). Saturation and simple extensions of models of Peano arithmetic. *Ann. Pure Appl. Logic*, **27**(2), 109–136. [24], [278], [292]

[69] Kaufmann, Matt & Schmerl, James H. (1987). Remarks on weak notions of saturation in models of Peano arithmetic. *J. Symbolic Logic*, **52**(1), 129–148. [24], [280], [292]

[70] Kaye, Richard (1991). Model-theoretic properties characterizing Peano arithmetic. *J. Symbolic Logic*, **56**(3), 949–963. [22], [195], [196], [201]

[71] Kaye, R. (1991). *Models of Peano arithmetic*, Volume 15 of *Oxford Logic Guides.* Clarendon Press, Oxford. [viii], [14], [18], [23], [201]

[72] Kaye, Richard (1991). On cofinal extensions of models of fragments of arithmetic. *Notre Dame J. Formal Logic*, **32**(3), 399–408. [201]

[73] Kaye, Richard (1992). The automorphism group of a countable recursively saturated structure. *Proc. London Math. Soc. (3)*, **65**(2), 225–244. [255]

[74] Kaye, Richard (1994). *A Galois correspondence for countable recursively saturated models of Peano arithmetic*, pp. 293–312. Oxford University Press, New York. [227], [229], [245], [291]

[75] Kaye, Richard, Kossak, Roman, & Kotlarski, Henryk (1991). Automorphisms of recursively saturated models of arithmetic. *Ann. Pure Appl. Logic*, **55**(1), 67–99. [23], [227–229], [255]

[76] Kaye, Richard & Kotlarski, Henryk (2000). On models constructed by means of the arithmetized completeness theorem. *MLQ Math. Logic. Quart.*, **46**(4), 505–516. [24]

[77] Kaye, Richard & Macpherson, Dugald (eds.) (1994). *Automorphisms of first-order structures.* The Clarendon Press, Oxford University Press, New York. [vii], [23], [229], [239], [254]

[78] Keisler, H. Jerome & Leth, Steven C. (1991). Meager sets on the hyperfinite time line. *J. Symbolic Logic*, **56**(1), 71–102. [288]

[79] Keisler, H. Jerome & Schmerl, James H. (1991). Making the hyperreal line both saturated and complete. *J. Symbolic Logic*, **56**(3), 1016–1025. [179], [279]

[80] Kirby, L. A. S (1977). Initial segments of models of arithmetic. Ph.D. thesis, University of Manchester. [201]

[81] Kirby, L. A. S. (1984). Ultrafilters and types on models of arithmetic. *Ann. Pure Appl. Logic*, **27**(3), 215–252. [88]

[82] Kirby, L. A. S. (1992). Book review: Richard Kaye. Models of Peano Arithmetic. *Notre Dame J. Formal Logic*, **33**(3), 461–463. [viii]

[83] Kirby, L. A. S. & Paris, J. B. (1977). *Initial segments of models of Peano's axioms*, Volume of 619 *Lecture Notes in Mathematics*, pp. 211–226. Springer, Berlin. [23], [184], [194], [201]

[84] Kirby, L. A. S. & Pillay, A. (1986). Weakly definable types. *Trans. Amer. Math. Soc.*, **295**(2), 547–563. [88]

[85] Knight, Julia & Nadel, Mark (1982). Models of arithmetic and closed ideals. *J. Symbolic Logic*, **47**(4), 833–840 (1983). [23], [48]

[86] Knight, Julia F. (1976). Hanf numbers for omitting types over particular theories. *J. Symbolic Logic*, **41**(3), 583–588. [157]

[87] Knight, Julia F. (1976). Omitting types in set theory and arithmetic. *J. Symbolic Logic*, **41**(1), 25–32. [48], [157]

[88] Knight, Julia F. (1999). *True approximations and models of arithmetic*, Volume 259 of *London Mathematical Society Lecture Note Series*, pp. 255–278. Cambridge University Press, Cambridge. [23]

[89] Kopytov, Valeriĭ M. & Medvedev, Nikolaĭ Ya. (1996). *Right-ordered groups.* Siberian School of Algebra and Logic. Consultants Bureau, New York. [157]

[90] Körner, Friederike (1998). Automorphisms moving all non-algebraic points and an application to NF. *J. Symbolic Logic*, **63**(3), 815–830. [226], [228]

[91] Kossak, Roman (1983). A certain class of models of Peano arithmetic. *J. Symbolic Logic*, **48**(2), 311–320. [48], [279]

[92] Kossak, Roman (1984). $L_{\infty\omega_1}$-elementary equivalence of ω_1-like models of PA. *Fund. Math.*, **123**(2), 123–131. [279]

[93] Kossak, Roman (1985). Recursively saturated ω_1-like models of arithmetic. *Notre Dame J. Formal Logic*, **26**(4), 413–422. [267], [279]

[94] Kossak, Roman (1990). On extensions of models of strong fragments of arithmetic. *Proc. Amer. Math. Soc.*, **108**(1), 223–232. [201], [229]

[95] Kossak, Roman (1991). A correction to: "On extensions of models of strong fragments of arithmetic" [*Proc. Amer. Math. Soc.* **108** (1990), no. 1, 223–232; MR 90d:03123]. *Proc. Amer. Math. Soc.*, **112**(3), 913–914. [201]

[96] Kossak, Roman (1991). *Exercises in "back-and-forth"*, Volume 112 of *Seminarberichte*. Humboldt University Berlin. [228]

[97] Kossak, Roman (1995). Four problems concerning recursively saturated models of arithmetic. *Notre Dame J. Formal Logic*, **36**(4), 519–530. Special issue: Models of arithmetic. [279], [292]

[98] Kossak, Roman (1997). Automorphisms of recursively saturated models of Peano arithmetic: fixed point sets. *Logic. J. IGPL*, **5**(6), 787–794 (electronic). [290]

[99] Kossak, Roman (2004). A note on a theorem of Kanovei. *Arch. Math. Logic*, **43**(4), 565–569. [179], [201]

[100] Kossak, Roman & Bamber, Nicholas (1996). On two questions concerning the automorphism groups of countable recursively saturated models of PA. *Arch. Math. Logic*, **36**(1), 73–79. [228]

[101] Kossak, R. & Kotlarski, H. (1988). Results on automorphisms of recursively saturated models of PA. *Fund. Math.*, **129**(1), 9–15. [228]

[102] Kossak, Roman & Kotlarski, Henryk (1996). On extending automorphisms of models of Peano arithmetic. *Fund. Math.*, **149**(3), 245–263. [228]

[103] Kossak, Roman, Kotlarski, Henryk, & Schmerl, James H. (1993). On maximal subgroups of the automorphism group of a countable recursively saturated model of PA. *Ann. Pure Appl. Logic*, **65**(2), 125–148. [88], [229]

[104] Kossak, Roman, Nadel, Mark, & Schmerl, James (1989). A note on the multiplicative semigroup of models of Peano arithmetic. *J. Symbolic Logic*, **54**(3), 936–940. [280]

[105] Kossak, Roman & Paris, Jeffrey B. (1992). Subsets of models of arithmetic. *Arch. Math. Logic*, **32**(1), 65–73. [22], [47], [201], [290]

[106] Kossak, Roman & Schmerl, James H. (1991). Minimal satisfaction classes with an application to rigid models of Peano arithmetic. *Notre Dame J. Formal Logic*, **32**(3), 392–398. [279], [280]

[107] Kossak, Roman & Schmerl, James H. (1995). Arithmetically saturated models of arithmetic. *Notre Dame J. Formal Logic*, **36**(4), 531–546. Special Issue: Models of arithmetic. [88], [228], [229], [254]

[108] Kossak, Roman & Schmerl, James H. (1995). The automorphism group of an arithmetically saturated model of Peano arithmetic. *J. London Math. Soc. (2)*, **52**(2), 235–244. [227], [254]

[109] Kotlarski, H. (1983). On cofinal extensions od models of arithmetic. *J. Symbolic Logic*, **48**, 253–262. [22], [24]

[110] Kotlarski, Henryk (1983). On elementary cuts in models of arithmetic. *Fund. Math.*, **115**(1), 27–31. [229]

[111] Kotlarski, Henryk (1984). On elementary cuts in recursively saturated models of Peano arithmetic. *Fund. Math.*, **120**(3), 205–222. [23], [227], [229]

[112] Kotlarski, Henryk (1986). The recursively saturated part of models of Peano arithmetic. *Z. Math. Logik Grundlag. Math.*, **32**(4), 365–370. [48]

[113] Kotlarski, Henryk (1995). Automorphisms of countable recursively saturated models of pa: a survey. *Notre Dame J. Formal Logic*, **36**(4), 505–518. Special issue: Models of arithmetic. [227]

[114] Kotlarski, Henryk & Kaye, Richard (1994). Automorphisms of models of true arithmetic: recognizing some basic open subgroups. *Notre Dame J. Formal Logic*, **35**(1), 1–14. [229], [254]

[115] Kotlarski, H., Krajewski, S., & Lachlan, A. H. (1981). Construction of satisfaction classes for nonstandard models. *Canad. Math. Bull.*, **24**(3), 283–293. [14]

[116] Kotlarski, Henryk & Piekart, Bożena (2000). Automorphisms of models of true arithmetic: more on subgroups which extend to a maximal one uniquely. *MLQ Math. Logic. Quart.*, **46**(1), 111–120. [229]

[117] Krajewski, S. (1976). *Non-standard satisfaction classes*, Volume 537 of *Lecture Notes in Mathematics*, pp. 121–144. Springer, Berlin. [24]

[118] Kueker, David W. (1981). $L_{\infty\omega_1}$ *elementarily equivalent models of power* ω_1, pp. 120–131. Springer, Berlin. [279]

[119] Lachlan, A. H. (1981). Full satisfaction classes and recursive saturation. *Canad. Math. Bull.*, **24**(3), 295–297. [14]

[120] Lascar, Daniel (1994). *The small index property and recursively saturated models of Peano arithmetic*, pp. 281–292. Oxford University Press, New York. [254]

[121] Lessan, Hamid (1978). Models of arithmetic. Ph.D. thesis, University of Manchester. [23], [48], [229]

[122] Lucchini, A. (1994). Intervals in subgroup lattices of finite groups. *Comm. Algebra*, **22**(2), 529–549. [134]

[123] Mac Dowell, R. & Specker, E. (1961). *Modelle der Arithmetik*, pp. 257–263. Pergamon, Oxford. [v], [47]

[124] Macintyre, Angus (1980). *Ramsey quantifiers in arithmetic*, Volume 834 of *Lecture Notes in Mathematics*, pp. 186–210. Springer, Berlin. [279]

[125] Macpherson, Dugald & Thomas, Simon (2005). Comeagre conjugacy classes and free products with amalgamation. *Discrete Math.*, **291**(1–3), 135–142. [244]

[126] Magidor, Menachem & Malitz, Jerome (1977). Compact extensions of $L(Q)$. Ia. *Ann. Math. Logic*, **11**(2), 217–261. [279]

[127] Manevitz, Larry Michael (1976). Internal end-extensions of Peano arithmetic and a problem of Gaifman. *J. Lond. Math. Soc. (2)*, **13**(1), 80–82. [48]

[128] McAloon, Kenneth (1978). Completeness theorems, incompleteness theorems and models of arithmetic. *Trans. Amer. Math. Soc.*, **239**, 253–277. [48]

[129] McKenzie, Ralph N., McNulty, George F., & Taylor, Walter F. (1987). *Algebras, lattices, varieties. Volume I.* The Wadsworth & Brooks/Cole Mathematics Series. Wadsworth & Brooks/Cole Advanced Books & Software, Monterey, CA. [132]

[130] Milliken, Keith R. (1979). A Ramsey theorem for trees. *J. Combin. Theory Ser. A*, **26**(3), 215–237. [171]

[131] Mills, George (1978). A model of Peano arithmetic with no elementary end extension. *J. Symbolic Logic*, **43**(3), 563–567. [133], [179], [290]

[132] Mills, George (1979). Substructure lattices of models of arithmetic. *Ann. Math. Logic*, **16**(2), 145–180. [179]

[133] Morgenstern, Carl (1982). On generalized quantifiers in arithmetic. *J. Symbolic Logic*, **47**(1), 187–190. [279]

[134] Morley, Michael (1965). Omitting classes of elements. In *Theory of Models* (*Proceedings of the 1963 International Symposium Berkeley, CA*), pp. 265–273. North-Holland, Amsterdam. [157]

[135] Mostowski, A. (1952). On models of axiomatic systems. *Fund. Math.*, **39**, 133–158 (1953). [19]

[136] Nadel, Mark E. (1981). The completeness of Peano multiplication. *Israel J. Math.*, **39**(3), 225–233. [280]

[137] Nešetřil, Jaroslav & Rödl, Vojtěch (1977). Partitions of finite relational and set systems. *J. Combin. Theory Ser. A*, **22**(3), 289–312. [157]

[138] Neumann, B. H. & Neumann, Hanna (1959). Embedding theorems for groups. *J. Lond. Math. Soc.*, **34**, 465–479. [252]

[139] Nurkhaidarov, Ermek S. Automorphism groups of arithmetically saturated models. *J. Symbolic Logic*, **71**(2006), 203–216. [254], [291]

[140] Odifreddi, Piergiorgio (1983). Forcing and reducibilities. *J. Symbolic Logic*, **48**(2), 288–310. [179]

[141] Pabion, J.-F. (1981). Saturation en arithmétique et en analyse. In *Proceedings of the Model Theory Meeting (University of Brussels, Brussels/ University of Mons, Mons, 1980)*, Volume 33, pp. 73–82. [288]

[142] Pabion, J.-F. (1982). Saturated models of Peano arithmetic. *J. Symbolic Logic*, **47**(3), 625–637. [viii], [288]

[143] Pálfy, P. P. & Pudlák, P. (1980). Congruence lattices of finite algebras and intervals in subgroup lattices of finite groups. *Algebra Universalis*, **11**, 22–27. [133]

[144] Paris, J. B. (1972). *On models of arithmetic*, Volume 255 of *Lecture Notes in Mathematics*, pp. 251–280. Springer, Berlin. [48], [133]

[145] Paris, J. B. (1977). Models of arithmetic and the 1-3-1 lattice. *Fund. Math.*, **95**(3), 195–199. [133]

[146] Paris, J. B. (1980). *A hierarchy of cuts in models of arithmetic*, Volume 834 of *Lecture Notes in Mathematics*, pp. 312–337. Springer, Berlin. [201]

[147] Paris, J. B. (1981). *Some conservation results for fragments of arithmetic*, Volume 890 of *Lecture Notes in Mathematics*, pp. 251–262. Springer, Berlin. [179]

[148] Paris, J. B. & Mills, G. (1979). Closure properties of countable nonstandard integers. *Fund. Math.*, **CIII**, 205–215. [32], [88], [201]

[149] Phillips, Robert G. (1974). A minimal extension that is not conservative. *Michigan Math. J.*, **21**, 27–32. [47]

[150] Phillips, R. G. (1974). *Omitting types in arithmetic and conservative extensions*, Volume 369 of *Lecture Notes in Mathematics*, pp. 195–202. Springer, Berlin. [47]

[151] Piekart, Bożena (1993). On closed elementary cuts in recursively saturated models of Peano arithmetic. *Notre Dame J. Formal Logic*, **34**(2), 223–230. [229]

[152] Potthoff, Klaus (1976/77). A simple tree lemma and its application to a counterexample of Philips. *Arch. Math. Logik Grundlag.*, **18**(1–2), 67–71. [47]

[153] Pudlák, P. (1976). A new proof of the congruence charaterization theorem. *Algebra Universalis*, **6**, 269–275. [132]

[154] Pudlák, P. & Tůma, J. (1976). Yeast graphs and fermentation of algebraic lattices. In *Lattice theory (Proceedings of the Colloquium, Szeged, 1974)*, pp. 301–341. Colloquium of the Mathematical Society, János Bolyai, Volume 14. North-Holland, Amsterdam. [134]

[155] Pudlák, Pavel & Tůma, Jiří (1980). Every finite lattice can be embedded in a finite partition lattice. *Algebra Universalis*, **10**(1), 74–95. [106]

[156] Rabin, Michael O. (1962). Diophantine equations and non-standard models of arithmetic. In *Logic, methodology and philosophy of science (Proc. 1960 Internat. Congr.)*, pp. 151–158. Stanford University Press, Stanford, CA. [48]

[157] Robinson, Abraham (1963). On languages which are based on non-standard arithmetic. *Nagoya Math. J.*, **22**, 83–117. [24]

[158] Rogers, Jr., Hartley (1987). *Theory of recursive functions and effective computability*. Second edition. MIT Press, Cambridge, MA. [45]

[159] Sauer, Norbert (2001). Hedetniemi's conjecture—a survey. *Discrete Math.*, **229**(1–3), 261–292. Combinatorics, graph theory, algorithms and applications. [254]

[160] Schlipf, John Stewart (1978). Toward model theory through recursive saturation. *J. Symbolic Logic*, **43**(2), 183–206. [23]

[161] Schmerl, James H. (1974). A partition property characterizing cardinals hyperinaccessible of finite type. *Trans. Amer. Math. Soc.*, **188**, 281–291. [48]

[162] Schmerl, J. H. (1978). Extending models of arithmetic. *Ann. Math. Logic*, **14**, 89–109. [133]

[163] Schmerl, J. H. (1981). Correction to: "Peano models with many generic classes" [Pacific J. Math. **46** (1973), 523–536; MR **50** #6831]. *Pacific J. Math.*, **92**(1), 195–198. [179]

[164] Schmerl, James H. (1981). *Recursively saturated, rather classless models of Peano arithmetic*, pp. 268–282. Springer, Berlin. [22], [48], [278]

[165] Schmerl, James H. (1985). Models of Peano arithmetic and a question of Sikorski on ordered fields. *Israel J. Math.*, **50**(1–2), 145–159. [279]

[166] Schmerl, James H. (1985). Recursively saturated models generated by indiscernibles. *Notre Dame J. Formal Logic*, **26**(2), 99–105. [157], [179]

[167] Schmerl, James H. (1986). *Substructure lattices of models of Peano arithmetic*, pp. 225–243. North-Holland, Amsterdam. [133]

[168] Schmerl, James H. (1992). End extensions of models of arithmetic. *Notre Dame J. Formal Logic*, **33**(2), 216–219. [48]

[169] Schmerl, James H. (1993). Finite substructure lattices of models of Peano arithmetic. *Proc. Amer. Math. Soc.*, **117**(3), 833–838. [133]

[170] Schmerl, James H. (1993). A weakly definable type which is not definable. *Arch. Math. Logic*, **32**(6), 463–468. [88]

[171] Schmerl, James H. (1995). PA(aa). *Notre Dame J. Formal Logic*, **36**(4), 560–569. Special issue: Models of arithmetic. [268], [280]

[172] Schmerl, J. H. (1998). *Rather classless, highly saturated models of Peano arithmetic*, pp. 237–246. Springer, Berlin. [279]

[173] Schmerl, James H. (2001). Closed normal subgroups. *MLQ Math. Logic. Quart.*, **47**(4), 489–492. [229]

[174] Schmerl, James H. (2002). Automorphism groups of models of Peano arithmetic. *J. Symbolic Logic*, **67**(4), 1249–1264. [88], [157], [212], [228], [291]

[175] Schmerl, James H. (2002). Moving intersticial gaps. *MLQ Math. Logic. Quart.*, **48**(2), 283–296. [227], [228]

[176] Schmerl, James H. (2002). Some highly saturated models of Peano arithmetic. *J. Symbolic Logic*, **67**(4), 1265–1273. [279]

[177] Schmerl, James H. (2004). Diversity in substructures. In *Nonstandard models of arithmetic and set theory*, Volume 361 of *Contemporary Mathematics*, pp. 145–161. American Mathematical Society, Providence, RI. [157]

[178] Schmerl, James H. (2005). Generic automorphisms and graph coloring. *Discrete Math.*, **291**(1–3), 235–242. [133], [254]

[179] Schmerl, James H. & Simpson, Stephen G. (1982). On the role of Ramsey quantifiers in first order arithmetic. *J. Symbolic Logic*, **47**(2), 423–435. [279]

[180] Scott, Dana (1962). *Algebras of sets binumerable in complete extensions of arithmetic*, pp. 117–121. American Mathematical Society, Providence, RI. [vi], [22]

[181] Seetapun, David & Slaman, Theodore A. (1995). On the strength of Ramsey's theorem. *Notre Dame J. Formal Logic*, **36**(4), 570–582. Special issue: Models of arithmetic. [201], [254]

[182] Serre, Jean-Pierre (1980). *Trees.* Springer-Verlag, Berlin. [244]

[183] Shelah, Saharon (1976). A two-cardinal theorem and a combinatorial theorem. *Proc. Amer. Math. Soc.*, **62**(1), 134–136 (1977). [179]

[184] Shelah, Saharon (1978). Models with second-order properties. II. Trees with no undefined branches. *Ann. Math. Logic*, **14**(1), 73–87. [258], [292]

[185] Shelah, Saharon (1994). Vive la différence. II. The Ax-Kochen isomorphism theorem. *Israel J. Math.*, **85**(1–3), 351–390. [viii], [288]

[186] Simpson, S. G. (1974). Forcing and models of arithmetic. *Proc. Amer. Math. Soc.*, **43**, 193–194. [179]

[187] Simpson, Stephen G. (1999). *Subsystems of second order arithmetic.* Springer-Verlag, Berlin. [194], [201]

[188] Skolem, Th. (1933). Über die Unmglichkeit einer vollstndigen Charakterisierung der Zahlenreihe mittels eines endlichen Axiomensystems. *Norsk Matematisk Forenings Skrifter*, **10**, 73–82. [v]

[189] Skolem, Th. (1934). Über die Nicht-charakterisierbarkeit der Zahlenreihe mittels endlich oder abzhlbar unendlich vieler Aussagen mit ausschliesslich Zahlenvariablen. *Fundam. Math.*, **23**, 150–161. [v]

[190] Smith, Stuart T. (1989). Nonstandard definability. *Ann. Pure Appl. Logic*, **42**(1), 21–43. [47]

[191] Smoryński, C. (1978). Nonstandard models of arithmetic. Preprint Nr. 153, University of Utrecht. [viii]

[192] Smoryński, C. (1981). Elementary extensions of recursively saturated models of arithmetic. *Notre Dame J. Formal Logic*, **22**(3), 193–203. [23], [279]

[193] Smoryński, C. (1981). Recursively saturated nonstandard models of arithmetic. *J. Symbolic Logic*, **46**(2), 259–286. [23], [48], [227]

[194] Smoryński, C. (1982). *Back-and-forth inside a recursively saturated model of arithmetic*, pp. 273–278. North-Holland, Amsterdam. [227]

[195] Smoryński, C. (1982). A note on initial segment constructions in recursively saturated models of arithmetic. *Notre Dame J. Formal Logic*, **23**(4), 393–408. [279]

[196] Smoryński, C. (1984). *Lectures on nonstandard models of arithmetic*, Volume 112 of *Studies in Logic and the Foundations of Mathematics*, pp. 1–70. North-Holland, Amsterdam. [22], [23]

[197] Smoryński, C. (1985). *Nonstandard models and related developments*, Volume 117 of *Studies in Logic and the Foundations of Mathematics*, pp. 179–229. North-Holland, Amsterdam. [22]

[198] Smoryński, Craig (1991). *Logical number theory. I.* Universitext. Springer-Verlag, Berlin. An introduction. [280]

[199] Smoryński, C. & Stavi, J. (1980). *Cofinal extension preserves recursive saturation*, pp. 338–345. Springer, Berlin. [23], [24]

[200] Snow, John W. (2003). Every finite lattice in $\mathcal{V}(\mathbf{M}_3)$ is representable. *Algebra Universalis*, **50**(1), 75–81. [134]

[201] Stone, M. H. (1936). The theory of representations for Boolean algebras. *Trans. Amer. Math. Soc.*, **40**(1), 37–111. [133]

[202] Surman, J. (2003). Back and forth. *Free and Equal, ECM*, **1802**. [228]

[203] Tanaka, Kazuyuki (1997). The self-embedding theorem of WKL_0 and a non-standard method. *Ann. Pure Appl. Logic*, **84**(1), 41–49. [24]

[204] Tardif, Claude & Wehlau, David. Chromatic numbers of products of graphs: the directed and undirected versions of the poljak-rödl function. To appear in *J. Graph Theory*. [254]

[205] Tennenbaum, Stanley (1959). Non-archimedean models for arithmetic. *Notices Amer. Math. Soc.*, **6**, 270.

[206] Thomas, Simon (1997). *Cofinalities of infinite permutation groups*, Volume 9 of *Algebra Logic Application*, pp. 101–120. Gordon and Breach, Amsterdam. [254]

[207] Vaught, R. L. (1974). Model theory before 1945. In *Proceedings of the Tarski Symposium (Proc. Sympos. Pure Math., Vol. XXV, University of California, Berkeley, CA, 1971)*, American Mathematical Society, Providence, RI, pp. 153–172. [v]

[208] Weiss, Michael (1980). Relatively diophantine predicates and existentially complete for arithmetic. Ph.D. thesis, Rutgers University. [22]

[209] Wilkie, Alex (1975). On models of arithmetic–answers to two problems raised by H. Gaifman. *J. Symbolic Logic*, **40**(1), 41–47. [23], [48]

[210] Wilkie, A. J. (1977). On models of arithmetic having non-modular substructure lattices. *Fund. Math.*, **95**, 223–237. [133]

[211] Wilkie, A. J. (1977). *On the theories of end-extensions of models of arithmetic*, Volume 619 of *Lecture Notes in Mathematics*, pp. 305–310. Springer, Berlin. [48]

[212] Wilkie, A. J. (1987). *On schemes axiomatizing arithmetic*, pp. 331–337. American Mathematical Society, Providence, RI. [201]

[213] Wilmers, George (1980). *Minimally saturated models*, Volume 834 of *Lecture Notes in Mathematics*, pp. 370–380. Springer, Berlin. [23]

INDEX OF NAMES

INDEX OF TERMS